Springer Proceedings in Mathematics

Volume 7

For further volumes:
http://www.springer.com/series/8806

Springer Proceedings in Mathematics

The book series will feature volumes of selected contributions from workshops and conferences in all areas of current research activity in mathematics. Besides an overall evaluation, at the hands of the publisher, of the interest, scientific quality, and timeliness of each proposal, every individual contribution will be refereed to standards comparable to those of leading mathematics journals. It is hoped that this series will thus propose to the research community well-edited and authoritative reports on newest developments in the most interesting and promising areas of mathematical research today.

Mounir Zili • Darya V. Filatova
Editors

Stochastic Differential Equations and Processes

SAAP, Tunisia, October 7-9, 2010

Springer

Editors
Mounir Zili
Preparatory Institute
to the Military Academies
Department of Mathematics
Avenue Marechal Tito
4029 Sousse
Tunisia
zilimounir@yahoo.fr

Darya V. Filatova
Jan Kochanowski University in Kielce
ul. Krakowska 11
25-029 Kielce
Poland
daria_filatova@interia.pl

ISSN 2190-5614
ISBN 978-3-642-22367-9 e-ISBN 978-3-642-22368-6
DOI 10.1007/978-3-642-22368-6
Springer Heidelberg Dordrecht London New York

Library of Congress Control Number: 2011938117

Mathematical Subject Classification (2010): 60HXX, 60GXX, 60FXX, 91BXX, 93BXX, 92DX

Cover design: deblik, Berlin

Printed on acid-free paper

Springer is part of Springer Science+Business Media (www.springer.com)

Preface

Stochastic analysis is currently undergoing a period of intensive research and various new developments, motivated in part by the need to model, understand, forecast, and control the behavior of many natural phenomena that evolve in time in a random way. Such phenomena appear in the fields of finance, telecommunications, economics, biology, geology, demography, physics, chemistry, signal processing, and modern control theory, to mention just a few.

Often, it is very convenient to use stochastic differential equations and stochastic processes to study stochastic dynamics. In such cases, research needs the guarantee of some theoretical properties, such as the existence and uniqueness of the stochastic equation solution. Without a deep understanding of the nature of the stochastic process this is seldom possible. The theoretical background of both stochastic processes and stochastic differential equations are therefore very important.

Nowadays, quite a few stochastic differential equations can be solved by means of exact methods. Even if this solution exists, it cannot necessarily be used for computer simulations, in which the continuous model is replaced by a discrete one. The problems of "ill-posed" tasks, the "stiffness" or "stability" of the system limit numerical approximations of the stochastic differential equation. As a result, new approaches for the numerical solution and, consequently, new numerical algorithms are also very important.

This volume contains 8 refereed papers dealing with these topics, chosen from among the contributions presented at the international conference on *Stochastic Analysis and Applied Probability* (SAAP 2010), which was held at Yasmine-Hammamet, Tunisia, from 7 to 9 October 2010. This conference was organized by the "Applied Mathematics & Mathematical Physics" research unit of the preparatory institute to the military academies of Sousse, Tunisia. It brought together some 60 researchers and PhD students, from 14 countries and 5 continents. Through lectures, communications, and posters, these researchers reported on theoretical, numerical, or application work as well as on significant results obtained for several topics within the field of stochastic analysis and probability, particularly for "Stochastic processes and stochastic differential equations." The conference program was planned by an international committee chaired by Mounir Zili (Preparatory Institute to the Military Academies of Sousse, Tunisia) and consisted of Darya Filatova (Jan Kochanowski University in Kielce, Poland), Ibtissem Hdhiri (Faculty of Sciences of Gabès, Tunisia), Ciprian A. Tudor (University of Lille, France), and Mouna Ayachi (Faculty of Sciences of Monastir, Tunisia).

As this book emphasizes the importance of numerical and theoretical studies of the stochastic differential equations and stochastic processes, it will be useful for a wide spectrum of researchers in applied probability, stochastic numerical and theoretical analysis and statistics, as well as for graduate students.

To make it more complete and accessible for graduate students, practitioners, and researchers, we have included a survey dedicated to the basic concepts of numerical analysis of the stochastic differential equations, written by Henri Schurz. This survey is valuable not only due to its excellent theoretical conception with respect to modern tendencies, but also with regard to its comprehensive concept of the dynamic consistency of numerical methods for the stochastic differential equations. In a second paper, motivated by its applications in econometrics, Ciprian Tudor develops an asymptotic theory for some regression models involving standard Brownian motion and the standard Brownian sheet. The result proved in this paper is an impressive example of convergence in distribution to a non-Gaussian limit. The paper "General shot noise processes and functional convergence to stable processes" by Wissem Jedidi, Jalal Almhana, Vartan Choulakian, and Robert McGorman also addresses the topic of stochastic processes, and the authors consider a model appropriate for the network traffic consisting of an infinite number of sources linked to a unique server. This model is based on a general Poisson shot noise representation, which is a generalization of a compound Poisson process. In the fourth paper of this volume, Charles El-Nouty deals with the lower classes of the sub-fractional Brownian motion, which has been introduced to model some self-similar Gaussian processes, with non-stationary increments. Then, in a paper by Mohamed Erraoui and Youssef Ouknine, the bounded variation of the flow of a stochastic differential equation driven by a fractional Brownian motion and with non-Lipschitz coefficients is studied. In the sixth paper, Antoine Ayache and Qidi Peng develop an extension of several probabilistic and statistical results for stochastic volatility models satisfying some stochastic differential equations for cases in which the fractional Brownian motion is replaced by the multifractional Brownian motion. The advantage of the multifractional stochastic volatility models is that they allow account variations with respect to time of volatility local roughness. The seventh paper was written by Archil Gulisashvili and Josep Vives and addresses two-sided estimates for the distribution density of standard models, perturbed by a double exponential law. The results obtained in this paper can especially be used in the study of distribution densities arising in some stochastic stock price models. And in the last paper in the volume, Mario Lefebvre explicitly solves the problem of maximizing a function of the time spent by a stochastic process by arriving at solutions of some particular stochastic differential equations.

All the papers presented in this book were carefully reviewed by the members of the SAAP 2010 Scientific Committee, a list of which is presented in the appendix.

We would like to thank the anonymous reviewers for their reports and many others who contributed enormously to the high standards for publication in these proceedings by carefully reviewing the manuscripts that were submitted.

Finally, we want to express our gratitude to Marina Reizakis for her invaluable help in the process of preparing this volume edition.

Tunisia
September 2011

Mounir Zili
Daria Filatova

Contents

Contributors

Jalal Almhana GRETI Group, University of Moncton, Moncton, NB E1A3E9, Canada, almhanaj@umoncton.ca

Antoine Ayache U.M.R. CNRS 8524, Laboratory Paul Painlevé, University Lille 1, 59655 Villeneuve d'Ascq Cedex, France, smith@smith.edu

Vartan Choulakian GRETI Group, University of Moncton, Moncton, NB E1A3E9, Canada, choulav@umoncton.ca

Charles El-Nouty UMR 557 Inserm/ U1125 Inra/ Cnam/ Université Paris XIII, SMBH-Université Paris XIII, 74 rue Marcel Cachin, 93017 Bobigny Cedex, France, c.el-nouty@uren.smbh.univ-paris13.fr

Mohamed Erraoui Faculté des Sciences Semlalia Département de Mathématiques, Université Cadi Ayyad BP 2390, Marrakech, Maroc, erraoui@ucam.ac.ma

Archil Gulisashvili Department of Mathematics, Ohio University, Athens, OH 45701, USA, gulisash@ohio.edu

Wissem Jedidi Department of Mathematics, Faculty of Sciences of Tunis, Campus Universitaire, 1060 Tunis, Tunisia, wissem.jedidi@fst.rnu.tn

Mario Lefebvre Département de Mathématiques et de Génie Industriel, École Polytechnique, C.P. 6079, Succursale Centre-ville, Montréal, PQ H3C 3A7, Canada, mlefebvre@polymtl.ca

Robert McGorman NORTEL Networks, 4001 E. Chapel Hill-Nelson Hwy, Research Triangle Park, NC 27709, USA, mcgorman@nortelnetworks.com

Youssef Ouknine Faculté des Sciences Semlalia Département de Mathématiques, Université Cadi Ayyad BP 2390, Marrakech, Maroc, ouknine@ucam.ac.ma

Qidi Peng U.M.R. CNRS 8524, Laboratory Paul Painlevé, University Lille 1, 59655 Villeneuve d'Ascq Cedex, France, smith@smith.edu

Henry Schurz Department of Mathematics, Southern Illinois University, 1245 Lincoln Drive, Carbondale, IL 62901, USA, hschurz@math.siu.edu

Ciprian A. Tudor Laboratoire Paul Painlevé, Université de Lille 1, F-59655 Villeneuve d'Ascq, France, tudor@math.univ-lille1.fr

Josep Vives Departament de Probabilitat, Lògica i Estadística, Universitat de Barcelona, Gran Via 585, 08007-Barcelona (Catalunya), Spain, josep.vives@ub.edu

Chapter 1
Basic Concepts of Numerical Analysis of Stochastic Differential Equations Explained by Balanced Implicit Theta Methods

Henri Schurz

Abstract We present the comprehensive concept of dynamic consistency of numerical methods for (ordinary) stochastic differential equations. The concept is illustrated by the well-known class of balanced drift-implicit stochastic Theta methods and relies on several well-known concepts of numerical analysis to replicate the qualitative behaviour of underlying continuous time systems under adequate discretization. This involves the concepts of consistency, stability, convergence, positivity, boundedness, oscillations, contractivity and energy behaviour. Numerous results from literature are reviewed in this context.

1.1 Introduction

Numerous monographs and research papers on numerical methods of stochastic differential equations are available. Most of them concentrate on the construction and properties of consistency. A few deal with stability and longterm properties. However, as commonly known, the replication of qualitative properties of numerical methods in its whole is the most important issue for modeling and real-world applications. To evaluate numerical methods in a more comprehensive manner, we shall discuss the concept of **dynamic consistency** of numerical methods for stochastic differential equations. For the sake of precise illustration, we will treat the example class of balanced implicit **outer Theta methods**. This class is defined by

H. Schurz (✉)
Southern Illinois University, Department of Mathematics, 1245 Lincoln Drive, Carbondale, IL 62901, USA
e-mail: hschurz@math.siu.edu

M. Zili and D.V. Filatova (eds.), *Stochastic Differential Equations and Processes*, Springer Proceedings in Mathematics 7, DOI 10.1007/978-3-642-22368-6_1,

$$X_{n+1} = \begin{cases} X_n + [\Theta_n a(t_{n+1}, X_{n+1}) + (I - \Theta_n) a(t_n, X_n)]\, h_n + \sum_{j=1}^{m} b^j(t_n, X_n) \Delta W_n^j \\ \qquad + \sum_{j=0}^{m} c^j(t_n, X_n)(X_n - X_{n+1}) |\Delta W_n^j| \end{cases} \tag{1.1}$$

with appropriate (bounded) matrices c^j with continuous entries, where I is the unit matrix in $\mathbb{R}^{d \times d}$ and

$$\Delta W_n^0 = h_n, \quad \Delta W_n^j = W^j(t_{n+1}) - W^j(t_n)$$

along partitions

$$0 = t_0 < t_1 < \ldots < t_n < t_{n+1} < \ldots < t_{n_T} = T < +\infty$$

of finite time-intervals $[0, T]$. These methods are discretizations of d-dimensional ordinary stochastic differential equations (SDEs), [3, 14, 32, 33, 81, 86, 102, 108]

$$dX(t) = a(t, X(t))dt + \sum_{j=1}^{m} b^j(t, X(t)) dW^j(t) \left(= \sum_{j=0}^{m} b^j(t, X(t)) dW^j(t) \right) \tag{1.2}$$

(with $b^0 = a$, $W^0(t) = t$), driven by i.i.d. Wiener processes W^j and started at adapted initial values $X(0) = x_0 \in \mathbb{R}^d$. The vector fields a and b^j are supposed to be sufficiently smooth throughout this survey. All stochastic processes are constructed on the complete probability basis $(\Omega, \mathcal{F}, (\mathcal{F}_t)_{t \geq 0}, \mathbb{P})$.

The aforementioned Theta methods (1.1) represent a first natural generalization of explicit and implicit Euler methods. Indeed, they are formed by a convex linear combinations of explicit and implicit Euler increment functions of the drift part, whereas the diffusion part is explicitly treated due to the problem of adequate integration within one and the same stochastic calculus. The balanced terms c^j are appropriate matrices and useful to control the pathwise (i.e. almost sure) behaviour and uniform boundedness of those approximations. The parameter matrices $(\Theta_n)_{n \in \mathbb{N}} \in \mathbb{R}^{d \times d}$ determine the degree of implicitness and simplective behaviour (energy- and area-preserving character) of related approximations. Most popular representatives are those with simple scalar choices $\Theta_n = \theta_n I$ where I denotes the unit matrix in $\mathbb{R}^{d \times d}$ and $\theta_n \in \mathbb{R}^1$. Originally, without balanced terms c^j, they were invented by Talay [138] in stochastics, who proposed $\Theta_n = \theta I$ with autonomous scalar choices $\theta \in [0, 1]$. This family with matrix-valued parameters $\Theta \in \mathbb{R}^{d \times d}$ has been introduced by Ryashko and Schurz [116] who also proved their mean square convergence with an estimate of worst case convergence rate 0.5. If $\theta = 0$ then its scheme reduces to the classical **(forward) Euler method** (see Maruyama [90], Golec et al. [35–38], Guo [39, 40], Gyöngy [41, 42], Protter and

Talay [109], Römisch & Wakolbinger [115], Tudor & Tudor [143] among others), if $\theta = 1$ to the **backward Euler method** which is also called *(drift-)implicit Euler method* (Hu [55]), and if $\theta = 0.5$ to the **(drift-implicit) trapezoidal method**, reducing to the scheme

$$X_{n+1} = X_n + \frac{1}{2}\left[a(t_{n+1}, X_{n+1}) + a(t_n, X_n)\right] h_n + \sum_{j=1}^{m} b^j(t_n, X_n)\Delta W_n^j \quad (1.3)$$

without balanced terms c^j. A detailed study of the qualitative dynamic behaviour of these methods can be found in Stuart and Peplow [136] in deterministic numerical analysis (in the sense of spurious solutions), and in Schurz [120] in stochastic numerical analysis.

A slightly different class of numerical methods is given by the balanced implicit **inner Theta methods**

$$X_{n+1} = \begin{cases} X_n + a\left(t_n + \theta_n h_n, \Theta_n X_{n+1} + (I - \Theta_n) X_n\right) h_n + \sum_{j=1}^{m} b^j(t_n, X_n)\,\Delta W_n^j \\ \qquad + \sum_{j=0}^{m} c^j(t_n, X_n)(X_n - X_{n+1})|\Delta W_n^j| \end{cases} \quad (1.4)$$

where $\theta_n \in \mathbb{R}$, $\Theta_n \in \mathbb{R}^{d\times d}$ such that local algebraic resolution can be guaranteed always. The most known representative of this class (1.4) with $\Theta_n = 0.5I$ and without balanced terms c^j is known as the **drift-implicit midpoint method** governed by

$$X_{n+1} = X_n + a\left(\frac{t_{n+1} + t_n}{2}, \frac{X_{n+1} + X_n}{2}\right)\Delta_n + \sum_{j=1}^{m} b^j(t_n, X_n)\,\Delta W_n^j. \quad (1.5)$$

This method is superior for the integration of conservation laws and Hamiltonian systems. Their usage seems to be very promising for the control of numerical stability, area-preservation and boundary laws in stochastics as well. The drawback for their practical implementation can be seen in the local resolution of nonlinear algebraic equations which is needed in addition to explicit methods. However, this fact can be circumvented by its practical implementation through predictor-corrector methods (PCMs), their linear- (LIMs) or partial-implicit (PIMs) derivates (versions). In passing, note that the **partitioned Euler methods** (cf. Strommen–Melbo and Higham [135]) are also a member of stochastic Theta methods (1.1) with the special choice of constant implicitness-matrix

$$\Theta_n = \begin{pmatrix} 0 & 0 \\ 1 & 1 \end{pmatrix}.$$

In passing, note that stochastic Theta methods (1.1) represent the simplest class of stochastic Runge-Kutta methods. Despite their simplicity, they are rich enough to cover many aspects of numerical approximations in an adequate manner.

The purpose of this survey is to compile some of the most important facts on representatives of classes (1.1) and (1.4). Furthermore, we shall reveal the goodness of these approximation techniques in view of their dynamic consistency. In the following sections we present and discuss several important key concepts of stochastic numerical analysis explained by Theta methods. At the end we finalize our presentation with a summary leading to the governing concept of dynamic consistency unifying the concepts presented before in a complex fashion. The paper is organized in 12 sections. The remaining part of our introduction reports on auxiliary tools to construct, derive, improve and justify consistency of related numerical methods for SDEs. Topics as consistency in Sect. 1.2, asymptotic stability in Sect. 1.3, convergence in Sect. 1.4, positivity in Sect. 1.5, boundedness in Sect. 1.6, oscillations in Sect. 1.7, energy in Sect. 1.8, order bounds in Sect. 1.9, contractivity in Sect. 1.10 and dynamic consistency in Sect. 1.11 are treated. Finally, the related references are listed alphabetically, without claiming to refer to all relevant citations in the overwhelming literature on those subjects. We recommend also to read the surveys of Artemiev and Averina [5], Kanagawa and Ogawa [66], Pardoux and Talay [106], S. [125] and Talay [140] in addition to our paper. A good introduction to related basic elements is found in Allen [1] and [73] too.

1.1.1 Auxiliary tool: Itô Formula (Itô Lemma) with Operators $\mathscr{L}^j$

Define linear partial differential operators

$$\mathscr{L}^0 = \frac{\partial}{\partial t} + < a(t,x), \nabla_x >_d + \frac{1}{2} \sum_{j=1}^{m} \sum_{i,k=1}^{d} b_i^j(t,x) b_k^j(t,x) \frac{\partial^2}{\partial x_k \partial x_i} \tag{1.6}$$

and $\mathscr{L}^j = < b^j(t,x), \nabla_x >_d$ where $j = 1, 2, \ldots, m$. Then, thanks to the fundamental contribution of Itô [56] and [57], we have the following lemma.

Lemma 1.1.1 (Stopped Itô Formula in Integral Operator Form). *Assume that the given deterministic mapping $V \in C^{1,2}([0,T] \times \mathbb{R}^d, \mathbb{R}^k)$. Let τ be a finite $\mathscr{F}_t$-adapted stopping time with $0 \le t \le \tau \le T$.*
Then, we have

$$V(\tau, X(\tau)) = V(t, X(t)) + \sum_{j=0}^{m} \int_t^{\tau} \mathscr{L}^j V(s, X(s))\, dW^j(s). \tag{1.7}$$

1.1.2 Auxiliary Tool: Derivation of Stochastic Itô-Taylor Expansions

By iterative application of Itô formula we gain the family of stochastic Taylor expansions. This idea is due to Wagner and Platen [144]. Suppose we have enough smoothness of V and of coefficients a, b^j of the Itô SDE. Remember, thanks to Itô's formula, for $t \geq t_0$

$$V(t, X(t)) = V(t_0, X(t_0)) + \int_{t_0}^{t} \mathscr{L}^0 V(s, X(s)) ds + \sum_{j=1}^{m} \int_{t_0}^{t} \mathscr{L}^j V(s, X(s))\, dW^j(s).$$

Now, take $V(t, x) = x$ at the first step, and set $b^0(t, x) \equiv a(t, x)$, $W^0(t) \equiv t$. Then one derives

$$\begin{aligned}
X(t) &= X(t_0) + \int_{t_0}^{t} a(s, X(s)) ds + \sum_{j=1}^{m} \int_{t_0}^{t} b^j(s, X(s))\, dW^j(s) \\
V &\equiv b^j \\
&= X(t_0) + \int_{t_0}^{t} \left[a(t_0, X(t_0)) + \sum_{k=0}^{m} \int_{t_0}^{s} \mathscr{L}^k a(u, X(u)) dW^k(u) \right] ds \\
&\quad + \sum_{j=1}^{m} \int_{t_0}^{t} \left[b^j(t_0, X(t_0)) + \sum_{k=0}^{m} \int_{t_0}^{s} \mathscr{L}^k a(u, X(u)) dW^k(u) \right] dW^j(s) \\
b^0 &\equiv a \\
&= X(t_0) + \underbrace{\sum_{j=0}^{m} b^j(t_0, X(t_0)) \int_{t_0}^{t} dW^j(s)}_{\textbf{Euler Increment}} \\
&\quad + \underbrace{\sum_{j,k=0}^{m} \int_{t_0}^{t} \int_{t_0}^{s} \mathscr{L}^k b^j(u, X(u))\, dW^k(u) dW^j(s)}_{\textbf{Remainder Term } \mathbf{R}_E} \\
V &\equiv \mathscr{L}^k b^j \\
&= X(t_0) \\
&\quad \underbrace{+ \sum_{j=0}^{m} b^j(t_0, X(t_0)) \int_{t_0}^{t} dW^j(s) + \sum_{j,k=1}^{m} \mathscr{L}^k b^j(t_0, X(t_0)) \int_{t_0}^{t} \int_{t_0}^{s} dW^k(u) dW^j(s)}_{\textbf{Increment of Milstein Method}}
\end{aligned}$$

$$+\sum_{j=1}^{m}\int_{t_0}^{t}\int_{t_0}^{s}\mathscr{L}^0 b^j(u,X(u))\,du\,dW^j(s)$$

$$+\sum_{k=1}^{m}\int_{t_0}^{t}\int_{t_0}^{s}\mathscr{L}^k a(u,X(u))\,dW^k(u)ds$$

$$+\underbrace{\sum_{j,k=1,l=0}^{m}\int_{t_0}^{t}\int_{t_0}^{s}\int_{t_0}^{u}\mathscr{L}^l\mathscr{L}^k b^j(z,X(z))\,dW^l(z)dW^k(u)dW^j(s)}_{\textbf{Remainder Term } \mathbf{R}_M}$$

$$\begin{aligned}V &\equiv \mathscr{L}^k b^j\\ &= X(t_0)\\ &\quad+\underbrace{\sum_{j=0}^{m}b^j(t_0,X(t_0))\int_{t_0}^{t}dW^j(s)+\sum_{j,k=0}^{m}\mathscr{L}^k b^j(t_0,X(t_0))\int_{t_0}^{t}\int_{t_0}^{s}dW^k(u)dW^j(s)}_{\textbf{Increment of 2nd order Taylor Method}}\\ &\quad+\underbrace{\sum_{j,k,l=0}^{m}\int_{t_0}^{t}\int_{t_0}^{s}\int_{t_0}^{u}\mathscr{L}^l\mathscr{L}^k b^j(z,X(z))\,dW^l(z)dW^k(u)dW^j(s)}_{\textbf{Remainder Term } \mathbf{R}_{TM2}}\end{aligned}$$

$$\begin{aligned}V &\equiv \mathscr{L}^r\mathscr{L}^k b^j\\ &= X(t_0)\\ &\quad+\underbrace{\sum_{j=0}^{m}b^j(t_0,X(t_0))\int_{t_0}^{t}dW^j(s)+\sum_{j,k=0}^{m}\mathscr{L}^k b^j(t_0,X(t_0))\int_{t_0}^{t}\int_{t_0}^{s}dW^k(u)dW^j(s)}_{\textbf{Increment of 3rd order Taylor Method}}\\ &\quad+\underbrace{\sum_{j,k,r=0}^{m}\mathscr{L}^r\mathscr{L}^k b^j(t_0,X(t_0))\int_{t_0}^{t}\int_{t_0}^{s}\int_{t_0}^{u}dW^r(v)dW^k(u)dW^j(s)}_{\textbf{Increment of 3rd order Taylor Method}}\\ &\quad+\underbrace{\sum_{j,k,r,l=0}^{m}\int_{t_0}^{t}\int_{t_0}^{s}\int_{t_0}^{u}\int_{t_0}^{v}\mathscr{L}^l\mathscr{L}^r\mathscr{L}^k b^j(z,X(z))dW^l(z)dW^r(v)dW^k(u)dW^j(s)}_{\textbf{Remainder Term } \mathbf{R}_{TM3}}\end{aligned}$$

....

This process can be continued under appropriate assumptions of smoothness and boundedness of the involved expressions. Thus, this is the place from which most numerical methods systematically originate, and where the main tool for consistency analysis is coming from. One has to expand the functionals in a hierarchical way, otherwise one would loose important order terms, and the implementation would be inefficient. Of course, for qualitative, smoothness and efficiency reasons we do not have to expand all terms in the Taylor expansions at the same time (e.g. cf. Milstein increment versus 2nd order Taylor increments). The *Taylor method* can be read down straight forward by truncation of stochastic Taylor expansion. *Explicit* and *implicit methods*, *Runge-Kutta methods*, *inner and outer Theta methods*, *linear-implicit* or *partially implicit* methods are considered as modifications of Taylor methods by substitution of derivatives by corresponding difference quotients, explicit expressions by implicit ones, respectively. However, it necessitates finding a more efficient form for representing stochastic Taylor expansions and hence Taylor-type methods. For this aim, we shall report on hierarchical sets, coefficient functions and multiple integrals in the subsection below.

In general, Taylor-type expansions are good to understand the systematic construction of numerical methods with certain orders. Moreover, they are useful to prove certain rates of local consistency of numerical methods. However, the rates of convergence (global consistency) of them are also determined by other complex dynamical features of numerical approximations, and "order bounds" and "practical modeling / simulation issues" may decisively limit their usage in practice. To fully understand this statement, we refer to the concept of "dynamic consistency" as developed in the following sections in this paper.

1.1.3 Auxiliaries: Hierarchical Sets, Coefficient Functions, Multiple Integrals

Kloeden and Platen [72] based on the original work of Wagner and Platen [144] have introduced a more compact, efficient formulation of stochastic Taylor expansions. For its statement, we have to formulate what is meant by multiple indices, hierarchical sets, remainder sets, coefficient functions and multiple integrals in the Itô sense.

Definition 1.1.1. A multiple index has the form $\alpha = (\alpha_1, \alpha_2, \ldots, \alpha_{l(\alpha)})$ where $l(\alpha) \in \mathbb{N}$ is called the *length* of the multiple index α, and $n(\alpha)$ is the *total number* of zero entries of α. The symbol ν denotes the *empty multiple index* with $l(\nu) = 0$. The operations $\alpha- = (\alpha_1, \ldots, \alpha_{l(\alpha)-1})$ and $-\alpha = (\alpha_2, \ldots, \alpha_{l(\alpha)})$ are called *right-* and *left-subtraction*, respectively (in particular, $(\alpha_1)- = -(\alpha_1) = \nu$). The *set of all multiple indices* is defined to be

$$\mathscr{M}_{k,m} = \{\alpha = (\alpha_1, \alpha_2, \ldots, \alpha_{l(\alpha)}) : \alpha_i \in \{k, k+1, \ldots, m\}, i = 1, 2, \ldots, l(\alpha), l(\alpha) \in \mathbb{N}\}.$$

A *hierarchical set* $Q \subset \mathscr{M}_{0,m}$ is any set of multiple indices $\alpha \in \mathscr{M}_{0,m}$ such that $\nu \in Q$ and $\alpha \in Q$ implies $-\alpha \in Q$. The hierarchical set Q_k denotes the set of all multiple indices $\alpha \in \mathscr{M}_{0,m}$ with length smaller than $k \in \mathbb{N}$, i.e.

$$Q_k = \{\alpha \in \mathscr{M}_{0,m} : l(\alpha) \leq k\}.$$

The set

$$R(Q) = \{\alpha \in \mathscr{M}_{0,m} \setminus Q : \alpha- \in Q\}$$

is called the *remainder set* $R(Q)$ of the hierarchical set Q. A *multiple (Itô) integral* $I_{\alpha,s,t}[V(.,.)]$ is defined to be

$$I_{\alpha,s,t}[V(.,.)] = \begin{cases} \int_s^t I_{-\alpha,s,u}[V(.,.)]\, dW^{\alpha_1}(u) \text{ if } & l(\alpha) > 1 \\ \int_s^t V(u, X_u)\, dW^{\alpha_{l(\alpha)}}(u) & \text{otherwise} \end{cases}$$

for a given process $V(t, X(t))$ where $V \in C^{0,0}$ and fixed $\alpha \in \mathscr{M}_{0,m} \setminus \{\nu\}$. A *multiple (Itô) coefficient function* $V_\alpha \in C^{0,0}$ for a given mapping $V = V(t,x) \in C^{l(\alpha),2l(\alpha)}$ is defined to be

$$V_\alpha(t,x) = \begin{cases} \mathscr{L}^{l(\alpha)} V_{\alpha-}(t,x) & \text{if} \quad l(\alpha) > 0 \\ V(t,x) & \text{otherwise} \end{cases}.$$

Similar notions can be introduced with respect to Stratonovich calculus (in fact, in general with respect to any stochastic calculus), see [72] for Itô and Stratonovich calculus.

1.1.4 Auxiliary Tool: Compact Formulation of Wagner-Platen Expansions

Now we are able to state a general form of Itô-Taylor expansions. Stochastic Taylor-type expansions for Itô diffusion processes have been introduced and studied by Wagner and Platen [144] (cf. also expansions in Sussmann [137], Arous [4], and Hu [54]). Stratonovich Taylor-type expansions can be found in Kloeden and Platen [72]. We will follow the original main idea of Wagner and Platen [144].

An Itô-Taylor expansion for an Itô SDE (1.2) is of the form

$$V(t, X(t)) = \sum_{\alpha \in Q} V_\alpha(s, X(s)) I_{\alpha,s,t} + \sum_{\alpha \in R(Q)} I_{\alpha,s,t}[V_\alpha(.,.)] \tag{1.8}$$

for a given mapping $V = V(t,x) : [0,T] \times \mathbb{R}^d \longrightarrow \mathbb{R}^k$ which is smooth enough, where $I_{\alpha,s,t}$ without the argument $[\cdot]$ is understood to be $I_{\alpha,s,t} = I_{\alpha,s,t}[1]$. Sometimes this formula is also referred to as *Wagner-Platen expansion*. Now, for completeness, let us restate the Theorem 5.1 of Kloeden and Platen [72].

Theorem 1.1.1 (Wagner-Platen Expansion). *Let ρ and τ be two $\mathcal{F}_t$-adapted stopping times with $t_0 \leq \rho \leq \tau \leq T < +\infty$ (a.s.). Assume $V : [0,T] \times \mathbb{R}^d \longrightarrow \mathbb{R}^k$. Take any hierarchical set $Q \in \mathcal{M}_{0,m}$.*

Then, each Itô SDE with coefficients a, b^j possesses a Itô-Taylor expansion (1.8) with respect to the hierarchical set Q, provided that all derivatives of V, a, b^j (related to Q) exist.

A proof is carried out in using the Itô formula and induction on the maximum length $\sup_{\alpha \in Q} l(\alpha) \in \mathbb{N}$. A similar expansion holds for Stratonovich SDEs.

1.1.5 Auxiliary Tool: Relations Between Multiple Integrals

The following lemma connects different multiple integrals. In particular, its formula can be used to express multiple integrals by other ones and to reduce the computational effort of their generation. The following lemma is a slightly generalized version of an auxiliary lemma taken from Kloeden and Platen [72], see proposition 5.2.3, p. 170.

Lemma 1.1.2 (Fundamental Lemma of Multiple Integrals). *Let $\alpha = (j_1, j_2, \ldots, j_{l(\alpha)}) \in \mathcal{M}_{0,m} \setminus \{v\}$ with $l(\alpha) \in \mathbb{N}$.*
Then, $\forall k \in \{0, 1, \ldots, m\}$ $\forall t, s : 0 \leq s \leq t \leq T$ we have

$$(W^k(t) - W^k(s)) I_{\alpha,s,t} = \sum_{i=0}^{l(\alpha)} I_{(j_1, j_2, \ldots, j_i, k, j_{i+1}, \ldots, j_{l(\alpha)}),s,t} \tag{1.9}$$

$$+ \sum_{i=0}^{l(\alpha)} \chi_{\{j_i = k \neq 0\}} I_{(j_1, j_2, \ldots, j_{i-1}, 0, j_{i+1}, \ldots, j_{l(\alpha)}),s,t}$$

$$= I_{(k, j_1, j_2, \ldots, j_{l(\alpha)}),s,t} + I_{(j_1, k, j_2, \ldots, j_{l(\alpha)}),s,t} + I_{(j_1, j_2, k, j_3, \ldots, j_{l(\alpha)}),s,t} + \ldots +$$

$$+ I_{(j_1, j_2, j_3, \ldots, j_{l(\alpha)}, k),s,t} + \sum_{i=0}^{l(\alpha)} \chi_{\{j_i = k \neq 0\}} I_{(j_1, j_2, \ldots, j_{i-1}, 0, j_{i+1}, \ldots, j_{l(\alpha)}),s,t}$$

where $\chi_{\{.\}}$ denotes the characteristic function of the subscribed set.

Hence, it obviously suffices to generate "minimal basis sets" of multiple integrals. In order to have a more complete picture on the structure of multiple integrals, we note the following assertion.

Lemma 1.1.3 (Hermite Polynomial Recursion of Multiple Integrals). *Assume that the multiple index α is of the form*

$$\alpha = (j_1, j_2, \ldots, j_{l(\alpha)}) \in \mathcal{M}_{0,m} \text{ with } j_1 = j_2 = \ldots = j_{l(\alpha)} = j \in 0, 1, \ldots, m$$

and its length $l(\alpha) \geq 2$.
Then, for all t with $t \geq s \geq 0$ we have

$$I_{\alpha,s,t} = \begin{cases} \dfrac{(t-s)^{l(\alpha)}}{l(\alpha)!}, & j = 0 \\ \dfrac{(W^j(t) - W^j(s))I_{\alpha-,s,t} - (t-s)I_{(\alpha-)-,s,t}}{l(\alpha)!}, & j \geq 1 \end{cases} \tag{1.10}$$

This lemma corresponds to a slightly generalized version of Corollary 5.2.4 (p. 171) in [72]. It is also interesting to note that this recursion formula for multiple Itô integrals coincides with the recursion formula for hermite polynomials. Let us conclude with a list of relations between multiple integrals which exhibit some consequences of Lemmas 1.1.2 and 1.1.3. For more details, see [72]. Take $j, k \in \{0, 1, \ldots, m\}$ and $0 \leq s \leq t \leq T$.

$$\begin{aligned} I_{(j),s,t} &= W^j(t) - W^j(s) \\ I_{(j,j),s,t} &= \frac{1}{2!}\left(I^2_{(j),s,t} - (t-s)\right) \\ I_{(j,j,j),s,t} &= \frac{1}{3!}\left(I^3_{(j),s,t} - 3(t-s)I_{(j),s,t}\right) \\ I_{(j,j,j,j),s,t} &= \frac{1}{4!}\left(I^4_{(j),s,t} - 6(t-s)I^2_{(j),s,t} + 3(t-s)^2\right) \\ I_{(j,j,j,j,j),s,t} &= \frac{1}{5!}\left(I^5_{(j),s,t} - 10(t-s)I^3_{(j),s,t} + 15(t-s)^2 I_{(j),s,t}\right) \\ & \ldots\ldots\ldots\ldots \\ (t-s)I_{(j),s,t} &= I_{(j,0),s,t} + I_{(0,j),s,t} \\ (t-s)I_{(j,k),s,t} &= I_{(j,k,0),s,t} + I_{(j,0,k),s,t} + I_{(0,j,k),s,t} \\ I_{(j),s,t}I_{(0,j),s,t} &= 2I_{(0,j,j),s,t} + I_{(j,0,j),s,t} + I_{(j,j,0),s,t} \\ I_{(j),s,t}I_{(j,0),s,t} &= I_{(0,j,j),s,t} + I_{(j,0,j),s,t} + 2I_{(j,j,0),s,t} \\ & \ldots\ldots\ldots\ldots. \end{aligned}$$

Some attempts has been made to approximate multiple stochastic integrals. For example, [72] use the technique of Karhunen-Loeve expansion (i.e. the Fourier series expansion of the Wiener process) or [30] exploit Box counting methods and related levy areas. A minimal basis set for multiple integrals is known, see [28, 29]. However, computationally more efficient approximation procedures of multiple stochastic integrals are still a challenge to be constructed and verified (especially in higher dimensions).

1.2 Local Consistency

Throughout this section, fix the time interval $[0, T]$ with finite and nonrandom terminal time T. Let $\|.\|_d$ be the Euclidean vector norm on $\mathbb{R}^d$ and $\mathcal{M}_p([s,t])$ the Banach space of $(\mathcal{F}_u)_{s\leq u\leq t}$-adapted, continuous, $\mathbb{R}^d$-valued stochastic processes X with finite norm $\|X\|_{\mathcal{M}_p} = (\sup_{s\leq u\leq t}\mathbb{E}||X(s)||_d^p)^{1/p} < +\infty$ where $p \geq 1$, $\mathcal{M}([0,s])$ the space of $(\mathcal{F}_s, \mathcal{B}(\mathbb{R}^d))$-measurable stochastic processes and $\mathcal{B}(S)$ the σ-algebra of Borel sets of inscribed set S.

Recall that every (one-step) numerical method Y (difference scheme) defined by

$$Y_{n+1} = Y_n + \Phi_n(Y)$$

with increment functional Φ_n has an associated continuous one-step representation

$$Y_{s,x}(t) = x + \Phi(t|s,x)$$

along partitions

$$0 = t_0 < t_1 < \ldots < t_n < t_{n+1} < \ldots < t_{n_T} = T < +\infty.$$

The continuity modulus of this one-step representation is the main subject of related consistency analysis. For this analysis, the auxiliary tools we presented in the introduction such as Itô formula and relations between multiple integrals are essential in deriving estimates of the one-step representation. For example, the **continuous time one-step representation** of stochastic Theta methods (1.1) is given by

$$\begin{aligned} Y_{s,x}(t) := {} & x + [\Theta a(t, Y_{s,x}(t)) + (I - \Theta)a(s,x)]\,(t-s) \\ & + \sum_{j=1}^{m} b^j(s,x)(W^j(t) \\ & - W^j(s)) + \sum_{j=0}^{m} c^j(s,x)(x - Y_{s,x}(t))|W^j(t) - W^j(s)| \end{aligned} \qquad (1.11)$$

driven by stochastic processes W^j, for all $t \geq s \geq 0$ and started at $x \in \mathbb{D}$ at time s.

Definition 1.2.1. A numerical method Y with one-step representation $Y_{s,y}(t)$ is said to be **mean consistent with rate** r_0 on $[0, T]$ iff $\exists$ Borel-measurable function $V : \mathbb{D} \to \mathbb{R}^1_+$ and $\exists$ real constants $K_0^C \geq 0, \delta_0 > 0$ such that $\forall (\mathcal{F}_s, \mathcal{B}(\mathbb{D}))$-measurable random variables $Z(s)$ with $Z \in \mathcal{M}([0,s])$ and $\forall s,t : 0 \leq t-s \leq \delta_0$

$$||\mathbb{E}[X_{s,Z(s)}(t) - Y_{s,Z(s)}(t)|\mathcal{F}_s]||_d \;\leq\; K_0^C\, V(Z(s))\,(t-s)^{r_0}. \qquad (1.12)$$

Remark 1.2.1. In the subsections below, we shall show that the balanced Theta methods (1.1) with uniformly bounded weights c^j and uniformly bounded parameters Θ_n are mean consistent with worst case rate $r_0 \geq 1.5$ and moment control function $V(x) = (1 + ||x||_d^2)^{1/2}$ for SDEs (1.2) with global Hölder-continuous and linear growth-bounded coefficients $b^j \in F \subset C^{1,2}([0,T] \times \mathbb{D})$ $(j = 0, 1, 2, \ldots, m)$.

Definition 1.2.2. A numerical method Y with one-step representation $Y_{s,y}(t)$ is said to be ***p*-th mean consistent with rate** r_2 on $[0,T]$ iff $\exists$ Borel-measurable function $V : \mathbb{D} \to \mathbb{R}_+^1$ and $\exists$ real constants $K_p^C \geq 0, \delta_0 > 0$ such that $\forall(\mathscr{F}_s, \mathscr{B}(\mathbb{D}))$-measurable random variables $Z(s)$ with $Z \in \mathscr{M}_p([0,s])$ and $\forall s,t$: $0 \leq t - s \leq \delta_0$

$$\left(\mathbb{E}[||X_{s,Z(s)}(t) - Y_{s,Z(s)}(t)||_d^p | \mathscr{F}_s]\right)^{1/p} \leq K_p^C \, V(Z(s)) \, (t-s)^{r_p}. \quad (1.13)$$

If $p = 2$ then we also speak of **mean square consistency** with local mean square rate r_2.

Remark 1.2.2. Below, we shall prove that the balanced Theta methods (1.1) are mean square consistent with worst case rate $r_2 \geq 1.0$ and moment control function $V(x) = (1 + ||x||^2)_d^{1/2}$ for SDEs (1.2) with global Lipschitz-continuous and linear growth-bounded coefficients $b^j \in F \subset C^{1,2}([0,T] \times \mathbb{D})$ $(j = 0, 1, 2, \ldots, m)$.

In the proofs of consistency of balanced Theta methods (1.1) below, it is crucial that one exploits the explicit identity

$$\begin{aligned} &Y_{s,x}(t) - x \\ &= M_{s,x}^{-1}(t)\,[\Theta a(t, Y_{s,x}(t)) - (I - \Theta) a(s,x)]\,(t-s) \\ &\quad + M_{s,x}^{-1}(t) \sum_{j=1}^{m} b^j(s,x)(W^j(t) - W^j(s)) \\ &= M_{s,x}^{-1}(t)\Theta\,[a(t, Y_{s,x}(t)) - a(s,x)] \int_s^t du + M_{s,x}^{-1}(t) \sum_{j=0}^{m} b^j(s,x) \int_s^t dW^j(u) \end{aligned}$$

where I is the $d \times d$ unit matrix in $\mathbb{R}^{d \times d}$, $b^0(s,x) = a(s,x)$, $W^0(t) = t$, $W^0(s) = s$ and

$$M_{s,x}(t) = I + \sum_{j=0}^{m} c^j(s,x)|W^j(t) - W^j(s)|.$$

1.2.1 Main Assumptions for Consistency Proofs

Let $||.||_{d \times d}$ denote a matrix norm on $\mathbb{R}^{d \times d}$ which is compatible to the Euclidean vector norm $||.||_d$ on $\mathbb{R}^d$, and $\langle .,. \rangle_d$ the Euclidean scalar product on $\mathbb{R}^d$.

Furthermore we have to assume that the coefficients a and b^j are Caratheodory functions such that a strong, unique solution $X = (X(t))_{0\le t\le T}$ exists. Recall that $\mathbb{D} \subseteq \mathbb{R}^d$ is supposed to be a nonrandom set. Let $\mathbb{D}$ be simply connected. To guarantee certain rates of consistency of the BTMs (and also its rates of convergence) the following conditions have to be satisfied:

(A0) $\forall s,t \in [0,T] : s < t \Longrightarrow \mathbb{P}\ (\{X(t) \in \mathbb{D} | X(s) \in \mathbb{D}\}) = \mathbb{P}\ (\{Y_{s,y}(t) \in \mathbb{D} | y \in \mathbb{D}\}) = 1$

(A1) $\exists$ constants $K_B = K_B(T), K_V = K_V(T) \ge 0$ such that

$$\forall t \in [0,T]\ \forall x \in \mathbb{D} : \sum_{j=0}^{m} ||b^j(t,x)||_d^2 \le (K_B)^2\,[V(x)]^2 \tag{1.14}$$

$$\sup_{0\le t\le T} \mathbb{E}[V(X(t))]^2 \le (K_V)^2 \mathbb{E}[V(X(0))]^2 < +\infty \tag{1.15}$$

with appropriate Borel-measurable function $V : \mathbb{D} \to \mathbb{R}^1_+$ satisfying

$$\forall x \in \mathbb{D} : ||x||_d \ \le\ V(x)$$

(A2) Hölder continuity of (a, b^j), i.e. $\exists$ real constants L_a and L_b such that

$$\forall s,t : 0 \le t-s \le \delta_0, \forall x,y \in \mathbb{D} : ||a(t,y) - a(s,x)||_d \le L_a(|t-s|^{1/2} + ||y-x||_d) \tag{1.16}$$

$$\sum_{j=1}^{m} ||b^j(t,y) - b^j(s,x)||_d^2 \le (L_b)^2(|t-s| + ||y-x||_d^2) \tag{1.17}$$

(A3) $\exists$ real constants $K_M = K_M(T) \ge 0$ such that, for the chosen weight matrices $c^j \in \mathbb{R}^{d\times d}$ of balanced Theta methods (1.1), we have

$$\forall s,t : 0 \le t-s \le \delta_0, \forall x \in \mathbb{D} : \exists M_{s,x}^{-1}(t) \text{ with } ||M_{s,x}^{-1}(t)||_{d\times d} \le K_M \tag{1.18}$$

(A4) $\exists$ real constants $K_{ca} = K_{ca}(T) \ge 0$ and $K_{cb} = K_{cb}(T) \ge 0$ such that, for the chosen weight matrices $c^j \in \mathbb{R}^{d\times d}$ of BTMs (1.1), we have

$$\forall s \in [0,T]\ \forall x \in \mathbb{D} : \sum_{j=0}^{m} ||c^j(s,x)a(s,x)||_d \ \le\ K_{ca}V(x) \tag{1.19}$$

$$\sum_{k=0}^{m}\sum_{j=0}^{m} ||c^k(s,x)b^j(s,x)||_d^2 \le K_{cb}^2[V(x)]^2 \tag{1.20}$$

(A5) $||\Theta||_{d\times d} \leq K_\Theta$, $||I-\Theta||_{d\times d} \leq K_{I-\Theta}$, and all step sizes $h_n \leq \delta_0$ are uniformly bounded by nonrandom quantity δ_0 such that

$$K_M K_B K_\Theta \delta_0 < 1.$$

Remark 1.2.3. Condition (A3) with uniform estimate (1.18) is guaranteed with the choice of positive semi-definite weight matrices c^j $(j = 0, 1, \ldots, m)$ in BTMs (1.1). In this case, we have $K_M \leq 1$. To control boundedness of moments and an appropriate constant K_M for invertible matrices M, it also suffices to take uniformly bounded weights c^0 and vanishing c^j for $j = 1, 2, \ldots, m$ together with sufficiently small step sizes h. Assumption (A5) ensures that the implicit expressions of Y are well-defined, together with the finiteness of some moments and Hölder-continuity (i.p. a guarantee of local resolution).

1.2.2 *Rate of Mean Consistency*

For simplicity, consider BTMs (1.1) with autonomous implicitness matrices $\Theta \in \mathbb{R}^{d\times d}$ (i.e. Θ is independent of time-variable n).

Theorem 1.2.1 (Mean Consistency of BTMs with Rate $r_0 \geq 1.5$). *Assume that the assumptions (A0)–(A5) are satisfied.*
Then, the BTMs (1.1) with autonomous implicitness matrices $\Theta \in \mathbb{R}^{d\times d}$ and nonrandom step sizes $h_n \leq \delta_0 < 1$ are mean consistent with worst case rate $r_0 \geq 1.5$.

Remark 1.2.4. The proof is based on auxiliary Lemmas 1.2.1 and 1.2.2 as stated and proved below.

Proof. First, rewrite the one-step representations of X and Y in integral form to

$$X_{s,x}(t) = x + \sum_{j=0}^{m} \int_s^t b^j(u, X(u))dW^j(u)$$

$$Y_{s,x}(t) = x + M_{s,x}^{-1}(t)[\Theta a(t, Y_{s,x}(t)) + (I-\Theta)a(s,x)] \int_s^t du$$
$$+ M_{s,x}^{-1}(t) b^j(s,x) \int_s^t dW^j(u).$$

Notice that

$$Y_{s,x}(t)$$
$$= x + \int_s^t a(s,x)du + (M_{s,x}^{-1}(t) - I)\int_s^t a(s,x)du + \sum_{j=1}^{m}\int_s^t b^j(s,x)dW^j(u)$$
$$+ M_{s,x}^{-1}(t)\Theta\int_s^t [a(t, Y_{s,x}(t)) - a(s,x)]du + \sum_{j=1}^{m}(M_{s,x}^{-1}(t) - I)\int_s^t b^j(s,x)dW^j(u).$$

Second, subtracting both representations gives

$$
\begin{aligned}
& X_{s,x}(t) - Y_{s,x}(t) \\
&= \int_s^t [a(u, X(u)) - a(s,x)]du + \sum_{j=1}^m \int_s^t [b^j(u, X(u)) - b^j(s,x)]dW^j(u) \\
&\quad + (M_{s,x}^{-1}(t) - I) \int_s^t a(s,x)du + M_{s,x}^{-1}(t)\Theta \int_s^t [a(t, Y_{s,x}(t)) - a(s,x)]du \\
&\quad + \sum_{j=1}^m (M_{s,x}^{-1}(t) - I) \int_s^t b^j(s,x)dW^j(u).
\end{aligned}
$$

Recall that the above involved stochastic integrals driven by W^j form martingales with vanishing first moment.
Third, pulling the expectation $\mathbb{E}$ over the latter identity and applying triangle inequality imply that

$$
\begin{aligned}
& ||\mathbb{E}[X_{s,x}(t) - Y_{s,x}(t)]||_d \\
&\leq \int_s^t \mathbb{E}||a(u, X(u)) - a(s,x)||_d du + \mathbb{E}||M_{s,x}^{-1}(t) - I||_d \int_s^t ||a(s,x)||_d du \\
&\quad + \mathbb{E}[||M_{s,x}^{-1}(t)\Theta||_d \int_s^t ||a(t, Y_{s,x}(t)) - a(s,x)||_d du] \\
&\quad + \sum_{j=1}^m \mathbb{E}[||M_{s,x}^{-1}(t) - I|| \cdot || \int_s^t b^j(s,x)dW^j(u)||_d] \\
&\leq L_a \int_s^t [|u-s|^{1/2} + (\mathbb{E}||X_{s,x}(u) - x||_d^2)^{1/2}]du \\
&\quad + K_M(t-s) \sum_{j=0}^m ||c^j(s,x)a(s,x)||_d \mathbb{E}[| \int_s^t dW^j(u)|] \\
&\quad + K_M(t-s)||\Theta||L_a[|t-s|^{1/2} + (\mathbb{E}||Y_{s,x}(t) - x||_d^2)^{1/2}].
\end{aligned}
$$

Note that we used the facts that

$$
M_{s,x}^{-1}(t) - I = -M_{s,x}^{-1}(t) \sum_{k=0}^m c^k(s,x) \left| \int_s^t dW^k(u) \right|
$$

and

$$
\mathbb{E}\left[\sum_{k=0}^m \sum_{j=1}^m M_{s,x}^{-1}(t) c^k(s,x) b^j(s,x) \left| \int_s^t dW^k(u) \right| \int_s^t dW^j(v) \right] = 0
$$

since all W^k are independent and symmetric about 0 (i.e. odd moments of them are vanishing to zero).
Fourth, we apply Lemma 1.2.2 in order to arrive at

$$
\begin{aligned}
||\mathbb{E}[X_{s,x}(t) - Y_{s,x}(t)]||_d &\leq L_a[1 + K_X^H V(x)] \int_s^t (u-s)^{1/2} du \\
&+ K_M(t-s) \sum_{j=0}^{m} ||c^j(s,x)a(s,x)||_d \left(\mathbb{E}\left[\,|\int_s^t dW^j(u)|^2 \right] \right)^{1/2} \\
&+ K_M K_\Theta L_a [1 + K_Y^H V(x)](t-s)^{3/2} \\
&\leq [\frac{2}{3} L_a + K_M(K_{ca} + K_\Theta L_a)](1 + K_X^H) V(x)(t-s)^{3/2} \\
&= O((t-s)^{3/2})
\end{aligned}
$$

under the assumptions (A0)–(A5). Consequently, this confirms the estimate of local rate $r_0 \geq 1.5$ of mean consistency of BTMs (1.1) with autonomous implicitness matrices $\Theta \in \mathbb{R}^{d \times d}$ along any nonrandom partitions of fixed time-intervals $[0, T]$ with step sizes $h < 1$, and hence the conclusion of Theorem 1.2.1 is verified. ◇

Remark 1.2.5. Moreover, returning to the proof and extracted from its final estimation process, the leading mean error constant K_0^C can be estimated by

$$K_0^C \leq [\frac{2}{3} L_a + K_M(K_{ca} + K_\Theta L_a)](1 + K_X^H)$$

where K_X^H is as in (1.23) (see also Remark 1.2.7). For uniformly Lipschitz-continuous coefficients b^j $(j = 0, 1, \ldots, m)$, the functional V of consistency is taken as

$$V(x) = (1 + ||x||_d^2)^{1/2}$$

which represents the functional of linear polynomial growth (as it is common in the case with globally Hölder-continuous coefficients b^j).

Lemma 1.2.1 (Local uniform boundedness of L^2-norms of X and Y). *Assume that*

$$\forall x \in \mathbb{D} : ||x||_d^2 \;\leq\; [V(x)]^2 \;\leq\; 1 + ||x||_d^2,$$

the assumptions (A1) and (A3) are satisfied, and both X and Y are $\mathbb{D}$-invariant for deterministic set $\mathbb{D} \subseteq \mathbb{R}^d$ (i.e. assumption (A0)). Furthermore, for well-definedness of Y, let (A5) hold, i.e.

$$K_M K_B K_\Theta \delta_0 < 1$$

with $||\Theta||_{d\times d} \le K_\Theta$ *and* $||I-\Theta||_{d\times d} \le K_{I-\Theta}$.
Then $\exists K_X^B, K_Y^B$ *(constants)* $\forall x \in \mathbb{D}\ \forall T \ge t \ge s \ge 0$ *with* $|t-s| \le \delta_0 < 1$

$$\mathbb{E}[||X_{s,x}(t)||_d^2] \le (K_X^B)^2[V(x)]^2 \tag{1.21}$$

$$\mathbb{E}[||Y_{s,x}(t)||_d^2] \le (K_Y^B)^2[V(x)]^2, \tag{1.22}$$

i.e. we may take $V(x) = (1+||x||_d^2)^{1/2}$ *in the estimates of local uniform boundedness of 2nd moments.*

Proof of Lemma 1.2.1. Let $t, s \ge 0$ such that $|t-s| \le 1$ and $x \in \mathbb{D} \subseteq \mathbb{R}^d$ for $\mathbb{D}$-invariant stochastic processes X and Y ($\mathbb{P}$ -a.s.).
First, consider the estimate for X. Recall the property of Itô isometry of stochastic integrals and independence of all processes W^j. We arrive at (by CBS- or Hölder-inequality)

$$\begin{aligned}
&\mathbb{E}||X_{s,x}(t)||_d^2\\
&= \mathbb{E}\left\| x + \int_s^t a(u,X(u))du + \sum_{j=1}^m \int_s^t b^j(u,X(u))dW^j(u)\right\|_d^2\\
&\le 3||x||_d^2 + 3\mathbb{E}\left\|\int_s^t a(u,X(u))du\right\|_d^2 + 3\mathbb{E}\left\|\sum_{j=1}^m \int_s^t b^j(u,X(u))dW^j(u)\right\|_d^2\\
&\le 3||x||_d^2 + 3(t-s)\mathbb{E}\left[\int_s^t ||a(u,X(u))||_d^2 du\right] + 3\mathbb{E}\left[\sum_{j=1}^m \int_s^t ||b^j(u,X(u))||_d^2 du\right]\\
&\le 3||x||_d^2 + 3\mathbb{E}\int_s^t \left[\sum_{j=0}^m ||b^j(u,X(u))||_d^2\right] du\\
&\le 3||x||_d^2 + 3(K_B)^2 \int_s^t \mathbb{E}[V(X(u))]^2 du\\
&\le 3||x||_d^2 + 3(K_B)^2(K_V)^2[V(x)]^2(t-s)\\
&\le 3[1+(K_B)^2(K_V)^2(t-s)][V(x)]^2
\end{aligned}$$

whenever $0 \le s \le t \le T$ are nonrandom and $|t-s| \le 1$. Hence, the uniform boundedness of second moments of X could be established.
Second, consider a similar estimation for Y. We obtain (apply CBS-inequality)

$$\begin{aligned}
&\mathbb{E}||Y_{s,x}(t)||_d^2\\
&= \mathbb{E}\Big\| x + M_{s,x}^{-1}\Big(\Theta a(t, Y_{s,x}(t)) + (I-\Theta)a(s,x)\Big)(t-s)
\end{aligned}$$

$$
\begin{aligned}
&+ \sum_{j=1}^{m} \int_s^t M_{s,x}^{-1}(t) b^j(s,x) dW^j(u) \Bigg\|_d^2 \\
&\leq 4||x||_d^2 + 4K_M^2 K_\Theta^2 \mathbb{E}||a(t, Y_{s,x}(t))||_d^2 (t-s)^2 \\
&\quad + 4K_M^2 K_{I-\Theta}^2 ||a(s,x)||_d^2 (t-s)^2 + 4K_M^2 \sum_{j=1}^{m} ||b^j(s,x)||_d^2 (t-s) \\
&\leq 4[V(x)]^2 + 4K_M^2 K_\Theta^2 (K_B)^2 \mathbb{E}[V(Y_{s,x}(t))]^2 (t-s)^2 \\
&\quad + 4K_M^2 (1 + K_{I-\Theta}^2) \sum_{j=1}^{m} ||b^j(s,x)||_d^2 (t-s) \\
&\leq 4[V(x)]^2 + 4K_M^2 K_\Theta^2 (K_B)^2 \mathbb{E}[V(Y_{s,x}(t))]^2 (t-s)^2 \\
&\quad + 4K_M^2 (1 + K_{I-\Theta}^2)(K_B)^2 [V(x)]^2 (t-s) \\
&\leq 4 \frac{1 + K_M^2 (K_B)^2 K_\Theta^2 (t-s)^2 + 1 + K_{I-\Theta}^2 (t-s)}{1 - K_M^2 (K_B)^2 K_\Theta^2 (\delta_0)^2} [V(x)]^2 \\
&\leq 4 \frac{1 + K_M^2 (K_B)^2 K_\Theta^2 (\delta_0)^2 + 1 + K_{I-\Theta}^2 \delta_0}{1 - K_M^2 (K_B)^2 K_\Theta^2 (\delta_0)^2} [V(x)]^2,
\end{aligned}
$$

where $[V(x)]^2 \leq 1 + ||x||_d^2$ and $K_M^2 (K_B)^2 K_\Theta^2 (\delta_0)^2 < 1$ are additionally supposed in the latter estimation. Therefore, the uniform boundedness of second moments of Y is verified. ◇

Remark 1.2.6. Indeed, while returning to the proof of Lemma 1.2.1, under the assumptions (A0)–(A1), (A3) and (A5) with $V(x) = (1 + ||x||_d^2)^{1/2}$ we find that K_X^B can be estimated by

$$
K_X^B \leq 3[1 + (K_B)^2 (K_V)^2].
$$

as long as $|t - s| \leq 1$. Similarly, K_Y^B is bounded by

$$
K_Y^B \leq 4 \frac{1 + K_M^2 (K_B)^2 K_\Theta^2 (\delta_0)^2 + 1 + K_{I-\Theta}^2 \delta_0}{1 - K_M^2 (K_B)^2 K_\Theta^2 (\delta_0)^2}
$$

whenever $K_M^2 (K_B)^2 K_\Theta^2 (\delta_0)^2 < 1$ and $|t - s| \leq \delta_0 \leq 1$, $||x||_d^2 \leq [V(x)]^2 \leq 1 + ||x||_d^2$ for all $x \in \mathbb{D}$.

Lemma 1.2.2 (Local mean square Hölder continuity of X and Y). *Assume that*

$$
\forall x \in \mathbb{D} : ||x||_d^2 \leq [V(x)]^2 \leq 1 + ||x||_d^2,
$$

the assumptions (A1)–(A3) and (A5) are satisfied, and both X and Y are $\mathbb{D}$-invariant for deterministic set $\mathbb{D} \subseteq \mathbb{R}^d$ (i.e. assumption (A0)).
Then $\exists K_X^H, K_Y^H$ (constants) $\forall T \geq t \geq s \geq 0$ with $|t-s| < 1$

$$\mathbb{E}[||X_{s,x}(t) - x||_d^2] \leq (K_X^H)^2 [V(x)]^2 (t-s) \tag{1.23}$$

$$\mathbb{E}[||Y_{s,x}(t) - x||_d^2] \leq (K_Y^H)^2 [V(x)]^2 (t-s) \tag{1.24}$$

for all nonrandom $x \in \mathbb{D} \subseteq \mathbb{R}^d$, i.e. we may take $V(x) \leq (1 + ||x||_d^2)^{1/2}$ in the estimates of Hölder-continuity modulus of 2nd moments.

Proof of Lemma 1.2.2. Let $t, s \geq 0$ such that $|t-s| \leq 1$ and $x \in \mathbb{D} \subseteq \mathbb{R}^d$ for $\mathbb{D}$-invariant stochastic processes X and Y ($\mathbb{P}$ -a.s.).
First, consider the estimate for X. Recall the property of Itô isometry of stochastic integrals and independence of all processes W^j. We arrive at

$$\begin{aligned}
\mathbb{E}||X_{s,x}(t) - x||_d^2 &= \mathbb{E}\left\| \int_s^t a(u, X(u))du + \sum_{j=1}^m \int_s^t b^j(u, X(u))dW^j(u) \right\|_d^2 \\
&= \mathbb{E}\left\| \int_s^t a(u, X(u))du \right\|^2 + \sum_{j=1}^m \mathbb{E}\left\| \int_s^t b^j(u, X(u))dW^j(u) \right\|_d^2 \\
&\leq \int_s^t \mathbb{E}\left[(t-s)||a(u, X(u))||_d^2 + \sum_{j=1}^m ||b^j(u, X(u))||_d^2 \right] du \\
&\leq \int_s^t \mathbb{E}\left[\sum_{j=0}^m ||b^j(u, X(u))||_d^2 \right] du \leq (K_B)^2 \int_s^t \mathbb{E}[V(X(u))]^2 du \\
&\leq (K_B)^2 (K_V)^2 \mathbb{E}[V(X_{s,x}(s))]^2 |t-s| = (K_B)^2 (K_V)^2 [V(x)]^2 (t-s)
\end{aligned}$$

whenever $0 \leq s \leq t \leq T$ are nonrandom and $|t-s| \leq 1$.
Second, consider a similar estimation for Y. We obtain (apply CBS-inequality)

$$\begin{aligned}
&\mathbb{E}||Y_{s,x}(t) - x||_d^2 \\
&= \mathbb{E}\Bigg\| M_{s,x}^{-1}\Big(\Theta a(t, Y_{s,x}(t)) + (I - \Theta)a(s, x)\Big)(t-s) \\
&\quad + \sum_{j=1}^m \int_s^t M_{s,x}^{-1}(t) b^j(s, x) dW^j(u) \Bigg\|_d^2 \\
&\leq 3K_M^2 K_\Theta^2 \mathbb{E}||a(t, Y_{s,x}(t))||_d^2 (t-s)^2 + 3K_M^2 K_{I-\Theta}^2 ||a(s, x)||_d^2 (t-s)^2
\end{aligned}$$

$$+3K_M^2\sum_{j=1}^{m}||b^j(s,x)||_d^2(t-s)$$
$$\leq 3K_M^2(K_B)^2(K_\Theta^2\mathbb{E}[V(Y_{s,x}(t))]^2$$
$$+K_{I-\Theta}^2[V(x)]^2)(t-s)^2+3K_M^2(K_B)^2[V(x)]^2(t-s)$$
$$\leq 3K_M^2(K_B)^2[K_\Theta^2(K_Y^B)^2+K_{I-\Theta}^2+1][V(x)]^2(t-s)$$

whenever $0 \leq s \leq t \leq T$ are nonrandom and $|t-s| \leq 1$. This completes the proof of Lemma 1.2.2. ◇

Remark 1.2.7. Indeed, while returning to the proof of Lemma 1.2.2, under the assumptions (A0)–(A5) with $V(x) = (1+||x||_d^2)^{1/2}$ we find that K_X^H can be estimated by

$$K_X^H \leq (K_B)^2(K_V)^2.$$

Similarly, K_Y^H is bounded by

$$K_Y^H \leq 3K_M^2(K_B)^2[K_\Theta^2(K_Y^B)^2+K_{I-\Theta}^2+1].$$

1.2.3 Mean Square Consistency

For simplicity, consider BTMs (1.1) with autonomous implicitness matrices $\Theta \in \mathbb{R}^{d\times d}$.

Theorem 1.2.2 (Mean Square Consistency of BTMs with Rate $r_2 \geq 1.0$). *Assume that the assumptions (A0) - (A5) are satisfied, together with*

$$\forall x \in \mathbb{D} : V(x) = (1+||x||_d^2)^{1/2}.$$

Then, the BTMs (1.1) with autonomous implicitness matrices $\Theta \in \mathbb{R}^{d\times d}$ and nonrandom step sizes $h_n \leq \delta_0 < 1$ are mean square consistent with worst case rate $r_2 \geq 1.0$ along V on $\mathbb{D} \subseteq \mathbb{R}^d$.

Remark 1.2.8. The proof is based on auxiliary Lemmas 1.2.1 and 1.2.2 as stated in previous subsection.

Proof. First, from the proof of Theorem 1.2.1, recall the difference of one-step representations of X and Y in integral form is given by

$$X_{s,x}(t)-Y_{s,x}(t)$$
$$=\int_s^t [a(u,X(u))-a(s,x)]du+\sum_{j=1}^{m}\int_s^t [b^j(u,X(u))-b^j(s,x)]dW^j(u)$$

$$+ (M_{s,x}^{-1}(t) - I) \int_s^t a(s,x)du + M_{s,x}^{-1}(t)\Theta \int_s^t [a(t, Y_{s,x}(t)) - a(s,x)]du$$
$$+ \sum_{j=1}^{m} (M_{s,x}^{-1}(t) - I) \int_s^t b^j(s,x)dW^j(u).$$

Second, take the square norm and apply Hölder-inequality (CBS-inequality) in order to encounter the estimation

$$\begin{aligned}
&\mathbb{E}||X_{s,x}(t) - Y_{s,x}(t)||_d^2 \\
&= \mathbb{E}\Bigg\| \int_s^t [a(u, X(u)) - a(s,x)]du + \sum_{j=1}^{m} \int_s^t [b^j(u, X(u)) - b^j(s,x)]dW^j(u) \\
&\qquad + (M_{s,x}^{-1}(t) - I) \int_s^t a(s,x)du + M_{s,x}^{-1}(t)\Theta \int_s^t [a(t, Y_{s,x}(t)) - a(s,x)]du \\
&\qquad + \sum_{j=1}^{m} (M_{s,x}^{-1}(t) - I) \int_s^t b^j(s,x)dW^j(u)\Bigg\|_d^2 \\
&\leq 5\mathbb{E}\left\| \int_s^t [a(u, X(u)) - a(s,x)]du \right\|_d^2 + 5\sum_{j=1}^{m} \mathbb{E}\left\| \int_s^t [b^j(u, X(u)) - b^j(s,x)]dW^j(u) \right\|_d^2 \\
&\quad + 5\mathbb{E}\left\| (M_{s,x}^{-1}(t) - I) \int_s^t a(s,x)du \right\|_d^2 + 5\mathbb{E}\left\| M_{s,x}^{-1}(t)\Theta \int_s^t [a(t, Y_{s,x}(t)) - a(s,x)]du \right\|_d^2 \\
&\quad + 5\sum_{j=1}^{m} \mathbb{E}\left\| (M_{s,x}^{-1}(t) - I) \int_s^t b^j(s,x)dW^j(u) \right\|_d^2 .
\end{aligned}$$

Third, recall Itô isometry and the facts that

$$M_{s,x}^{-1}(t) - I = -M_{s,x}^{-1}(t) \sum_{k=0}^{m} c^k(s,x) \left| \int_s^t dW^k(u) \right|$$

and

$$\mathbb{E}\left[\sum_{k=0}^{m} \sum_{j=1}^{m} M_{s,x}^{-1}(t) c^k(s,x) b^j(s,x) \left| \int_s^t dW^k(u) \right| \int_s^t dW^j(v) \right] = 0$$

since all W^k are independent and symmetric about 0 (i.e. odd moments of them are vanishing to zero). Therefore, we may simplify our latter estimation to

$$
\begin{aligned}
&\mathbb{E}||X_{s,x}(t) - Y_{s,x}(t)||_d^2 \\
&\leq 5(t-s)\mathbb{E}\left[\int_s^t ||a(u,X(u)) - a(s,x)||_d^2 du\right] \\
&\quad +5\sum_{j=1}^{m}\mathbb{E}\left[\int_s^t ||b^j(u,X(u)) - b^j(s,x)||_d^2 du\right] \\
&\quad +5\mathbb{E}\left[\left\|M_{s,x}^{-1}(t)\sum_{k=0}^{m} c^k(s,x)a(s,x)\right\|_d^2 \left|\int_s^t dW^k(u)\right|\right](t-s) \\
&\quad +5\mathbb{E}\left[||M_{s,x}^{-1}(t)\Theta||_{d\times d}^2 \int_s^t ||a(t,Y_{s,x}(t)) - a(s,x)||_d^2 du\right](t-s) \\
&\quad +5\sum_{j=1}^{m}\sum_{k=0}^{m}\mathbb{E}\left\|M_{s,x}^{-1}(t)\left|\int_s^t dW^k(u)\right|\int_s^t c^k(s,x)b^j(s,x)dW^j(u)\right\|_d^2 \\
&\leq 10L_a^2(t-s)\int_s^t [(u-s) + \mathbb{E}||X_{s,x}(u) - x||_d^2]du + 5L_b^2\int_s^t [(u-s) + \mathbb{E}||X_{s,x}(u) - x||_d^2]du \\
&\quad +5K_M^2(K_{ca})^2[V(x)]^2(t-s)^2 \\
&\quad +10K_M^2K_\Theta^2L_a^2\int_s^t [(t-s) + \mathbb{E}||Y_{s,x}(t) - x||_d^2]du(t-s) \\
&\quad +5K_M^2(K_{cb})^2[V(x)]^2(t-s)^2 \\
&\leq 10L_a^2(t-s)\int_s^t [1 + K_X^H[V(x)]^2](u-s)du + 5L_b^2\int_s^t [1 + K_X^H[V(x)]^2](u-s)du \\
&\quad +5K_M^2(K_{ca})^2[V(x)]^2(t-s)^2 \\
&\quad +10K_M^2K_\Theta^2L_a^2\int_s^t [2 + K_Y^H[V(x)]^2](u-s)du(t-s) \\
&\quad +5K_M^2(K_{cb})^2[V(x)]^2(t-s)^2 \\
&\leq 5L_a^2[1 + K_X^H[V(x)]^2](t-s)^3 + \frac{5}{2}L_b^2[1 + K_X^H[V(x)]^2](t-s)^2 \\
&\quad +5K_M^2(K_{ca})^2[V(x)]^2(t-s)^2 \\
&\quad +10K_M^2K_\Theta^2L_a^2[1 + K_Y^H[V(x)]^2](t-s)^3 \\
&\quad +5K_M^2(K_{cb})^2[V(x)]^2(t-s)^2 \\
&\leq 5\left[(L_a^2 + \frac{1}{2}L_b^2)(1 + K_X^H) + K_M^2[(K_{ca})^2 + 2K_\Theta^2L_a^2(1 + K_Y^H) + (K_{cb})^2]\right][V(x)]^2(t-s)^2
\end{aligned}
$$

whenever $|t-s| \leq \delta_0 < 1$ and $[V(x)]^2 = 1 + ||x||_d^2$. Thus, the rate of mean square consistency of BTMs (1.1) can be estimated by $r_2 \geq 1.0$. This completes the proof of Theorem 1.2.2. ◇

Remark 1.2.9. Indeed, while returning to the proof of Theorem 1.2.2, we may extract an estimate for the leading mean square consistency coefficient K_2^C by setting

$$K_2^C = \sqrt{5\Big[(L_a^2 + \frac{1}{2}L_b^2)(1 + K_X^H) + K_M^2[(K_{ca})^2 + 2K_\Theta^2 L_a^2(1 + K_Y^H) + (K_{cb})^2]\Big]}$$

along functional $V(x) = (1 + ||x||_d^2)^{1/2}$ under the assumptions (A0)–(A5).

1.2.4 P-th Mean Consistency

Similarly, to the previous proof of mean square consistency, one can verify estimates on the rates r_p of p-th mean consistency with $p \geq 1$.

Theorem 1.2.3 (p-th mean consistency rate $r_p \geq 1.0$ of BTMs (1.1)). *Assume that the assumptions (A0)–(A5) with $V(x) = (1 + ||x||_d^p)^{1/p}$ on $\mathbb{D} \subseteq \mathbb{R}^d$ are satisfied with general exponent $p \geq 1$ instead of $p = 2$.*
Then, the BTMs (1.1) with autonomous implicitness matrices $\Theta \in \mathbb{R}^{d \times d}$ and nonrandom step sizes $h_n \leq \delta_0 < 1$ are p-th mean consistent with worst case rate $r_p \geq 1.0$.

The details of the proof are left to the reader (just apply analogous steps as in proof of Theorem 1.2.2, exploiting Hölder-type and Burkholder inequalities together with Itô isometry relation and the symmetry of Gaussian moments. For the subcase $1 \leq p < 2$, one may also apply Lyapunov inequality (see Shiryaev [134]) to conclude $r_p = 1.0$. For p-th mean consistency of Euler methods, see also Kanagawa [61–65]. In fact, the error distribution caused by Euler approximations is studied in Bally et al. [6–9], in Jacod & Protter [58, 59] and in Kohatsu-Higa & Protter [77].

1.3 Asymptotic Stability

Asymptotic stability and long-term behavior of numerical methods has been investigated by numerous authors, see [1, 2, 10, 16–19, 45–47, 53, 73, 78–80, 87, 88, 97, 107, 110, 111, 113, 114, 117].

For the sake of simplicity, we will mostly separate our stability analysis applied to purely **drift-implicit BTMs**

$$Y_{n+1} = Y_n + [\theta a(t_{n+1}, Y_{n+1}) + (1-\theta)a(t_n, Y_n)]h_n + \sum_{j=1}^{m} b^j(t_n, Y_n)\Delta W_n^j \quad (1.25)$$

with nonrandom scalar implicitness $\theta \in [0, 1]$ and **linear-implicit BIMs**

$$Y_{n+1} = Y_n + \sum_{j=0}^{m} b^j(t_n, Y_n)\Delta W_n^j + \sum_{j=0}^{m} c^j(t_n, Y_n)(Y_n - Y_{n+1})|\Delta W_n^j| \quad (1.26)$$

with appropriate weight matrices c^j.

1.3.1 Weak *V*-Stability of Numerical Methods

Definition 1.3.1. A numerical method Y with one-step representation $Y_{s,y}(t)$ is said to be **weakly** V**-stable** with real constant $K_S = K_S(T)$ on $[0,T]$ iff $V:\mathbb{R}^d \to \mathbb{R}^1_+$ is Borel-measurable and $\exists$ real constant $\delta_0 > 0$ such that $\forall(\mathscr{F}_s, \mathscr{B}(\mathbb{R}^d))$-measurable random variables $Z(s)$ and $\forall s,t : 0 \le t - s \le \delta_0 \le 1$

$$\mathbb{E}[V(Y_{s,Z(s)}(t))|\mathscr{F}_s] \le \exp(K_S(t-s))\, V(Z(s)). \quad (1.27)$$

A weakly V-stable numerical method Y is called **weakly exponential** V**-stable** iff $K_S < 0$. A numerical method Y is **exponential** p**-th mean stable** iff Y is weakly exponential V-stable with $V(x) = ||x||_d^p$ and $K_S < 0$. In the case $p = 2$, we speak of **exponential mean square stability**.

Remark 1.3.1. Exponential stability is understood with respect to the trivial solution (equilibrium or fixed point 0) throughout this paper.

Theorem 1.3.1. *Assume that the numerical method Y started at a $(\mathscr{F}_0, \mathscr{B}(\mathbb{R}^d))$-measurable Y_0 and constructed along any $(\mathscr{F}_t)$-adapted time-discretization of $[0,T]$ with maximum step size $h_{max} \le \delta_0$ is weakly V-stable with δ_0 and stability constant K_S on $[0,T]$. Then*

$$\forall t \in [0,T]: \quad \mathbb{E}[V(Y_{0,Y_0}(t))] \le \exp(K_S t)\mathbb{E}[V(Y_0)], \quad (1.28)$$

$$\sup_{0\le t\le T} \mathbb{E}[V(Y_{0,Y_0}(t))] \le \exp([K_S]_+ T)\mathbb{E}[V(Y_0)] \quad (1.29)$$

where $[.]_+$ denotes the positive part of the inscribed expression.

Proof. Suppose that $t_k \le t \le t_{k+1}$ with $h_k \le \delta_0$. If $\mathbb{E}[V(Y_0)] = +\infty$ then nothing is to prove. Now, suppose that $\mathbb{E}[V(Y_0)] < +\infty$. Using elementary properties of conditional expectations, we estimate

$$\begin{aligned}
\mathbb{E}[V(Y_{0,Y_0}(t))] &= \mathbb{E}[\mathbb{E}[V(Y_{t_k,Y_k}(t)|\mathscr{F}_{t_k}]] \\
&\le \exp(K_S(t-t_k)) \cdot \mathbb{E}[V(Y_k)] \\
&= \exp(K_S(t-t_k)) \cdot \mathbb{E}[V(Y_{t_{k-1},Y_{k-1}}(t_k))] \le \dots \\
&\le \exp(K_S t) \cdot \mathbb{E}[V(Y_0)] \le \exp([K_S]_+ t) \cdot \mathbb{E}[V(Y_0)]
\end{aligned}$$

$$\leq \exp([K_S]_+ T) \cdot \mathbb{E}[V(Y_0)]$$

by induction. Hence, taking the supremum confirms the claim of Theorem 1.3.1. $\diamond$

Remark. Usually V plays the role of a Lyapunov functional for controlling the stability of the numerical method Y.

Theorem 1.3.2. *Assume that $(A1)$ and $(A3)$ with $V(x) = \rho^2 + ||x||_d^2$ ($\rho \in \mathbb{R}^1$ some real constant) hold. Then the BIMs (1.26) with $h_{max} \leq \delta_0 \leq \min(1,T)$ are weakly V-stable with stability constant*

$$K_S \ \leq \ K_M \cdot K_B \cdot (2 + K_M \cdot K_B) \tag{1.30}$$

and they satisfy global weak V-stability estimates (1.28) and (1.29).

Proof. Suppose that $(A1)$ and $(A3)$ hold with $V(x) = \rho^2 + ||x||^2$. Recall that $0 \leq t - s \leq \delta_0 \leq 1$. Let $Z(s)$ be any $(\mathscr{F}_s, \mathscr{B}(\mathbb{R}^d))$-measurable random variable. Then

$$\begin{aligned}
\mathbb{E}[\rho^2 + ||Y_{s,Z(s)}||_d^2|\mathscr{F}_s] &= \mathbb{E}[\rho^2 + ||Z(s) + M_{s,Z(s)}^{-1}(t)\sum_{j=0}^{m} b^j(s,Z(s))(W^j(t) \\
&\quad - W^j(s))||_d^2|\mathscr{F}_s] \\
&= \mathbb{E}[\rho^2 + ||Z(s) + M_{s,Z(s)}^{-1}(t)a(s,Z(s))(t-s) \\
&\quad + M_{s,Z(s)}^{-1}(t)\sum_{j=1}^{m} b^j(s,Z(s))(W^j(t) - W^j(s))||_d^2|\mathscr{F}_s] \\
&= \rho^2 + \frac{1}{2}\mathbb{E}[||z + M_{s,z}^{-1}(t)a(s,z)(t-s) \\
&\quad + M_{s,z}^{-1}(t)\sum_{j=1}^{m} b^j(s,z)(W^j(t) - W^j(s))||_d^2]\Big|_{z=Z(s)} \\
&\quad + \frac{1}{2}\mathbb{E}[||z + M_{s,z}^{-1}(t)a(s,z)(t-s) \\
&\quad - M_{s,z}^{-1}(t)\sum_{j=1}^{m} b^j(s,z)(W^j(t) - W^j(s))||_d^2]\Big|_{z=Z(s)} \\
&= \rho^2 + \mathbb{E}[||z + M_{s,z}^{-1}(t)a(s,z)(t-s)||_d^2]\Big|_{z=Z(s)} \\
&\quad + \mathbb{E}[||M_{s,z}^{-1}(t)\sum_{j=1}^{m} b^j(s,z)(W^j(t) - W^j(s))||_d^2]\Big|_{z=Z(s)} \\
&= \rho^2 + ||Z(s)||_d^2 + 2\,[\mathbb{E}<z, M_{s,z}^{-1}(t)a(s,z)>_d]\Big|_{z=Z(s)}(t-s) \\
&\quad + \mathbb{E}[||M_{s,z}^{-1}(t)a(s,z)||_d^2]\Big|_{z=Z(s)}(t-s)^2
\end{aligned}$$

$$+\sum_{j=1}^{m} \mathbb{E}[||M_{s,z}^{-1}(t)b^j(s,z)||_d^2 (W^j(t)-W^j(s))^2]\big|_{z=Z(s)}$$
$$\leq (1+[2K_M K_B + K_M^2 K_B^2](t-s)) \cdot (\rho^2 + ||Z(s)||_d^2)$$
$$\leq \exp([2K_M K_B + K_M^2 K_B^2](t-s)) \cdot (\rho^2 + ||Z(s)||_d^2),$$

hence the BIMs (1.26) are weakly V-stable with $V(x) = \rho^2 + ||x||_d^2$. It obviously remains to apply Theorem 1.3.1 in order to complete the proof. ◇

Remark 1.3.2. Interestingly, by setting $\rho = 0$, we gain also a result on numerical mean square stability. However, for results on asymptotic mean square stability of BTMs and BIMs, see below or [120].

1.3.2 Asymptotic Mean Square Stability for Bilinear Test Equations

Consider (autonomous) linear systems of Itô SDEs with stationary solution $X_\infty = 0$

$$dX(t) \;=\; AX(t)\,dt + \sum_{j=1}^{m} B^j X(t) dW^j(t) \tag{1.31}$$

started in $(\mathscr{F}_0, \mathscr{B}(\mathbb{R}^d))$-measurable initial data $X(0)$ at time $t = 0$. Let $p \in (0, +\infty)$ be a nonrandom constant.

Definition 1.3.2. Assume $X \equiv 0$ is an equilibrium (fixed point, steady state). The trivial solution $X \equiv 0$ of SDE (1.2) is called **globally (asymptotically) p-th mean stable** iff $\forall X_0 : \mathbb{E}\|X(0)\|_d^p < +\infty \Longrightarrow \lim_{t\to+\infty} \mathbb{E}\|X(t)\|_d^p = 0$. In the case $p = 2$ we speak of **global (asymptotic) mean square stability**. Moreover, the trivial solution $X \equiv 0$ of SDE (1.2) is said to be **locally (asymptotically) p-th mean stable** iff $\forall \varepsilon > 0 \exists \delta : \forall X_0 : \mathbb{E}\|X(0)\|_d^p < \delta \Longrightarrow \forall t > 0 : \mathbb{E}\|X(t)\|_d^p < \varepsilon$. In the case $p = 2$ we speak of **local (asymptotic) mean square stability**.

Remark 1.3.3. To recall some well-known facts from general stability theory, an exponentially p-th mean stable process is also asymptotically p-th mean stable. However, not vice versa in general. Furthermore, a globally (asymptotically) p-th mean stable trivial solution is also locally (asymptotically) p-th mean stable. For nonlinear systems of SDEs (1.2), the concepts of local and global stability do not coincide. For linear autonomous systems, these concepts coincide.

Let Y_0 be $(\mathscr{F}_0, \mathscr{B}(\mathbb{R}^d))$-measurable in all statements in what follows.

Definition 1.3.3. Assume $Y \equiv 0$ is an equilibrium (fixed point, steady state). The trivial solution $Y \equiv 0$ of BTMs (1.1) is called **globally (asymptotically) p-th mean stable** iff $\forall Y_0 : \mathbb{E}\|Y_0\|_d^p < +\infty \Longrightarrow \lim_{n\to+\infty} \mathbb{E}\|Y_n\|_d^p = 0$. In the case $p = 2$

we speak of **global (asymptotic) mean square stability** of Y. Moreover, the trivial solution $Y \equiv 0$ of BTMs (1.1) is said to be **locally (asymptotically) p-th mean stable** iff $\forall \varepsilon > 0 \exists \delta : \forall Y_0 : \mathbb{E}\|Y_0\|_d^p < \delta \implies \forall n > 0 : \mathbb{E}\|Y_n\|_d^p < \varepsilon$. In the case $p = 2$ we speak of **local (asymptotic) mean square stability** of Y.

Now, consider the family of drift-implicit Theta methods

$$Y_{n+1} = Y_n + (\theta A Y_{n+1} + (1-\theta) A Y_n) h_n + \sum_{j=1}^{m} B^j Y_n \Delta W_n^j \tag{1.32}$$

applied to autonomous systems of SDEs (1.31) with scalar implicitness parameter $\theta \in \mathbb{R}^1$ (i.e. $\Theta = \theta I$, all $c^j = 0$ in (1.1)).

Theorem 1.3.3 (Asymptotic m.s. stability of BTMs for linear SDEs [118, 120]). *For all equidistant approximations with $h > 0$, the following (global and local) asymptotic properties hold:*

(1) For $\theta = 0.5$, $X \equiv 0$ mean square stable $\iff$ $Y \equiv 0$ mean square stable.
(2) $X \equiv 0$ mean square stable and $\theta \geq 0.5 \implies Y \equiv 0$ mean square stable.
(3) $X \equiv 0$ mean square unstable and $\theta \leq 0.5 \implies Y \equiv 0$ mean square unstable.
(4) $Y^{\theta_1} \equiv 0$ mean square stable with $\theta_1 \leq \theta_2 \implies Y^{\theta_2} \equiv 0$ mean square stable with θ_2.

Proof. Use spectral theory of monotone positive operators $\mathscr{L}$ and some knowledge on Lyapunov equation

$$AM + MA^T + \sum_{j=1}^{m} B^j M {B^j}^T = -C$$

(see concept of mean square operators $\mathscr{L}$ as introduced by S. [118, 120], cf. remarks below). ◇

Remark 1.3.4. (i) Systematic stability analysis of systems of discrete random mappings is possible by the study of sequences of positive operators (*mean square operators*) related to each numerical method. For example, the family of mean square stability operators $\mathscr{L}_n$ related to drift-implicit Theta-Euler methods (1.32) is given by the mappings.

$$S \in \mathscr{S}_+^{d\times d} \mapsto \mathscr{L}_n S = (I_d - \theta h_n A)^{-1}\Big((I_d + (1-\theta)h_n A) S (I_d + (1-\theta) h_n A)^T$$
$$+ h_n \sum_{j=1}^{m} B^j S (B^j)^T\Big)((I_d - \theta h_n A)^{-1})^T$$

where I_d represents the $d \times d$ unit matrix, and, for $n \in \mathbb{N}$ and nonrandom step sizes h_n, we have

$$\mathbb{E}[Y_{n+1}Y_{n+1}^T] = \mathscr{L}_n\left(\mathbb{E}[Y_nY_n^T]\right) = \prod_{k=0}^{n} \mathscr{L}_k\left(\mathbb{E}[Y_0Y_0^T]\right)$$

for the related evolution of all 2nd moments (similar for random step sizes).

(ii) The system formulation in terms of positive operators is needed since **the problem of stochastic test equation is not solved** and one cannot reduce the general case of SDEs to simple one-dimensional test equations within non-anticipating stochastic calculus in sharp contrast to the situation in deterministic calculus (due to the presence of non-commutative operators in general). One-dimensional test equations are only relevant for stability of systems with complete commutative structure (This truly striking fact is commonly not recognized in the literature correctly).

(iii) Stochastic A-, AN-stability and monotone nesting of stability domains can be established, see [118, 120].

(iv) More general, there is a systematic study of stochastic dissipative and monotone systems to carry over ideas for linear systems to nonlinear systems ones (cf. [120, 123, 124]). One needs to distinguish between systems with additive and multiplicative noise in related stability analysis.

An interesting, illustrative and simple complex-valued test equation is given by the **stochastic Kubo oscillator** perturbed by multiplicative white noise in Stratonovich sense

$$dX(t) = iX(t)dt + i\rho X(t) \circ dW(t) \tag{1.33}$$

where $\rho \in \mathbb{R}^1, i^2 = -1$. This equation describes rotations on the circle with radius $||X(0)||_{\mathbb{C}} = |X_0|$. S. [118] has studied this example and shown that the corresponding discretization of drift-implicit Milstein methods explodes for any step size selection, whereas the lower order drift-implicit trapezoidal method with the same implicitness parameter θ or appropriately balanced implicit methods (BIMs) could stay very close to the circle of the exact solution even for large integration times! This is a test equation which manifests that "stochastically coherent" (i.e. asymptotically exact) numerical methods are needed and the search for efficient higher order convergent methods is somehow restricted even under linear boundedness and infinitely smooth assumptions on drift and diffusion coefficients. So Dahlquist barriers for consistent and stable linear multi-step methods are bounded by order 1.0 in stochastic settings.

In passing, note that Stratonovich-type equations (1.33) are equivalent to Itô-type SDE

$$dX(t) = (i - \frac{1}{2}\rho^2)X(t)\,dt + i\rho X(t)dW(t),$$

thereby SDEs (1.33) belong to following subclass (1.34). A more general bilinear test equation is studied in the following example.

Example of one-dimensional complex-valued test SDE. Many authors (e.g. Mitsui and Saito [98] or Schurz [118]) have studied the SDE

$$dX(t) \;=\; \lambda X(t)\,dt + \gamma X(t)\,dW(t), \tag{1.34}$$

with $X(0) = x_0, \lambda, \gamma \in \mathbb{C}^1$, representing a test equation for the class of completely commutative systems of linear SDEs with multiplicative white noise. This stochastic process has the unique exact solution $X(t) = X(0)\cdot\exp((\lambda - \gamma^2/2)t + \gamma W(t))$ with second moment

$$\begin{aligned}\mathbb{E}[X(t)X^*(t)] &= \mathbb{E}|X(t)|^2 = \mathbb{E}\exp(2(\lambda - \gamma^2/2)_r t + 2\gamma_r W(t))\cdot|x_0|^2\\ &= |x_0|^2\cdot\exp(2(\lambda_r - \gamma_r^2/2 + \gamma_i^2/2)t + 2\gamma_r^2 t)\\ &= |x_0|^2\cdot\exp((2\lambda_r + |\gamma|^2)t)\end{aligned}$$

where $x_0 \in \mathbb{C}^1$ is nonrandom (z_r is the real part, z_i the imaginary part of $z \in \mathbb{C}^1$) and $*$ denotes the complex conjugate value. The trivial solution $X \equiv 0$ of (1.34) is mean square stable for the process $\{X(t) : t \geq 0\}$ iff $2\lambda_r + |\gamma|^2 < 0$. Now, let us compare the numerical approximations of families of (drift-)implicit Theta and Milstein methods. Applied to (1.34), the drift-implicit Milstein and drift-implicit Theta methods are given by

$$Y_{n+1}^{(M)} = \frac{1 + (1-\theta)\lambda h + \gamma\xi_n\sqrt{h} + \gamma^2(\xi_n^2 - 1)h/2}{1 - \theta\lambda h}\cdot Y_n^{(M)} \tag{1.35}$$

and

$$Y_{n+1}^{(E)} = \frac{1 + (1-\theta)\lambda h + \gamma\xi_n\sqrt{h}}{1 - \theta\lambda h}\cdot Y_n^{(E)}\,, \tag{1.36}$$

respectively. Their second moments $P_n^{(E/M)} = \mathbb{E}[Y_n^{(E/M)} Y_n^{(E/M)^*}]$ satisfy

$$\begin{aligned}P_{n+1}^{(M)} &= \left(\mathbb{E}|\frac{1 + (1-\theta)\lambda h + \gamma\xi_n\sqrt{h}}{1 - \theta\lambda h}|^2 + \mathbb{E}|\frac{\gamma^2(\xi_n^2 - 1)}{1 - \theta\lambda h}|^2\cdot h^2/4\right)\cdot P_n^{(M)}\\ &= P_0^{(M)}\cdot\left(\frac{|1 + (1-\theta)\lambda h|^2 + |\gamma|^2 h + |\gamma|^4 h^2/2}{|1 - \theta\lambda h|^2}\right)^{n+1}\\ &> P_0^{(E)}\cdot\left(\frac{|1 + (1-\theta)\lambda h|^2 + |\gamma|^2 h}{|1 - \theta\lambda h|^2}\right)^{n+1} \;=\; P_{n+1}^{(E)} \qquad (n = 0, 1, 2, \ldots),\end{aligned}$$

provided that $P_0^{(M)} = \mathbb{E}[Y_0^{(M)} Y_0^{(M)^*}] \geq \mathbb{E}[Y_0^{(E)} Y_0^{(E)^*}] = P_0^{(E)}$, and

$$P_{n+1}^{(M)} = P_{n+1}^{(E)} \cdot \left(\frac{|1+(1-\theta)\lambda h|^2 + |\gamma|^2 h + |\gamma|^4 h^2/2}{|1+(1-\theta)\lambda h|^2 + |\gamma|^2 h} \right)^{n+1}$$

$$= P_{n+1}^{(E)} \cdot \left(1 + \frac{|\gamma|^4 h^2/2}{|1+(1-\theta)\lambda h|^2 + |\gamma|^2 h} \right)^{n+1}.$$

while assuming identical initial values $P_0^{(M)} = P_0^{(E)}$. Hence, if the drift-implicit Milstein method (1.35) possesses a mean square stable null solution then the corresponding drift-implicit Theta method (1.36) possesses it too. The mean square stability domain of (1.35) is smaller than that of (1.36) for any implicitness $\theta \in [0,1]$. Besides, the drift-implicit Theta method (1.36) has a mean square stable null solution if $\theta \geq \frac{1}{2}$ and $2\lambda_r + |\gamma|^2 < 0$. The latter condition coincides with the necessary and sufficient condition for the mean square stability of the null solution of SDE (1.34). Thus, the drift-implicit Theta method (1.36) with implicitness $\theta = 0.5$ is useful to indicate mean square stability of the equilibrium solution of (1.34). More general theorems concerning the latter observations can be found in S. [118], [120].

1.3.3 Mean Square A-Stability of Drift-Implicit Theta Methods

A-stability is one of the most desired properties of numerical algorithms. We could distinguish between the linear A-stability and nonlinear A-stability concepts, depending on the corresponding linear and nonlinear test classes of dissipative SDEs. However, one may find a unified treatment of the classical AN- and A-stability concepts in moment sense. Following S. [118, 120, 124, 125], one can introduce the following meaningful definitions, motivated basically by the fundamental works of Dahlquist [23] in deterministic numerical analysis. Fix $p \in [1, +\infty)$.

Definition 1.3.4. The numerical sequence $Y = (Y_n)_{n \in \mathbb{N}}$ (method, approximation, etc.) is called **p-th mean A-stable** iff it has an asymptotically (numerically) p-th mean stable equilibrium solution $Y \equiv 0$ for all autonomous SDEs (1.34) with asymptotically p-th mean stable equilibrium solution $X \equiv 0$, while using any admissible step size sequence h_n with

$$0 < \inf_{n \in \mathbb{N}} h_n \leq h_k \leq \sup_{n \in \mathbb{N}} h_n < +\infty.$$

The numerical sequence $Y = (Y_n)_{n \in \mathbb{N}}$ (method, approximation, etc.) is said to be **p-th mean AN-stable** iff it has an asymptotically (numerically) p-th mean stable equilibrium solution $Y \equiv 0$ for all SDEs (1.2) with asymptotically p-th mean stable equilibrium solution $X \equiv 0$, while using any admissible step size sequence h_n with

$$0 < \inf_{n \in \mathbb{N}} h_n \leq h_k \leq \sup_{n \in \mathbb{N}} h_n < +\infty.$$

In case of $p = 2$, we speak of **mean square A-** and **mean square AN-stability**.

Remark 1.3.5. Recall, from literature, that the bilinear real-valued test SDE (1.34) possesses an asymptotically mean square stable trivial solution $X \equiv 0$ iff

$$2\lambda + |\gamma|^2 < 0.$$

Theorem 1.3.4 (M.s. A-stability of drift-implicit Theta methods [120]). *Assume that $Y_0 = X_0$ is independent of all σ-algebras $\sigma(W^j(t) : t \geq 0)$ and $\mathbb{E}[X_0]^2 < \infty$. Then the drift-implicit Theta method (1.25) applied to SDEs (1.34) with $\theta \geq 0.5$ provides mean square A-stable numerical approximations (i.e. when $p = 2$).*

Proof. For simplicity, let $\lambda, \gamma \in \mathbb{R}^1$. Suppose that $(h_n)_{n\in\mathbb{N}}$ is nonrandom and admissible with

$$0 < \inf_{n\in\mathbb{N}} h_n \leq h_k \leq \sup_{n\in\mathbb{N}} h_n < +\infty.$$

First, note that the drift-implicit Theta methods (1.25) applied to SDEs (1.34) with scalar $\theta \in [0,1]$ is governed by the scheme (with i.i.d. ξ_n with $\mathbb{E}[\xi_n] = 0$ and $\mathbb{E}[\xi_n]^2 = 1$)

$$Y_{n+1}^\theta = \frac{1 + (1-\theta)\lambda h_n + \gamma \xi_n \sqrt{h_n}}{1 - \theta\lambda h_n} \cdot Y_n^\theta.$$

Second, its temporal second moment evolution $P_n^\theta = \mathbb{E}[Y_n^\theta]^2$ can be rewritten to as

$$\begin{aligned} P_{n+1}^\theta &= P_0^\theta \cdot \prod_{k=0}^{n} \frac{|1 + (1-\theta)\lambda h_k|^2 + |\gamma|^2 h_k}{|1 - \theta\lambda h_k|^2} \\ &= P_0^\theta \cdot \prod_{k=0}^{n} \left(1 + \frac{2\lambda + |\gamma|^2 + (1-2\theta)|\lambda|^2 h_k}{|1-\theta\lambda h_k|^2} h_k\right) \\ &\leq P_0^\theta \cdot \prod_{k=0}^{n} \left(1 + \frac{2\lambda + |\gamma|^2}{|1-\theta\lambda h_k|^2} h_k\right) \\ &\leq P_0^\theta \cdot \prod_{k=0}^{n} \left(1 + \frac{2\lambda + |\gamma|^2}{|1-\theta\lambda \sup_{l\in\mathbb{N}} h_l|^2} h_k\right) \end{aligned}$$

where $P_0^\theta = \mathbb{E}[Y_0]^2$, $\theta \geq 0.5$ and $2\lambda + |\gamma|^2 \leq 0$. Third, one shows that there are real constants c_1 and c_2 such that

$$\exp\left(c_1 \sum_{k=0}^{n} h_k\right) P_0^\theta = \exp(c_1 t_{n+1}) P_0^\theta \leq P_{n+1}^\theta \leq \exp(c_2 t_{n+1}) P_0^\theta$$

$$= \exp\left(c_2 \sum_{k=0}^{n} h_k\right) P_0^\theta \tag{1.37}$$

for all $n \in \mathbb{N}$. Fourth, it remains to verify that $c_2 < 0$ whenever $2\lambda + \gamma^2 < 0$ and $\theta \geq 0.5$. In fact, one finds that

$$P_{n+1}^\theta \leq P_0^\theta \cdot \prod_{k=0}^{n} \exp\left(\frac{2\lambda + |\gamma|^2}{|1 - \theta\lambda \sup_{l \in \mathbb{N}} h_l|^2} h_k\right)$$

$$= P_0^\theta \cdot \exp\left(\frac{2\lambda + |\gamma|^2}{|1 - \theta\lambda \sup_{l \in \mathbb{N}} h_l|^2} \sum_{k=0}^{n} h_k\right)$$

$$= P_0^\theta \cdot \exp\left(\frac{2\lambda + |\gamma|^2}{|1 - \theta\lambda \sup_{l \in \mathbb{N}} h_l|^2} t_{n+1}\right),$$

hence

$$c_2 \leq \frac{2\lambda + \gamma^2}{(1 - \theta\lambda \sup_{n \in \mathbb{N}} h_n)^2}$$

since $2\lambda + \gamma^2 < 0$ and $\theta \geq 0.5$. Consequently, $\lim_{n \to +\infty} P_n^\theta = 0$ for all P_0 with adapted $Y_0 \in L^2(\omega, \mathscr{F}_0, \mathbb{P})$ and for all admissible step sizes h_n if $2\lambda + \gamma^2 < 0$ and $\theta \geq 0.5$. This fact establishes mean square A-stability of drift-implicit Theta methods (1.25) with $\theta \geq 0.5$, hence the proof is complete. ◇

A warning. The drift-implicit Theta methods with $\theta > 0.5$ such as backward Euler methods are on the "sure numerically stable" side. However, we must notice that they provide so-called "superstable" numerical approximations. Superstability is a property which may lead to undesired stabilization effects of numerical dynamics whereas the underlying SDE does not have stable behaviour. Of course, it would be **better to make use of asymptotically exact numerical methods** such as **midpoint-type or trapezoidal methods** (cf. explanations in sections below).

Remark 1.3.6. Artemiev and Averina [5], Mitsui and Saito [98], and Higham [47], [48] have also established mean square A-stability, however exclusively for equidistant partitions only. They analyze the more standard concept of stability function which is only adequate for equidistant discretizations and test classes of linear SDEs (for stability regions for 1D linear test SDEs, see also [15]). There is also an approach using the concept of **weak A-stability**, i.e. the A-stability of related deterministic parts of numerical dynamics discretizing linear SDEs. However, this concept does not lead to new insights into the effects of stochasticity with respect to stability (cf. [95] or [72]).

1.3.4 Asymptotic Mean Square Stability for Nonlinear Test Equations

Let $X_{s,x}(t)$ denote the value of the stochastic process X at time $t \geq s$, provided that it has started at the value $X_{s,x}(s) = x$ at prior time s. x and y are supposed to be adapted initial values. Let Π denote an ordered time-scale (discrete ($\Pi = \mathbb{N}$) or continuous ($\Pi = [0, +\infty)$)) and $p \neq 0$ be a nonrandom constant.

Definition 1.3.5. A stochastic process $X=(X(t))_{t\in\Pi}$ with basis $(\Omega, \mathcal{F}, (\mathcal{F}_t)_{t\in\Pi}, \mathbb{P})$ is said to be uniformly **p-th mean (forward) dissipative** on $\mathbb{R}^d$ iff $\exists K_D^X \in \mathbb{R}\ \forall t \geq s \in \Pi\ \forall x \in \mathbb{R}^d$

$$\mathbb{E}\Big[||X_{s,x}(t)||_d^p \Big| \mathcal{F}_s\Big] \leq \exp\Big(pK_D^X(t-s)\Big)||x||_d^p \tag{1.38}$$

with **p-th mean dissipativity constant** K_D^X. X is said to be a process with **p-th mean non-expansive norms** iff $\forall t \geq s \in \Pi\ \forall x \in \mathbb{R}^d$

$$\mathbb{E}\Big[||X_{s,x}(t)||_d^p \Big| \mathcal{F}_s\Big] \leq ||x||_d^p. \tag{1.39}$$

If $p = 2$ then we speak of **mean square dissipativity** and **mean square non-expansive norms**, resp.

For dissipative processes with adapted initial data, one can re-normalize p-th power norms to be uniformly bounded and find uniformly bounded p-th moment Lyapunov exponents by K_D^X. Hence, its p-th mean longterm dynamic behaviour is under some control. Processes with non-expansive p-th mean norms have uniformly bounded L^p-norms even without renormalization along the entire time-scale Π. These concepts are important for the uniform boundedness and stability of values of numerical methods. They are also meaningful to test numerical methods while applying to SDEs with uniformly p-th norm bounded coefficient systems.

Let $p > 0$ be a nonrandom constant.

Definition 1.3.6. A coefficient system (a, b^j) of SDEs (1.2) and its SDE are said to be **strictly uniformly p-th mean bounded** on $\mathbb{R}^d$ iff $\exists K_{OB} \in \mathbb{R} \forall t \in \mathbb{R} \forall x \in \mathbb{R}^d$

$$< a(t,x), x >_d + \frac{1}{2}\sum_{j=1}^{m} ||b^j(t,x)||_d^2 + \frac{p-2}{2}\sum_{j=1}^{m} \frac{< b^j(t,x), x >_d^2}{||x||_d^2} \leq K_{OB}\ ||x||_d^2. \tag{1.40}$$

If $p = 2$ then we speak of **mean square boundedness**.

Lemma 1.3.1 (Dissipativity of SDE (1.2), S.[120, 124]). *Assume that X satisfies SDE (1.2) with p-th mean bounded coefficient system (a, b^j).*
Then X is p-th mean dissipative for all $p \geq 2$ and its p-th mean dissipativity

constant K_D^X can be estimated by

$$K_D^X \leq K_{OB}.$$

This lemma can be proved by Dynkin's formula (averaged Itô formula). Let us discuss the possible "worst case effects" on the temporal evolution of p-th mean norms related to numerical methods under condition (1.40) with $p = 2$. It turns out that the drift-implicit backward Euler methods are mean square dissipative under this condition and mean square stable if additionally $K_{OB} < 0$.

Theorem 1.3.5 (Dissipativity + expo. m.s. stability of backward EM[120, 124]). *Assume that*

(i) $\theta_n = 1$.
(ii) $0 < \inf_{n\in I\!N} h_n \leq \sup_{n\in I\!N} h_n < +\infty$, *all h_n nonrandom (i.e. only admissible step sizes).*
(iii) $\exists K_a \leq 0 \forall x \in R^d \; \forall t \geq 0 \; : < a(t,x), x > \leq K_a ||x||_d^2$.
(iv) $\exists K_b \forall x \in R^d \; \forall t \geq 0 \; : \; \sum_{j=1}^m ||b^j(t,x)||_d^2 \leq K_b ||x||_d^2$.

Then, the drift-implicit Euler methods (1.1) with scalar implicitness $\theta_n = 1$ and vanishing $c^j = 0$ are mean square dissipative when applied to SDEs (1.2) with mean square bounded coefficients (a, b^j) with dissipativity constant

$$K_D^X = \sup_{n\in I\!N} \frac{2K_a + K_b}{1 - 2h_n K_a}. \tag{1.41}$$

If additionally $2K_a + K_b \leq 0$ then they are mean square non-expansive and

$$K_D^X = \frac{2K_a + K_b}{1 - 2K_a \sup_{n\in I\!N} h_n} \tag{1.42}$$

and hence exponentially mean square stable if $2K_a + K_b < 0$

Proof. Rearrange the scheme (1.1) for the drift-implicit Theta methods with nonrandom scalar implicitness $(\Theta_n) = \theta_n I$ to separate implicit from explicit part such that

$$X_{n+1} - \theta_n h_n a(t_{n+1}, X_{n+1}) = X_n + (1 - \theta_n) h_n a(t_n, X_n) + \sum_{j=1}^m b^j(t_n, X_n) \Delta W_n^j. \tag{1.43}$$

Recall that X denotes the value of the iteration scheme (1.1) started at values $X_0 \in L^2(\Omega, \mathcal{F}_0, I\!P)$. Now, take the square of Euclidean norms on both sides. By taking the expectation on both sides we arrive at

$$\mathbb{E}||X_{n+1}||_d^2 - 2\theta_n h_n \mathbb{E} < X_{n+1}, a(t_{n+1}, X_{n+1}) >_d + \theta_n^2 h_n^2 \mathbb{E}||a(t_{n+1}, X_{n+1})||_d^2$$
$$= \mathbb{E}||X_n||_d^2 + 2(1-\theta_n)h_n \mathbb{E} < X_n, a(t_n, X_n) >_d + (1-\theta_n)^2 h_n^2 \mathbb{E}||a(t_n, X_n)||_d^2$$
$$+ h_n \sum_{j=1}^{m} \mathbb{E}||b^j(t_n, X_n)||_d^2.$$

Under the assumption (iii) we have

$$-2\theta_n h_n < a(t,x), x >_d \geq -2\theta_n h_n K_a ||x||_d^2 \geq 0$$

for all $x \in \mathbb{R}^d$ and $t \geq 0$. Consequently, under (iii) and (iv), we may estimate

$$(1 - 2\theta_n h_n K_a)\mathbb{E}||X_{n+1}||_d^2$$
$$\leq [1 + (2(1-\theta_n)K_a + K_b)h_n]_+ \mathbb{E}||X_n||_d^2 + (1-\theta_n)^2 h_n^2 \mathbb{E}||a(t_n, X_n)||_d^2.$$

for all $n \in \mathbb{N}$. This leads to the estimate

$$\mathbb{E}||X_{n+1}||_d^2 \leq \frac{[1 + (2(1-\theta_n)K_a + K_b)h_n]_+}{1 - 2\theta_n h_n K_a} \mathbb{E}||X_n||_d^2$$
$$= \left(1 + \frac{(2K_a + K_b)h_n}{1 - 2\theta_n h_n K_a}\right) \mathbb{E}||X_n||_d^2 + \frac{(1-\theta_n)^2 h_n^2}{1 - 2\theta_n h_n K_a} \mathbb{E}||a(t_n, X_n)||_d^2$$
$$\leq \exp\left(\frac{(2K_a + K_b)h_n}{1 - 2\theta_n h_n K_a}\right) \mathbb{E}||X_n||_d^2 + \frac{(1-\theta_n)^2 h_n^2}{1 - 2\theta_n h_n K_a}$$
$$\mathbb{E}||a(t_n, X_n)||_d^2 \tag{1.44}$$

since $1 + z \leq \exp(z)$ for $z \geq -1$. Now, set all parameters $\theta_n = 1$ in the above inequality. In this case one encounters

$$\mathbb{E}||X_{n+1}||_d^2 \leq \exp\left(\frac{2K_a + K_b}{1 - 2h_n K_a} h_n\right) \mathbb{E}||X_n||_d^2.$$

Therefore, the drift-implicit backward Euler methods are mean square dissipative with dissipativity constant

$$K_D^X = \sup_{n \in \mathbb{N}} \frac{2K_a + K_b}{1 - 2h_n K_a}$$
$$= \frac{2K_a + K_b}{1 - 2 \sup_{n \in \mathbb{N}} h_n K_a} \text{ if } 2K_a + K_b \leq 0.$$

If additionally $2K_a + K_b \leq 0$, then their norms are mean square non-expansive. Exponential mean square stability follows from estimate (1.44) with setting $\theta_n = 1.0$ under the hypothesis $2K_a + K_b < 0$. This completes the proof. ◇

Theorem 1.3.6 (Local m.s. stability of drift-implicit BTMs for dissipative SDEs). *Assume that:*

(i) $\theta_n \geq 0.5$.
(ii) $0 < \inf_{n\in\mathbb{N}} h_n \leq \sup_{n\in\mathbb{N}} h_n < +\infty$, *all* h_n *nonrandom (i.e. only admissible step sizes),* $(\theta_n h_n)_{n\in\mathbb{N}}$ *is non-increasing in* n.
(iii) $\exists K_a \leq 0 \forall x \in R^d \ \forall t \geq 0 \ : < a(t,x), x >\leq K_a ||x||_d^2$.
(iv) $\exists K_b \forall x \in R^d \ \forall t \geq 0 \ : \ \sum_{j=1}^m ||b^j(t,x)||_d^2 \leq K_b ||x||_d^2$.
(v) $2K_a + K_b \leq 0$.
(vi) $a(0,0) = 0$ *and* $a(0,\cdot) \in C^0(N(0), \mathbb{R}^d)$ *for a neighborhood* $N(0)$ *around* $0 \in \mathbb{R}^d$.

Then, the drift-implicit BTMs (1.25) with scalar implicitness $\theta_n \geq 0.5$ *(and vanishing* $c^j = 0$*) are locally (asymptotically) mean square stable when applied to SDEs (1.2) with mean square bounded coefficients* (a, b^j).

Proof. As before, rearrange the scheme (1.1) for the drift-implicit Theta methods with nonrandom scalar implicitness $(\Theta_n) = \theta_n I$ to separate implicit from explicit part such that

$$X_{n+1} - \theta_n h_n a(t_{n+1}, X_{n+1}) = X_n + (1-\theta_n) h_n a(t_n, X_n) + \sum_{j=1}^m b^j(t_n, X_n)\Delta W_n^j. \tag{1.45}$$

Recall that X denotes the value of the iteration scheme (1.1) started at values $X_0 \in L^2(\Omega, \mathscr{F}_0, \mathbb{P})$. Now, take the square of Euclidean norms on both sides. By taking the expectation on both sides we arrive at

$$\begin{aligned}\mathbb{E}||X_{n+1}||_d^2 &- 2\theta_n h_n \mathbb{E} < X_{n+1}, a(t_{n+1}, X_{n+1}) >_d + \theta_n^2 h_n^2 \mathbb{E}||a(t_{n+1}, X_{n+1})||_d^2 \\ &= \mathbb{E}||X_n||_d^2 + 2(1-\theta_n) h_n \mathbb{E} < X_n, a(t_n, X_n) >_d + (1-\theta_n)^2 h_n^2 \mathbb{E}||a(t_n, X_n)||_d^2 \\ &+ h_n \sum_{j=1}^m \mathbb{E}||b^j(t_n, X_n)||_d^2.\end{aligned}$$

Under the assumption (iii) we have

$$-2\theta_n h_n < a(t,x), x >_d \geq -2\theta_n h_n K_a ||x||_d^2 \geq 0$$

for all $x \in \mathbb{R}^d$ and $t \geq 0$. Consequently, under the requirements (i)–(iv), we may estimate

$$\begin{aligned}&(1 - 2\theta_n h_n K_a)\mathbb{E}||X_{n+1}||_d^2 + \theta_n^2 h_n^2 \mathbb{E}||a(t_{n+1}, X_{n+1})||_d^2 \\ &\leq [1 + (2(1-\theta_n)K_a + K_b)h_n]_+ \mathbb{E}||X_n||_d^2 + (1-\theta_n)^2 h_n^2 \mathbb{E}||a(t_n, X_n)||_d^2\end{aligned}$$

$$
\begin{aligned}
&= [1 + (2(1-\theta_n)K_a + K_b)h_n]_+ \mathbb{E}||X_n||_d^2 + (1-2\theta_n)h_n^2 \mathbb{E}||a(t_n, X_n)||_d^2 \\
&\quad + \theta_n^2 h_n^2 \mathbb{E}||a(t_n, X_n)||_d^2 \\
&\leq [1 + (2(1-\theta_n)K_a + K_b)h_n]_+ \mathbb{E}||X_n||_d^2 + \theta_{n-1}^2 h_{n-1}^2 \mathbb{E}||a(t_n, X_n)||_d^2 .
\end{aligned}
$$

for all $n \in \mathbb{N}$. This leads to the estimate

$$
\begin{aligned}
&\mathbb{E}||X_{n+1}||_d^2 + \frac{\theta_n^2 h_n^2}{1-2\theta_n h_n K_a} \mathbb{E}||a(t_{n+1}, X_{n+1})||_d^2 \\
&\leq \frac{[1 + (2(1-\theta_n)K_a + K_b)h_n]_+}{1-2\theta_n h_n K_a} \mathbb{E}||X_n||_d^2 + \frac{\theta_{n-1}^2 h_{n-1}^2}{1-2\theta_n h_n K_a} \mathbb{E}||a(t_n, X_n)||_d^2 \\
&\leq \frac{[1 + (2(1-\theta_n)K_a + K_b)h_n]_+}{1-2\theta_n h_n K_a} \mathbb{E}||X_n||_d^2 + \frac{\theta_{n-1}^2 h_{n-1}^2}{1-2\theta_{n-1} h_{n-1} K_a} \mathbb{E}||a(t_n, X_n)||_d^2 \\
&= \left(1 + \frac{(2K_a + K_b)h_n}{1-2\theta_n h_n K_a}\right) \mathbb{E}||X_n||_d^2 + \frac{\theta_{n-1}^2 h_{n-1}^2}{1-2\theta_{n-1} h_{n-1} K_a} \mathbb{E}||a(t_n, X_n)||_d^2 \\
&\leq \exp\left(\frac{(2K_a + K_b)h_n}{1-2\theta_n h_n K_a}\right) \mathbb{E}||X_n||_d^2 + \frac{\theta_{n-1}^2 h_{n-1}^2}{1-2\theta_{n-1} h_{n-1} K_a} \mathbb{E}||a(t_n, X_n)||_d^2 \\
&\leq \mathbb{E}||X_n||_d^2 + \frac{\theta_{n-1}^2 h_{n-1}^2}{1-2\theta_{n-1} h_{n-1} K_a} \mathbb{E}||a(t_n, X_n)||_d^2
\end{aligned}
$$

since $1 + z \leq \exp(z)$ for $z \geq -1$, $K_a \leq 0$, $\theta_n \geq 0.5$, $\theta_n h_n > 0$ non-increasing in n and (v). Now, it remains to apply the principle of complete induction in order to conclude that

$$
\begin{aligned}
&\mathbb{E}||X_{n+1}||_d^2 + \frac{\theta_n^2 h_n^2}{1-2\theta_n h_n K_a} \mathbb{E}||a(t_{n+1}, X_{n+1})||_d^2 \\
&\leq \mathbb{E}||X_0||_d^2 + \frac{\theta_{-1}^2 h_{-1}^2}{1-2\theta_{-1} h_{-1} K_a} \mathbb{E}||a(t_0, X_0)||_d^2 \\
&= \mathbb{E}||X_0||_d^2 + \frac{\theta_0^2 h_0^2}{1-2\theta_0 h_0 K_a} \mathbb{E}||a(0, X_0)||_d^2
\end{aligned}
$$

for all $n \in \mathbb{N}$ (recall the convention $t_0 = 0$, $h_{-1} := h_0$ and $\theta_{-1} = \theta_0$). Eventually, the continuity hypothesis (vi) on a w.r.t. x in the neighborhood of 0 together with $a(0,0) = 0$ guarantees that the drift-implicit BTMs (1.25) are locally (asymptotically) mean square stable for nonlinear SDEs (1.2) with p-th mean bounded coefficient systems (a, b^j), provided that (i)–(v). This conclusion completes the proof. ◇

In contrast to backward Euler methods and drift-implicit Theta methods with $\theta \geq 0.5$, we may establish expansivity of p-th norms for norm-expansive nonlinear SDEs. To see this fact, consider the concept of expansive norms in what follows.

Definition 1.3.7. A stochastic process $X=(X(t))_{t\in\Pi}$ with basis $(\Omega, \mathscr{F}, (\mathscr{F}_t)_{t\in\Pi}, \mathbb{P})$ is said to be **p-th mean (forward) expansive** on $\mathbb{R}^d$ iff $\forall t \geq s \in \Pi$ $\forall x \in \mathbb{R}^d$ (adapted)

$$\mathbb{E}\Big[||X_{s,x}(t)||_d^p \Big| \mathscr{F}_s\Big] \geq ||x||_d^p. \tag{1.46}$$

X is said to be a process with **strictly p-th mean expansive norms** iff $\forall t > s \in \Pi$ $\forall x \in \mathbb{R}^d (x \neq 0)$ (adapted)

$$\mathbb{E}\Big[||X_{s,x}(t)||_d^p \Big| \mathscr{F}_s\Big] > ||x||_d^p. \tag{1.47}$$

If $p=2$ then we speak of **mean square expansivity** and **strict mean square expansivity**, respectively.

For expansive processes, p-th mean norms of the initial data may significantly enlarge as time t advances. The L^p-norms of adapted initial data of strict expansive processes may even show an non-controllable or exponential growth, hence explosions as t tends to $+\infty$. These concepts are important for testing and control of the temporal evolution of L^p-norms through numerical methods applied to SDEs with expansive coefficient systems.

Let $p > 0$ be a nonrandom constant.

Definition 1.3.8. A coefficient system (a, b^j) of SDEs (1.2) and its SDE are said to be **(uniformly) p-th mean expansive** on $\mathbb{R}^d$ iff $\exists K_{OB} \geq 0 \in \mathbb{R}$ $\forall t \in \mathbb{R}$ $\forall x \in \mathbb{R}^d$

$$< a(t,x), x >_d + \frac{1}{2}\sum_{j=1}^{m} ||b^j(t,x)||_d^2 + \frac{p-2}{2}\sum_{j=1}^{m} \frac{< b^j(t,x), x >_d^2}{||x||_d^2}$$
$$\geq K_{OB}\, ||x||_d^2. \tag{1.48}$$

If $K_{OB} > 0$ in (1.48) then the coefficient system (a, b^j) is said to be **strictly p-th mean expansive** and its SDE has p-th mean expansive norms. Moreover, if $p = 2$ then we speak of **mean square expansivity** and **strictly mean square expansive** norms and systems, respectively.

Lemma 1.3.2. *Assume that X satisfies SDE (1.2) with p-th mean expansive coefficient system (a, b^j).*
Then X has p-th mean expansive norms for all $p \geq 2$. If additionally $K_{OB} > 0$ in (1.48) then X possesses strict p-th mean expansive norms.

This lemma can be proved by Dynkin's formula (averaged Itô formula). Let us discuss the possible "worst case effects" on L^p-norms of numerical methods under condition (1.48) with $p = 2$. It turns out that the drift-implicit forward Euler methods have mean square expansive norms under this condition and may even possess strict mean square expansive norms.

Theorem 1.3.7 (Mean square expansivity of forward Euler methods). *Assume that*

(i) $\theta_n = 0$.
(ii) $0 < \inf_{n \in \mathbb{N}} h_n \leq \sup_{n \in \mathbb{N}} h_n < +\infty$, *all* h_n *nonrandom (i.e. only admissible step sizes).*
(iii) $\exists K_a \; \forall x \in R^d \; \forall t \geq 0 \; : < a(t,x), x >_d \geq \; K_a ||x||_d^2$.
(iv) $\exists K_b \; \forall x \in R^d \; \forall t \geq 0 \; : \; \sum_{j=1}^m ||b^j(t,x)||_d^2 \; \geq \; K_b ||x||_d^2$.

Then, the drift-implicit (forward) Euler methods (1.1) with scalar implicitness $\theta_n = 0$ *(and vanishing* $c^j = 0$*) have mean square expansive norms when applied to SDEs (1.2) with mean square expansive coefficients* (a, b^j) *satisfying* $2K_a + K_b \geq 0$. *If additionally* $2K_a + K_b > 0$ *then their norms are strictly mean square expansive.*

Proof. Consider the scheme (1.1) for the drift-implicit Theta methods with nonrandom scalar implicitness $(\Theta_n) = \theta_n I$ and separate implicit from explicit part such that

$$X_{n+1} - \theta_n h_n a(t_{n+1}, X_{n+1}) = X_n + (1-\theta_n) h_n a(t_n, X_n) + \sum_{j=1}^m b^j(t_n, X_n) \Delta W_n^j. \tag{1.49}$$

Recall that X denote the value of the scheme (1.1) started at values $X_0 \in L^2(\Omega, \mathcal{F}_0, \mathbb{P})$. Now, take the square of Euclidean norms on both sides. By taking the expectation on both sides we arrive at

$$\begin{aligned} &\mathbb{E}||X_{n+1}||_d^2 - 2\theta_n h_n \mathbb{E} < X_{n+1}, a(t_{n+1}, X_{n+1}) >_d + \theta_n^2 h_n^2 \mathbb{E}||a(t_{n+1}, X_{n+1})||_d^2 \\ &= \mathbb{E}||X_n||_d^2 + 2(1-\theta_n) h_n \mathbb{E} < X_n, a(t_n, X_n) >_d + (1-\theta_n)^2 h_n^2 \mathbb{E}||a(t_n, X_n)||_d^2 \\ &\quad + h_n \sum_{j=1}^m \mathbb{E}||b^j(t_n, X_n)||_d^2. \end{aligned}$$

Under the assumption (iii) and $\theta_n \leq 1$ we have

$$2(1-\theta_n) h_n < a(t,x), x >_d + h_n \sum_{j=1}^m ||b^j(t,x)||_d^2 \geq [2(1-\theta_n) K_a + K_b] h_n ||x||_d^2$$

for all $x \in \mathbb{R}^d$ and $t \geq 0$. Consequently, under (ii)–(iv), $\theta_n \leq 1$ and $2K_a + K_b \geq 0$, we may estimate

$$\begin{aligned} &(1 - 2\theta_n h_n K_a) \mathbb{E}||X_{n+1}||_d^2 + \theta_n^2 h_n^2 \mathbb{E}||a(t_{n+1}, X_{n+1})||_d^2 \\ &\geq [1 + (2(1-\theta_n) K_a + K_b) h_n] \mathbb{E}||X_n||_d^2 + (1-\theta_n)^2 h_n^2 \mathbb{E}||a(t_n, X_n)||_d^2. \end{aligned}$$

for all $n \in \mathbb{N}$. Now, set $\theta_n = 0$. This leads to the estimate

$$\mathbb{E}||X_{n+1}||_d^2 \geq [1 + (2K_a + K_b)h_n]\mathbb{E}||X_n||_d^2 \ \geq \ \mathbb{E}||X_n||_d^2.$$

Therefore, the forward Euler methods have mean square expansive norms under the condition $2K_a + K_b \geq 0$. After returning to the latter inequality above, one clearly recognizes that, if additionally $2K_a + K_b > 0$, then their norms are strictly mean square expansive. ◇

1.3.5 *Asymptotic Almost Sure Stability for Bilinear Systems in $\mathbb{R}^d$*

In the following two subsections we discuss the almost sure stability behavior of numerical methods (and sequences of random variables) with both constant h and variable step sizes h_n with respect to the trivial equilibrium $0 \in \mathbb{R}^d$. Let $\|.\|_d$ be a vector norm of $\mathbb{R}^d$ which is compatible with the matrix norm $\|.\|_{d\times d}$ of $\mathbb{R}^{d\times d}$.

Definition 1.3.9. A random sequence $Y = (Y_n)_{n\in\mathbb{N}}$ of real-valued random variables $Y_n : (\Omega, \mathscr{F}_n, \mathbb{P}) \to (\mathbb{R}^d, \mathscr{B}(\mathbb{R}^d))$ is called **(globally) asymptotically stable with probability one** (or **(globally) asymptotically a.s. stable**) iff

$$\lim_{n\to+\infty} \|Y_n\|_d \ = \ 0 \ (a.s.)$$

for all $Y_0 = y_0 \in \mathbb{R}^d \backslash \{0\}$, where $y_0 \in \mathbb{R}^d$ is nonrandom, otherwise **asymptotically a.s. unstable**.

Remark 1.3.7. One may introduce a similar definition of local asymptotic stability with probability one (i.e. a.s.). However, we omit this here, and we leave this and related investigations to the interested reader. We shall follow the concept of global asymptotic a.s. stability in what follows.

Throughout this subsection, consider the real-valued system of bilinear stochastic Theta methods

$$Y_{n+1} = Y_n + \Big(\theta_n A_{n+1} Y_{n+1} + (1-\theta_n) A_n Y_n\Big) h_n$$
$$+\sqrt{h_n} \sum_{j=1}^{m} B_{j,n} Y_n \xi_n^j \in \mathbb{R}^d, n \in \mathbb{N} \qquad (1.50)$$

driven by independent random variables ξ_n with centered first moments $\mathbb{E}[\xi_n] = 0$ and finite second moments $\sigma_n^2 = \mathbb{E}[\xi_n]^2 < +\infty$. Here, A_n and $B_{j,n}$ are nonrandom matrices in $\mathbb{R}^{d\times d}$. Recall that $h_n = t_{n+1} - t_n > 0$ represents the current step size of numerical integration. We suppose that the initial value $Y_0 \in \mathbb{R}^d$ is independent of all ξ_n^j. Let parameter $\theta_n \in [0,1]$ (which controls the degree of local implicitness). These schemes serve as discretizations of (non-autonomous) Itô-type SDEs

$$dX(t) = A(t)X(t)dt + \sum_{j=1}^{m} B^j(t)X(t)dW^j(t) \tag{1.51}$$

driven by standard i.i.d. Wiener processes W^j, where $A_n = A(t_n)$ and $B_n^j = B^j(t_n)$ along time-partitions $(t_n)_{n\in\mathbb{N}}$.

We are interested in studying the almost sure stability and almost sure convergence behaviour of these methods as t_n tends to infinity (i.e. as $n \to +\infty$) in $\mathbb{R}^d$ (d any dimension). So the main attention is drawn to the study of limits $\lim_{n\to+\infty} Y_n$ (a.s.) and whether its value is equal to 0. In passing, we note that the case of dimension $d = 1$ has been considered in [112].

Less is known about almost sure stability of multi-dimensional stochastic systems without a tedious calculation of related top Lyapunov exponents. We shall exploit a "semi-martingale convergence approach" by appropriate decompositions in order to investigate almost sure asymptotic convergence and stability of linear systems of stochastic Theta methods (1.50) in any dimension $d \in \mathbb{N}$.

For our further analysis and the sake of abbreviation, consider the equivalent real-valued system of linear stochastic difference equations

$$X_{n+1} = X_n + \tilde{A}_n X_n h_n + \sqrt{h_n} \sum_{j=1}^{m} \tilde{B}_{j,n} X_n \xi_n^j \in \mathbb{R}^d, n \in \mathbb{N} \tag{1.52}$$

driven by independent random variables ξ_n^j with centered first moments $\mathbb{E}[\xi_n^j] = 0$ and finite second moments $\sigma_n^2 = \mathbb{E}[\xi_n^j]^2 < +\infty$. Here, $\tilde{A}_n$ and $\tilde{B}_{j,n}$ are nonrandom matrices in $\mathbb{R}^{d\times d}$, defined by

$$\tilde{A}_n := (I - h_n\theta_n A_{n+1})^{-1}\Big(\theta_n A_{n+1} + (1-\theta_n)A_n\Big), \tag{1.53}$$

$$\tilde{B}_{j,n} := (I - h_n\theta_n A_{n+1})^{-1} B_{j,n}. \tag{1.54}$$

Throughout the presentation, we assume that $h_n\theta_n$ are chosen such that the inverses of $I - h_n\theta_n A_{n+1}$ exists for all $n \in \mathbb{N}$ (indeed, this is not very restrictive since always possible for small $h_n\theta_n$). It is obvious that the explicit system (1.52) is equivalent to the implicit system (1.50) under the aforementioned assumption.

The following theorem shall represent our major tool to establish almost sure stability of the trivial solution of stochastic Theta methods (1.50) and is proved below for the sake of relative self-contained character of this paper. Recall that $\mathcal{B}$ denotes the σ-algebra of all Borel subsets of $\mathbb{R}^1$.

Theorem 1.3.8 (Semi-martingale Convergence Theorem). *Let $Z = (Z_n)_{n\in\mathbb{N}}$ be an a.s. non-negative stochastic sequence of $(\mathcal{F}_n, \mathcal{B})$-measurable random vari-*

ables Z_n on probability space $(\Omega, \mathscr{F}, (\mathscr{F}_n)_{n\in\mathbb{N}}, \mathbb{P})$. Assume that Z permits the decomposition

$$Z_n \le Z_0 + A_n^1 - A_n^2 + M_n, n \in \mathbb{N} \tag{1.55}$$

where $A^1 = (A_n^1)_{n\in\mathbb{N}}$ and $A^2 = (A_n^2)_{n\in\mathbb{N}}$ are two non-decreasing, predictable processes with $A_0^i = 0$ $(i = 1,2)$, $M = (M_n)_{n\in\mathbb{N}}$ is a local $(\mathscr{F}_n)_{n\in\mathbb{N}}$-martingale with $M_0 = 0$ on $(\Omega, \mathscr{F}, (\mathscr{F}_n)_{n\in\mathbb{N}}, \mathbb{P})$. Then, the requirement of $\exists \lim_{n\to+\infty} A_n^1 < +\infty$ ($\mathbb{P}$-a.s.) implies that

$$\exists \limsup_{n\to+\infty} Z_n < +\infty, \quad \exists \lim_{n\to+\infty} A_n^2 < +\infty \quad (\mathbb{P}-a.s.). \tag{1.56}$$

If additionally the inequality (1.55) renders to be an equation for all $n \in \mathbb{N}$, then the finite existence of limit $\lim_{n\to+\infty} A_n^1 < +\infty$ ($\mathbb{P}$-a.s.) guarantees the finite existence of

$$\exists \lim_{n\to+\infty} Z_n < +\infty \qquad (\mathbb{P}-a.s.).$$

Proof. Suppose that the random variables $Z_n : (\Omega, \mathscr{F}_n, \mathbb{P}) \to (\mathbb{R}^1_+, \mathscr{B}^1_+)$ are governed by the inequality (1.55) with non-decreasing, predictable processes A^i with $A_0^i = 0$ and M forms a local $(\mathscr{F}_n)_{n\in\mathbb{N}}$-martingale with $M_0 = 0$. Without loss of generality, we may suppose that M is a L^2-integrable martingale (due to well-known localization procedures). First, note that:

(a) $\forall n \ge 0 : \mathbb{E}[A_{n+1}^i - A_n^i|\mathscr{F}_n] \ge 0 (i = 1,2)$, hence the processes A^i must be predictable $(\mathscr{F}_n)_{n\in\mathbb{N}}$-sub-martingales.
(b) A^i $(i = 1,2)$ are non-negative since by telescoping

$$A_n^i = A_0^i + \sum_{k=1}^{n} \Delta A_k^i = \sum_{k=1}^{n} \Delta A_k^i \ge 0$$

with $A_0^i = 0$ and $\Delta A_k^i = A_k^i - A_{k-1}^i$.
(c) M is also a generalized $(\mathscr{F}_n)_{n\in\mathbb{N}}$-martingale with $\mathbb{E}[M_{n+1} - M_n|\mathscr{F}_n] = 0$ and $\mathbb{E}[|M_{n+1}||\mathscr{F}_n] < +\infty$ for all $n \ge 0$.
(d) The process $Z_0 + M = (Z_0 + M_n)_{n\in\mathbb{N}}$ is a local $(\mathscr{F}_n)_{n\in\mathbb{N}}$-martingale too.

Second, we may estimate

$$Z_0 + M_n \ge Z_n - A_n^1 + A_n^2 \ge -\sup_{n\in\mathbb{N}} A_n^1 = -\limsup_{n\to+\infty} A_n^1 = -\lim_{n\to+\infty} A_n^1 > -\infty \tag{1.57}$$

since the limit $\lim_{n\to+\infty} A_n^1$ exists and is finite (a.s.) by the hypothesis of Theorem 1.3.8, and the non-decreasing A^i are non-negative. Therefore, the process $Z_0 + M$ represents a bounded $(\mathscr{F}_n)_{n\in\mathbb{N}}$-martingale. From the well-known Doob's martingale convergence theorem, we know that every bounded martingale has a finite almost sure limit. Hence, it follows that

$$\exists \lim_{n\to+\infty} Z_0 + M_n < +\infty(\mathbb{P} - a.s.)$$

and

$$\exists M_\infty := \lim_{n\to+\infty} M_n < +\infty(\mathbb{P} - a.s.)$$

with $\mathbb{E}[|M_{+\infty}|] < +\infty$. Consequently, the sequence $Y = (Y_n)_{n\in\mathbb{N}}$ of bounded, non-negative random variables Y_n satisfying

$$0 \le Y_n := Z_n + A_n^2 \le Z_0 + A_n^1 + M_n \tag{1.58}$$

for all $n \in \mathbb{N}$ must have a finite limit $Y_{+\infty} := \limsup_{n\to+\infty} Y_n$ which is bounded by

$$\limsup_{n\to+\infty} Y_n \le Z_0 + \limsup_{n\to+\infty} A_n^1 + \limsup_{n\to+\infty} M_n = Z_0 + \lim_{n\to+\infty} A_n^1$$
$$+ \lim_{n\to+\infty} M_n < +\infty \tag{1.59}$$

($\mathbb{P}$ -a.s.) (just pull the limit lim sup as $n \to +\infty$ over the inequality (1.58) and use the fact that the limits of both A^1 and M exist, and are finite a.s.). Furthermore, note that

$$0 \le Z_n \le Y_n, \quad 0 \le A_n^2 \le Y_n, \quad n \in \mathbb{N}.$$

This fact implies that the non-decreasing process A^2 must have a finite limit and

$$0 \le \limsup_{n\to+\infty} Z_n \le \limsup_{n\to+\infty} Y_n < +\infty, 0 \le \lim_{n\to+\infty} A_n^2 \le \limsup_{n\to+\infty} Y_n < +\infty$$

by applying the Bounded Monotone Sequence Theorem from calculus. Eventually, the identity

$$Z_n = Z_0 + A_n^1 - A_n^2 + M_n$$

for all $n \in \mathbb{N}$ yields that

$$Y_n = Z_n + A_n^2 = Z_0 + A_n^1 + M_n$$

for all $n \in \mathbb{N}$. Hence, the process $Z = (Y_n - A_n^2)_{n\in\mathbb{N}}$ satisfies

$$0 \le \liminf_{n\to+\infty} Y_n = Z_0 + \liminf_{n\to+\infty} A_n^1 + \liminf_{n\to+\infty} M_n = Z_0 + \lim_{n\to+\infty} A_n^1 + \lim_{n\to+\infty} M_n$$
$$= Z_0 + A_{+\infty}^1 + M_{+\infty} = \limsup_{n\to=\infty} Y_n < +\infty$$

in view of the identity in (1.59), Consequently, the process $Z = (Y_n - A_n^2)_{n\in\mathbb{N}}$ converges (a.s.) too since the finite limits of Y and A^2 exist (a.s.) from previous argumentation. This completes the proof of Theorem 1.3.8. ◇

Remark 1.3.8. Recall that all semi-martingales Z can be decomposed into

$$Z_n = Z_0 + BV_n + M_n$$

where BV is a process of bounded variation with $BV_0 = 0$ and M a local martingale (c.f. Doob-Meyer decomposition) with $M_0 = 0$. Moreover, each process BV of bounded variation possesses the representation

$$BV_n = A_n^1 - A_n^2$$

where A^i are non-decreasing processes. Therefore, Theorem 1.3.8 represents a theorem on the almost sure asymptotic behaviour (existence of limits) of non-negative semi-martingales.

Now we are in the position to state and verify our major findings. Let $\lambda(S)$ denote the eigenvalue of the inscribed matrix S and S^T be the transpose of matrix S.

Theorem 1.3.9 (Existence of a.s. Limits). *Assume that:*

(i) $Y_0 = X_0 \in \mathbb{R}^d$ *is independent of all* ξ_n^j.
(ii) All ξ_n^j *are independent random variables with* $\mathbb{E}[\xi_n^j] = 0$ *and* $\mathbb{E}[\xi_n^j]^2 = \sigma_j^2(n) < +\infty$.
(iii) All eigenvalues $\lambda(S_n) \leq 0$ *with*

$$S_n = \tilde{A}_n^T + \tilde{A}_n + \sum_{j=1}^{m} \sigma_j^2(n) \tilde{B}_{j,n}^T \tilde{B}_{j,n} + h_n \tilde{A}_n^T \tilde{A}_n.$$

Then, for system (1.52), the following limits exist and are finite ($\mathbb{P}$ $- a.s.$)

$$\lim_{n\to+\infty} X_n,$$

$$-\sum_{n=0}^{+\infty} X_n^T \left[\tilde{A}_n^T + \tilde{A}_n + \sum_{j=1}^{m} \sigma_j^2(n) \tilde{B}_{j,n}^T \tilde{B}_{j,n} + h_n \tilde{A}_n^T \tilde{A}_n \right] X_n h_n = A_{+\infty}^2 \quad \textit{and}$$

$$\sum_{n=0}^{+\infty} X_n^T \left(\sum_{j=1}^{m} \left(\tilde{B}_{j,n}^T + \tilde{B}_{j,n} + h_n \tilde{B}_{j,n}^T \tilde{A}_n + h_n \tilde{A}_n^T \tilde{B}_{j,n} \right) \xi_n^j \sqrt{h_n} \right.$$

$$\left. + \sum_{j,k=1}^{m} \tilde{B}_{j,n}^T \tilde{B}_{k,n} (\xi_n^j \xi_n^k - \mathbb{E}[\xi_n^j \xi_n^k]) h_n \right) X_n.$$

Proof. Suppose that $(X_n)_{n\in\mathbb{N}}$ satisfies (1.52) under $(i) - -(ii)$. Define the functional process $Z = (Z_n)_{n\in\mathbb{N}}$ by the square of the Euclidean norm

$$Z_n = ||X_n||_d^2 = X_n^T X_n$$

for all $n \in \mathbb{N}$. Calculate

$$
\begin{aligned}
Z_{n+1} &= X_{n+1}^T X_{n+1} \\
&= \left((I + h_n \tilde{A}_n + \sqrt{h_n} \sum_{j=1}^{m} \tilde{B}_{j,n} \xi_n^j) X_n \right)^T \left(I + h_n \tilde{A}_n + \sqrt{h_n} \sum_{k=1}^{m} \tilde{B}_{k,m} \xi_n^k \right) X_n \\
&= X_n^T \Bigg(I + h_n \tilde{A}_n^T + \sqrt{h_n} \sum_{j=1}^{m} \tilde{B}_{j,n}^T \xi_n^j + h_n \tilde{A}_n + h_n^2 \tilde{A}_n^T \tilde{A}_n + h_n^{3/2} \sum_{j=1}^{m} \tilde{B}_{j,n}^T \tilde{A}_n \xi_n^j \\
&\quad + \sqrt{h_n} \sum_{k=1}^{m} \tilde{B}_{k,n} \xi_n^k + h_n^{3/2} \sum_{k=1}^{m} \tilde{A}_n^T \tilde{B}_{k,n} \xi_n^k + h_n \sum_{j,k=1}^{m} \tilde{B}_{j,n}^T \tilde{B}_{k,n} \xi_n^j \xi_n^k \Bigg) X_n \\
&= X_n^T X_n + X_n^T \left[\tilde{A}_n^T + \tilde{A}_n + \sum_{j,k=1}^{m} \tilde{B}_{j,n}^T \tilde{B}_{k,n} \mathbb{E}[\xi_n^j \xi_n^k] + h_n \tilde{A}_n^T \tilde{A}_n \right] X_n h_n \\
&\quad + X_n^T \left[\sum_{j=1}^{m} \tilde{B}_{j,n}^T \xi_n^j + \sum_{k=1}^{m} \tilde{B}_{k,n} \xi_n^k + \sum_{j=1}^{m} \tilde{B}_{j,n}^T \tilde{A}_n \xi_n^j h_n + \sum_{k=1}^{m} \tilde{A}_n^T \tilde{B}_{k,n} \xi_n^k h_n \right] X_n \sqrt{h_n} \\
&\quad + \sum_{j,k=1}^{m} X_n^T \tilde{B}_{j,n}^T \tilde{B}_{k,n} X_n (\xi_n^j \xi_n^k - \mathbb{E}[\xi_n^j \xi_n^k]) h_n \\
&= X_n^T X_n + X_n^T \left[\tilde{A}_n^T + \tilde{A}_n + \sum_{j=1}^{m} \tilde{B}_{j,n}^T \tilde{B}_{j,n} \mathbb{E}[\xi_n^j]^2 + h_n \tilde{A}_n^T \tilde{A}_n \right] X_n h_n \\
&\quad + X_n^T \left[\sum_{j=1}^{m} \left(\tilde{B}_{j,n}^T + \tilde{B}_{j,n} + h_n \tilde{B}_{j,n}^T \tilde{A}_n + h_n \tilde{A}_n^T \tilde{B}_{j,n} \right) \right] X_n \sqrt{h_n} \xi_n^j \\
&\quad + \sum_{j,k=1}^{m} X_n^T \tilde{B}_{j,n}^T \tilde{B}_{k,n} X_n (\xi_n^j \xi_n^k - \mathbb{E}[\xi_n^j \xi_n^k]) h_n \\
&= Z_n + \Delta A_n^1 - \Delta A_n^2 + \Delta M_n
\end{aligned} \tag{1.60}
$$

where $\Delta A_n^i = A_{n+1}^i - A_n^i$, and we set $A_n^1 = 0$ (for all $n \in \mathbb{N}$), $M_0 = 0$, $A_0^2 = 0$ and

$$
A_n^2 = -\sum_{l=0}^{n-1} X_l^T \left[\tilde{A}_l^T + \tilde{A}_l + \sum_{j=1}^{m} \sigma_j^2(l) \tilde{B}_{j,l}^T \tilde{B}_{j,l} + h_l \tilde{A}_l^T \tilde{A}_l \right] X_l h_l = \sum_{l=0}^{n-1} \Delta A_l^2 \tag{1.61}
$$

$$M_n = \sum_{l=0}^{n-1} X_l^T \left(\sum_{j=1}^{m} \Big(\tilde{B}_{j,l}^T + \tilde{B}_{j,l} + \tilde{B}_{j,l}^T \tilde{l} h_l + \tilde{A}_l^T \tilde{B}_{j,l} h_l \Big) \xi_l^j \sqrt{h_l} \right.$$

$$\left. + \sum_{j,k=1}^{m} \tilde{B}_{j,l}^T \tilde{B}_{k,l} (\xi_l^j \xi_l^k - \mathbb{E}[\xi_l^j \xi_l^k]) h_l \right) X_l \tag{1.62}$$

for all $n \geq 1$. Recall the definitions of differences $\Delta A_n^j = A_{n+1}^j - A_n^j$ and $\Delta M_n = M_{n+1} - M_n$ for all $n \in \mathbb{N}$. Next, telescoping and summing up over n in (1.60) implies that

$$Z_n = Z_0 + A_n^1 - A_n^2 + M_n$$

with $A^1 \equiv 0$. Now, recall the assumption (iii). This guarantees that $A_n^2 \geq 0$ for all values of x_l and A_n^2 is non-decreasing in n. Consequently, all assumptions of Theorem 1.3.8 with $A^1 \equiv 0$ are fulfilled. Hence, a direct application of Theorem 1.3.8 to the non-negative sequence $Z = (Z_n)_{n \in \mathbb{N}}$ yields that

$$\exists \lim_{n \to +\infty} Z_n = \lim_{n \to +\infty} ||X_n||_d^2 < +\infty \quad (\mathbb{P} - a.s.).$$

From the continuity of the Euclidean norm $||.||_d$, we conclude that

$$-\infty < \exists \lim_{n \to +\infty} X_n < +\infty \quad (\mathbb{P} - a.s.)$$

– a fact which confirms the main assertion of Theorem 1.3.9 together with the bounded martingale inequality $Z_0 + M_n \geq 0$ (a.s.) for all $n \in \mathbb{N}$. ◇

Theorem 1.3.10 (A.s. Stability). *Assume that:*

(i) $x_0 \in \mathbb{R}^d$ *is independent of all* ξ_n^j.
(ii) All ξ_n^j *are independent random variables with* $\mathbb{E}[\xi_n^j] = 0$, $\mathbb{E}[\xi_n^j]^2 = \sigma_j^2(n) < +\infty$.
(iii) $\exists c > 0$ *constant such that* $\forall n \in \mathbb{N}$ *we have all eigenvalues* $\lambda(\hat{S}_n) \leq -c < 0$ *with*

$$\hat{S}_n = \left(\tilde{A}_n^T + \tilde{A}_n + \sum_{j=1}^{m} \sigma_j^2(n) \tilde{B}_{j,n}^T \tilde{B}_{j,n} + h_n \tilde{A}_n^T \tilde{A}_n \right) h_n.$$

Then, $\lim_{n \to +\infty} X_n = 0$, *i.e. the trivial solution* $x = 0$ *of the system (1.52) is globally a.s. asymptotically stable.*

Proof. First, apply previous Theorem 1.3.9 to get the a.s. convergence and existence of the limit $\lim_{n \to +\infty} X_n$ since all eigenvalues also satisfy $\lambda(\hat{S}_n) \leq 0$ under (iii). From the proof of Theorem 1.3.9 based on Theorem 1.3.8, we also know that

$$\lim_{n\to+\infty} \tilde{A}_n^2 = -\sum_{l=0}^{+\infty} X_l^T \left[\tilde{A}_l^T + \tilde{A}_l + \sum_{j=1}^{m} \sigma_j^2(l) \tilde{B}_{j,l}^T \tilde{B}_{j,l} + h_l \tilde{A}_l^T \tilde{A}_l \right] X_l h_l = \sum_{l=0}^{+\infty} \Delta A_l^2.$$

From calculus of series, the convergence of a series requires that each series member ΔA_l^2 must converge to zero ($\mathbb{P}$ -a.s.). Therefore, we arrive at the estimation

$$\begin{aligned} 0 \le c \lim_{n\to+\infty} ||X_n||_d^2 &\le \lim_{n\to+\infty} \min -\lambda(\hat{S}_n)||X_n||_d^2 = -\lim_{n\to+\infty} \max \lambda(\hat{S}_n)||X_n||_d^2 \\ &\le \lim_{n\to+\infty} \Delta \tilde{A}_n^2 = 0. \end{aligned} \tag{1.63}$$

Consequently, squeezing yields that

$$\lim_{n\to+\infty} ||X_n||_d^2 = 0 \quad (\mathbb{P} - a.s.).$$

This leads us to the conclusion that

$$\lim_{n\to+\infty} X_n = 0 \quad (\mathbb{P} - a.s.),$$

thanks to the continuity and definiteness of the Euclidean norm in $\mathbb{R}^d$. ◇

Note that above (see (1.63)) we have used the known facts from linear algebra that all symmetric matrices $S \in \mathbb{R}^{d\times d}$ have real eigenvalues, a complete set of eigenvalues which form an orthogonal basis in $\mathbb{R}^d$ and, for such matrices, we have the uniform estimation

$$\min \lambda(S)||x||_d^2 \le x^T S x \le \max \lambda(S)||x||_d^2$$

where $\lambda(S)$ are the eigenvalues of inscribed matrix S.

A two-dimensional example (Systems with diagonal-noise terms). Consider the two-dimensional system of stochastic difference equations

$$\begin{aligned} x_{n+1} &= x_n + [\theta_n(a_{n+1}x_{n+1} + b_{n+1}y_{n+1}) + (1-\theta_n)(a_n x_n + b_n y_n)]h_n \\ &\quad + \sigma_n^1 \sqrt{h_n} x_n \xi_n \end{aligned} \tag{1.64}$$

$$\begin{aligned} y_{n+1} &= y_n + [\theta_n(c_{n+1}x_{n+1} + d_{n+1}y_{n+1}) + (1-\theta_n)(c_n x_n + d_n y_n)]h_n \\ &\quad + \sigma_n^2 \sqrt{h_n} y_n \xi_n \end{aligned} \tag{1.65}$$

where $a_n, b_n, c_n, \sigma_n^i$ are real constants and ξ_n are independent standardized random variables with moments $\mathbb{E}[\xi_n] = 0$ and $\mathbb{E}[\xi_n]^2 = 1$ for all $n \in \mathbb{N}$. (e.g. take i.i.d. $\xi_n \in \mathcal{N}(0,1)$, but the property of being Gaussian is not essential for the validity of our findings, however the fact of independence and properties of finite first two moments does). Obviously, system (1.64)–(1.65) has the form (1.50) with $d = 2$, $m = 1$, and matrices

$$A_n = \begin{pmatrix} a_n & b_n \\ c_n & d_n \end{pmatrix}, \quad B_n = \begin{pmatrix} \sigma_n^1 & 0 \\ 0 & \sigma_n^2 \end{pmatrix}. \tag{1.66}$$

To apply Theorem 1.3.9, we need to calculate $S_n = \tilde{A}_n^T + \tilde{A}_n + \tilde{B}_n^T \tilde{B}_n + h_n \tilde{A}_n^T \tilde{A}_n$. For this purpose, we study the equivalent system (1.52) with matrices

$$\tilde{A}_n = \begin{pmatrix} p_n & q_n \\ r_n & s_n \end{pmatrix},$$
$$\tilde{B}_n = \begin{pmatrix} -\dfrac{(-1+\theta_n h_n d_{n+1})\sigma_n^1}{det_n} & \dfrac{\theta_n h_n b_{n+1}\sigma_n^2}{det_n} \\ \dfrac{\theta_n h_n c_{n+1}\sigma_n^1}{det_n} & -\dfrac{(-1+\theta_n h_n a_{n+1})\sigma_n^2}{det_n} \end{pmatrix}. \tag{1.67}$$

where

$$det_n = 1-\theta_n h_n d_{n+1} - \theta_n h_n a_{n+1} + \theta_n^2 h_n^2 a_{n+1} d_{n+1} - \theta_n^2 h_n^2 b_{n+1} c_{n+1}, \tag{1.68}$$

$$p_n = \frac{-(-1+\theta_n h_n d_{n+1})(\theta_n a_{n+1} + (1-\theta_n) a_n) + \theta_n h_n b_{n+1}(\theta_n c_{n+1} + (1-\theta_n) c_n)}{1-\theta_n h_n d_{n+1} - \theta_n h_n a_{n+1} + \theta_n^2 h_n^2 a_{n+1} d_{n+1} - \theta_n^2 h_n^2 b_{n+1} c_{n+1}},$$

$$q_n = \frac{-(-1+\theta_n h_n d_{n+1})(\theta_n b_{n+1} + (1-\theta_n) b_n) + \theta_n h_n b_{n+1}(\theta_n d_{n+1} + (1-\theta_n) d_n)}{1-\theta_n h_n d_{n+1} - \theta_n h_n a_{n+1} + \theta_n^2 h_n^2 a_{n+1} d_{n+1} - \theta_n^2 h_n^2 b_{n+1} c_{n+1}},$$

$$r_n = \frac{\theta_n h_n c_{n+1}(\theta_n a_{n+1} + (1-\theta_n) a_n) - (-1+\theta_n h_n a_{n+1})(\theta_n c_{n+1} + (1-\theta_n) c_n)}{1-\theta_n h_n d_{n+1} - \theta_n h_n a_{n+1} + \theta_n^2 h_n^2 a_{n+1} d_{n+1} - \theta_n^2 h_n^2 b_{n+1} c_{n+1}},$$

$$s_n = \frac{\theta_n h_n c_{n+1}(\theta_n b_{n+1} + (1-\theta_n) b_n) - (-1+\theta_n h_n a_{n+1})(\theta_n d_{n+1} + (1-\theta_n) d_n)}{1-\theta_n h_n d_{n+1} - \theta_n h_n a_{n+1} + \theta_n^2 h_n^2 a_{n+1} d_{n+1} - \theta_n^2 h_n^2 b_{n+1} c_{n+1}}.$$

Thus, one obtains

$$S_n = \begin{pmatrix} u_n & w_n \\ w_n & v_n \end{pmatrix} \tag{1.69}$$

with

$$u_n = -2\frac{(-1+\theta_n h_n d_{n+1})(\theta_n a_{n+1} + (1-\theta_n) a_n)}{1-\theta_n h_n d_{n+1} - \theta_n h_n a_{n+1} + \theta_n^2 h_n^2 a_{n+1} d_{n+1} - \theta_n^2 h_n^2 b_{n+1} c_{n+1}} \tag{1.70}$$
$$+2\frac{\theta_n h_n b_{n+1}(\theta_n c_{n+1} + (1-\theta_n) c_n)}{1-\theta_n h_n d_{n+1} - \theta_n h_n a_{n+1} + \theta_n^2 h_n^2 a_{n+1} d_{n+1} - \theta_n^2 h_n^2 b_{n+1} c_{n+1}}$$

$$
\begin{aligned}
&+\frac{(-1+\theta_n\, h_n d_{n+1})^2\,(\sigma_n^1)^2}{\left(1-\theta_n\, h_n d_{n+1}-\theta_n\, h_n a_{n+1}+\theta_n^2 h_n^2 a_{n+1}\, d_{n+1}-\theta_n^2 h_n^2 b_{n+1}\, c_{n+1}\right)^2}\\
&+\frac{\theta_n^2 h_n^2 c_{n+1}^2 (\sigma_n^1)^2}{\left(1-\theta_n\, h_n d_{n+1}-\theta_n\, h_n a_{n+1}+\theta_n^2 h_n^2 a_{n+1}\, d_{n+1}-\theta_n^2 h_n^2 b_{n+1}\, c_{n+1}\right)^2}\\
&+h_n\left(p_n^2+r_n^2\right)\\
w_n =& \frac{\theta_n\, h_n c_{n+1}\,(\theta_n\, a_{n+1}+(1-\theta_n)\, a_n)}{1-\theta_n\, h_n d_{n+1}-\theta_n\, h_n a_{n+1}+\theta_n^2 h_n^2 a_{n+1}\, d_{n+1}-\theta_n^2 h_n^2 b_{n+1}\, c_{n+1}}\\
&-\frac{(-1+\theta_n\, h_n a_{n+1})\,(\theta_n\, c_{n+1}+(1-\theta_n)\, c_n)}{1-\theta_n\, h_n d_{n+1}-\theta_n\, h_n a_{n+1}+\theta_n^2 h_n^2 a_{n+1}\, d_{n+1}-\theta_n^2 h_n^2 b_{n+1}\, c_{n+1}}\\
&-\frac{(-1+\theta_n\, h_n d_{n+1})\,(\theta_n\, b_{n+1}+(1-\theta_n)\, b_n)}{1-\theta_n\, h_n d_{n+1}-\theta_n\, h_n a_{n+1}+\theta_n^2 h_n^2 a_{n+1}\, d_{n+1}-\theta_n^2 h_n^2 b_{n+1}\, c_{n+1}}\\
&+\frac{\theta_n\, h_n b_{n+1}\,(\theta_n\, d_{n+1}+(1-\theta_n)\, d_n)}{1-\theta_n\, h_n d_{n+1}-\theta_n\, h_n a_{n+1}+\theta_n^2 h_n^2 a_{n+1}\, d_{n+1}-\theta_n^2 h_n^2 b_{n+1}\, c_{n+1}}\\
&-\frac{(-1+\theta_n\, h_n d_{n+1})\,\sigma_n^1\,\theta_n\, h_n {-} b_{n+1}\,\sigma_n^2}{\left(1-\theta_n\, h_n d_{n+1}-\theta_n\, h_n a_{n+1}+\theta_n^2 h_n^2 a_{n+1}\, d_{n+1}-\theta_n^2 h_n^2 b_{n+1}\, c_{n+1}\right)^2}\\
&-\frac{\theta_n\, h_n c_{n+1}\,\sigma_n^1\,(-1+\theta_n\, h_n a_{n+1})\,\sigma_n^2}{\left(1-\theta_n\, h_n d_{n+1}-\theta_n\, h_n a_{n+1}+\theta_n^2 h_n^2 a_{n+1}\, d_{n+1}-\theta_n^2 h_n^2 b_{n+1}\, c_{n+1}\right)^2}\\
&+h_n\,(p_n q_n+r_n s_n)\\
v_n =& \,2\,\frac{\theta_n\, h_n c_{n+1}\,(\theta_n\, b_{n+1}+(1-\theta_n)\, b_n)}{1-\theta_n\, h_n d_{n+1}-\theta_n\, h_n a_{n+1}+\theta_n^2 h_n^2 a_{n+1}\, d_{n+1}-\theta_n^2 h_n^2 b_{n+1}\, c_{n+1}}\\
&-2\,\frac{(-1+\theta_n\, h_n a_{n+1})\,(\theta_n\, d_{n+1}+(1-\theta_n)\, d_n)}{1-\theta_n\, h_n d_{n+1}-\theta_n\, h_n a_{n+1}+\theta_n^2 h_n^2 a_{n+1}\, d_{n+1}-\theta_n^2 h_n^2 b_{n+1}\, c_{n+1}}\\
&+\frac{\theta_n^2 h_n^2 b_{n+1}^2 (\sigma_n^2)^2}{\left(1-\theta_n\, h_n d_{n+1}-\theta_n\, h_n a_{n+1}+\theta_n^2 h_n^2 a_{n+1}\, d_{n+1}-\theta_n^2 h_n^2 b_{n+1}\, c_{n+1}\right)^2}\\
&+\frac{(-1+\theta_n\, h_n a_{n+1})^2\,(\sigma_n^2)^2}{\left(1-\theta_n\, h_n d_{n+1}-\theta_n\, h_n a_{n+1}+\theta_n^2 h_n^2 a_{n+1}\, d_{n+1}-\theta_n^2 h_n^2 b_{n+1}\, c_{n+1}\right)^2}\\
&+h_n\left(q_n^2+s_n^2\right) \qquad (1.71)
\end{aligned}
$$

where p_n, q_n, r_n and s_n as in (1.68) above. It is hard to evaluate the eigenvalues λ of this matrix S_n analytically (i.e. depending on all possible parameters θ_n, h_n, etc.). Packages such as MAPLE return several pages of long expressions for all eigenvalues in terms of all parameters. However, it is fairly easy to check the sign of all eigenvalues with the help of symbolic packages such as MAPLE, MATHEMATICA, MATLAB, etc. for given fixed sets of parameters. For example, for the constant parameter set (i.e. the trapezoidal and midpoint methods)

$$a_n = -1, b_n = 1, c_n = -5, d_n = -5, \sigma_n^1 = \sigma_n^2 = 0.01, \theta_n = 0.5, h_n = 0.01,$$

the eigenvalues are both negative, namely

$$\lambda_n = -.342381027965132034\ldots, -11.0035987828310819\ldots$$

which guarantees us almost sure asymptotic stability and convergence to 0.

A convenient way to evaluate the eigenvalues of symmetric matrices is to apply the following lemma. Notice that the entries u_n, v_n and w_n of S_n depend on all system-parameters, above all on h_n and θ_n. This dependence-relation is important in order to control the qualitative behaviour of numerical discretizations (1.50) of SDEs (1.51). For a simplification of its practical evaluation, we present Lemma 1.3.3.

Lemma 1.3.3. *The eigenvalues of 2D matrices*

$$S_n = \begin{pmatrix} u_n & w_n \\ w_n & v_n \end{pmatrix} \tag{1.72}$$

can be easily computed by

$$\lambda_{1/2}(S_n) = \frac{u_n + v_n}{2} \pm \sqrt{\left(\frac{u_n - v_n}{2}\right)^2 + w_n^2} \tag{1.73}$$

and estimated by

$$\lambda_{1/2}(S_n) \leq \max\{u_n, v_n\} + |w_n| \tag{1.74}$$

where u_n, v_n, w_n are as in (1.70) for $n \in I\!N$.

Proof. Note that the characteristic polynomial of S_n is

$$p(\lambda) = det(S_n - \lambda I) = (u_n - \lambda)(v_n - \lambda) - w_n^2$$

with roots

$$\begin{aligned} \lambda_{1/2}(S_n) &= \frac{u_n + v_n}{2} + \sqrt{\left(\frac{u_n - v_n}{2}\right)^2 + w_n^2} \\ &\leq \frac{u_n + v_n}{2} + \frac{|u_n - v_n|}{2} + |w_n| \quad \text{(while using } \sqrt{x^2 + y^2} \leq |x| + |y|\text{)} \\ &= \max\{u_n, v_n\} + |w_n|. \end{aligned}$$

This confirms the assertion of Lemma 1.3.3.

Therefore, the direct evaluation of these eigenvalues by Lemma 1.3.3 and an application of Theorem 1.3.9 provide the following immediate consequence under the assumptions of this subsection.

Corollary 1.3.1 (Corollary to Theorem 1.3.9). *Assume that, for all $n \in \mathbb{N}$, we have*

$$\max\{u_n, v_n\} + |w_n| \leq 0$$

where u_n, v_n and w_n are defined as in (1.70).

Then, for all adapted initial values $(x_0 y_0)^T$, the following limits for system (1.64)-(1.65) with $\tilde{A}_n$, $\tilde{B}_n$ defined by (1.67) exist and are finite ($\mathbb{P}$ -a.s.)

$$\limsup_{n\to+\infty} x_n, \limsup_{n\to+\infty} y_n, \liminf_{n\to+\infty} x_n, \liminf_{n\to+\infty} y_n,$$

$$\lim_{n\to+\infty} \begin{pmatrix} x_n & y_n \end{pmatrix} \left[\tilde{A}_n^T + \tilde{A}_n + \tilde{B}_n^T \tilde{B}_n + h_n \tilde{A}_n^T \tilde{A}_n\right] \begin{pmatrix} x_n \\ y_n \end{pmatrix} = 0,$$

$$\lim_{n\to+\infty} \begin{pmatrix} x_n & y_n \end{pmatrix} \Big[(\tilde{B}_n^T + \tilde{B}_n + h_n \tilde{B}_n^T \tilde{A}_n + h_n \tilde{A}_n^T \tilde{B}_n)\xi_n \sqrt{h_n}$$

$$+ \tilde{B}_n^T \tilde{B}_n (\xi_n^2 - 1) h_n\Big] \begin{pmatrix} x_n \\ y_n \end{pmatrix} = 0.$$

Note that the main assumption of Corollary 1.3.1 requires that at least $u_n + v_n \leq 0$ for all $n \in \mathbb{N}$, where u_n and v_n are defined by (1.70). As an immediate consequence of Theorem 1.3.10 applied to $\hat{S}_n = S_n h_n$ and Lemma 1.3.3, we are able to find the following corollary on a.s. asymptotic stability.

Corollary 1.3.2 (Corollary to Theorem 1.3.10). *Assume that $\exists c = constant > 0$ such that $\forall n \in \mathbb{N}$*

$$(\max\{u_n, v_n\} + |w_n|)\, h_n \leq -c < 0$$

where u_n, v_n and w_n are defined by (1.70).

Then, for all adapted initial values $(x_0 y_0)^T$, the limits satisfy

$$\lim_{n\to+\infty} x_n = 0, \quad \lim_{n\to+\infty} y_n = 0 \quad (\mathbb{P} - a.s.),$$

i.e. global almost sure asymptotic stability of the trivial solution for system (1.64)-(1.65) is observed.

Remark 1.3.9. Moreover, the conditions of Corollaries 1.3.1 and 1.3.2 need to be fulfilled only for all $n \geq n_0$ where $n_0 \in \mathbb{N}$ is sufficiently large (due to the asymptotic character of main assertion).

Remark 1.3.10. To summarize our findings, an evaluation of eigenvalues λ in our example shows a strong dependence of λ on the determinants $\det_n$ determined by (1.68). Stabilizing effects of these determinants on the fluctuation of λ can be observed and maximized whenever

$$d_n \leq 0, \quad a_n \leq 0, \quad b_n c_n \leq 0 \tag{1.75}$$

for all $n \geq n_0$. In this case, one can show that $det_n > 1$ for all $n \geq n_0$ and "fluctuations of eigenvalues" λ are also under some control and the choice of step sizes h_n is not very sensitive for qualitative behavior of Theta methods. Moreover, an increase of implicitness $\theta \geq 0$ leads to stabilization of numerical dynamics under (1.75).

Remark 1.3.11. In order to avoid "unwanted spurious" oscillations through Theta-methods in our example we recommend to use only step sizes h_n and implicitness θ_n which guarantee that

$$\theta_n h_n \max\{d_{n+1}, a_{n+1}\} < 1$$

for all $n \in \mathbb{N}$. There are some hints to prefer the choice $\theta_n = 0.5$ to circumvent the problem of over- or under-stabilization through Theta-methods (cf. [118,120–122]).

This subsection has shown how to exploit semi-martingale convergence theorems (cf. Theorem 1.3.8) to discuss the existence of finite almost sure limits and global asymptotic stability for stochastic Theta methods applied to systems of stochastic differential equations. Sufficient conditions (cf. Theorems 1.3.9, 1.3.10) for almost sure stability and convergence of their solutions could be verified in terms of all parameters of non-autonomous systems such as variable step sizes h_n and variable implicitness θ_n. These conditions are applicable (see Sect. 1.3.5) and can be easily evaluated by computer packages such as MAPLE, MATHEMATICA, MATLAB, etc. It would be interesting to extend the main results to other type of noise sources (i.e. non-martingale-type or noise with infinite moments). Our conditions on the noise are fairly general (since we only asked for existence of first and second moments). However, a key assumption is the complete independence of all noise terms ξ_n^j. The relaxation of this major assumption is a potential subject for further research. It is also possible to extend the presented results to other numerical methods or to systems of nonlinear stochastic differential equations under appropriate conditions (recall that balanced Theta methods (1.1) represent the simplest parametric family of stochastic Runge-Kutta methods), thanks to the generality of Semimartingale Convergence Theorem 1.3.8).

1.3.6 Asymptotic Almost Sure Stability for Pure Diffusion Processes

This subsection is devoted to the analysis of discretizations of pure diffusion processes as the simplest class of stochastic test equations in $\mathbb{R}^1$.

Theorem 1.3.11. *Let* $V = (V(n))_{n \in \mathbb{N}}$ *be a sequence of non-negative random variables* $V(n) : (\Omega, \mathcal{F}_n, \mathbb{P}) \to \mathbb{R}^1_+$ *with* $V(0) > 0$ *satisfying the recursive scheme*

$$V(n+1) = V(n)G(n) \tag{1.76}$$

where $G(n) : (\Omega, \mathcal{F}_n, \mathbb{P}) \to \mathbb{R}^1_+$ *are i.i.d. random variables with the moment property* $\mathbb{E}|\ln[G(n)]| < +\infty$. *Then*

$$V \text{ (globally) asymptotically a.s. stable} \quad \text{iff} \quad \mathbb{E}\ln[G(0)] < 0\,.$$

Proof. This result is already found in [48] and in [125]. The main idea is to use the strong law of large numbers (SLLN) in conjunction with the law of iterated logarithm (LIL). Note that V possesses the explicit representation

$$V(n+1) = \left(\prod_{k=0}^{n} G(k)\right) V(0) \tag{1.77}$$

for all $n \in \mathbb{N}$. Now, suppose that $G(k)$ are i.i.d. random variables and define

$$\mu_0 := \mathbb{E}[\ln(G(0))], \quad S_n := \sum_{k=0}^{n-1} \ln G(k),$$

hence $V(n+1) = \exp(S_{n+1})V(0)$ and $\mathbb{E}[S_n] = n\mu$ for $n \in \mathbb{N}$. By SLLN, conclude that

$$\lim_{n \to +\infty} \frac{S_n}{n} = \mu_0 \ (a.s.)$$

thanks to the $\mathbb{P}$ -integrability of $G(k)$. This fact implies that if $\mu_0 < 0$ then $S_n \to -\infty$, i.e. $V(n) \to 0$ as n tends to $+\infty$ and if $\mu > 0$ then $S_n \to +\infty$, i.e. $V(n) \to +\infty$ as n tends to $+\infty$. Moreover, in the case $\mu_0 = 0$, we may use LIL (at first, under $\sigma^2 = Var(\ln G(k)) = \mathbb{E}[\ln G(k) - \mathbb{E}\ln G(k)]^2 < +\infty$, later we may drop $\sigma^2 < +\infty$ by localization procedures) to get

$$\liminf_{n \to +\infty} \frac{S_n}{\sqrt{2n \ln\ln n}} = -|\sigma|, \quad \limsup_{n \to +\infty} \frac{S_n}{\sqrt{2n \ln\ln n}} = |\sigma|,$$

hence S_n oscillates with growing amplitude and $\lim_{n \to +\infty} S_n$ does not exist. Therefore

$$\lim_{n \to +\infty} V(n) = \lim_{n \to +\infty} \exp(S_n)V(0)$$

does not exist either (a.s.). Thus, $\lim_{n \to +\infty} V(n) \neq 0$ and the proof is complete. ◇

Now, consider the one-dimensional test class of pure diffusion equations (a Girsanov SDE of Itô-type)

$$dX(t) = \sigma X(t)\, dW(t) \tag{1.78}$$

discretized by BIMs

$$Y_{n+1} = Y_n + \sigma Y_n \Delta W_n + (c^0 h_n + c^1 |\Delta W_n|)(Y_n - Y_{n+1}) \qquad (1.79)$$

as suggested by Milstein, Platen and Schurz [96]. Then, the following result provides a mathematical evidence that their numerical experiments for BIMs (1.79) led to the correct observation of numerical stability due to its asymptotic a.s. stability. It extends results which are found in [48, 120] and [125].

Theorem 1.3.12 (A.s. asymptotic stability of BIMs with constant h, [127]). *The BIMs (1.79) with scalar weights $c^0 = 0$ and $c^1 = |\sigma|$ applied to martingale test (1.78) for any parameter $\sigma \in I\!R^1 \setminus \{0\}$ with any equidistant step size h provide (globally) asymptotically a.s. stable sequences $Y = (Y_n)_{n \in I\!N}$.*

Proof. Suppose $|\sigma| > 0$. Then, the proof is an application of Theorem 1.3.11. For this purpose, consider the sequence $V = (V(n))_{n \in \mathbb{N}} = (|Y_n|)_{n \in \mathbb{N}}$. Note that $V(n+1) = G(n)V(n)$, $\mathbb{E}|\ln G(n)| < +\infty$ and $\mathbb{E}[\ln G(n)] < 0$ since

$$\mathbb{E}[|\ln G(n)|] \leq (\mathbb{E}[\ln G(n)]^2)^{1/2} \leq \ln(2) + |\sigma|\sqrt{h} \quad \text{and}$$

$$\begin{aligned}
\mathbb{E}[\ln G(n)] &= \mathbb{E}\left[\ln\left|\frac{1 + |\sigma \Delta W_n| + \sigma \Delta W_n}{1 + |\sigma \Delta W_n|}\right|\right] = \mathbb{E}\left[\ln\left|1 + \frac{\sigma \Delta W_n}{1 + |\sigma \Delta W_n|}\right|\right] \\
&= \frac{1}{2}\mathbb{E}\left[\ln\left|1 + \frac{\sigma \Delta W_n}{1 + |\sigma \Delta W_n|}\right|\right] + \frac{1}{2}\mathbb{E}\left[\ln\left|1 - \frac{\sigma \Delta W_n}{1 + |\sigma \Delta W_n|}\right|\right] \\
&= \frac{1}{2}\mathbb{E}\left[\ln\left|1 - \left(\frac{\sigma \Delta W_n}{1 + |\sigma \Delta W_n|}\right)^2\right|\right] < -\frac{1}{2}\mathbb{E}\left[\left(\frac{|\sigma \Delta W_n|}{1 + |\sigma \Delta W_n|}\right)^2\right] < 0
\end{aligned}$$

with independently identically Gaussian distributed increments $\Delta W_n \in \mathcal{N}(0, h)$ (In fact, note that, for all $\sigma \neq 0$ and Gaussian ΔW_n, we have

$$0 < 1 - \left(\frac{\sigma \Delta W_n}{1 + |\sigma \Delta W_n|}\right)^2 < 1$$

with probability one, hence, that ΔW_n has a non-degenerate probability distribution with nontrivial support is essential here!). Therefore, the assumptions of Theorem 1.3.11 are satisfied and an application of Theorem 1.3.11 yields the claim of Theorem 1.3.12. Thus, the proof is complete. ◇

Remark 1.3.12. The increments $\Delta W_n \in \mathcal{N}(0, h_n)$ can also be replaced by multi-point discrete probability distributions such as

$$\begin{aligned}
&\mathbb{P}\left\{\Delta W_n = \pm\sqrt{h_n}\right\} = \frac{1}{2} \\
\text{or} \quad &\mathbb{P}\{\Delta W_n = 0\} = \frac{2}{3}, \ \mathbb{P}\left\{\Delta W_n = \pm\sqrt{3h_n}\right\} = \frac{1}{6}
\end{aligned}$$

as commonly met in weak approximations. In this case, the almost sure stability of the BIMs as chosen by Theorem 1.3.12 is still guaranteed, as seen by our proof above (due to the inherent symmetry of ΔW_n with respect to 0).

For variable step sizes, we can also formulate and prove a general assertion with respect to asymptotic a.s. stability. Let $Var(Z)$ denote the variance of the inscribed random variable Z.

Theorem 1.3.13 (A.s. asymptotic stability of linear recursions, S. [127]). *Let* $V = (V(n))_{n\in\mathbb{N}}$ *be a sequence of non-negative random variables* $V(n) : (\Omega, \mathcal{F}_n, \mathbb{P}) \to \mathbb{R}^1_+$ *with* $V(0) > 0$ *satisfying the recursive scheme*

$$V(n+1) \;=\; V(n)G(n), \tag{1.80}$$

where $G(n) : (\Omega, \mathcal{F}_n, \mathbb{P}) \to \mathbb{R}^1_+$ *are independent random variables such that* $\exists$ *nonrandom sequence* $b = (b_n)_{n\in\mathbb{N}}$ *with* $b_n \to +\infty$ *as* $n \to +\infty$

$$\sum_{k=0}^{+\infty} \frac{Var(\ln(G(k)))}{b_k^2} \;<\; +\infty, \qquad \lim_{n\to+\infty} \frac{\sum_{k=0}^{n-1} \mathbb{E}\ln(G(k))}{b_n} \;<\; 0. \tag{1.81}$$

Then $V = (V(n))_{n\to+\infty}$ *is (globally) asymptotically a.s. stable sequence, i.e. we have* $\lim_{n\to+\infty} V(n) \;=\; 0$ *(a.s.).*
Moreover, if

$$\sum_{k=0}^{+\infty} \frac{Var(\ln(G(k)))}{b_k^2} \;<\; +\infty, \qquad \lim_{n\to+\infty} \frac{\sum_{k=0}^{n-1} \mathbb{E}\ln(G(k))}{b_n} \;>\; 0 \tag{1.82}$$

then $V = (V(n))_{n\to+\infty}$ *is (globally) asymptotically a.s. unstable sequence, i.e. we have* $\lim_{n\to+\infty} V(n) \;=\; +\infty$ *(a.s.) for all nonrandom* $y_0 \neq 0$.

Proof. The main idea is to apply Kolmogorov's SLLN, see Shiryaev [134] (p. 389). Recall that V possesses the explicit representation (1.77). Now, define

$$S_n := \sum_{k=0}^{n-1} \ln G(k),$$

hence $V(n+1) \;=\; \exp(S_{n+1})V(0)$ for $n \in \mathbb{N}$. By Kolmogorov's SLLN we may conclude that

$$\lim_{n\to+\infty} \frac{S_n}{b_n} = \lim_{n\to+\infty} \frac{\mathbb{E}S_n}{b_n} = \lim_{n\to+\infty} \frac{\sum_{k=0}^{n-1} \mathbb{E}\ln(G(k))}{b_n} < 0(a.s.)$$

thanks to the assumptions (1.81) of $\mathbb{P}$ -integrability of $G(k)$. This fact together with $b_n \to +\infty$ implies that $S_n \to -\infty$ (a.s.), i.e. $V(n) \to 0$ as n tends to $+\infty$.

The reverse direction under (1.82) is proved analogously to previous proof-steps. Thus, the proof is complete. ◇

Now, let us apply this result to BIMs (1.79) applied to test (1.78). For $k = 0, 1, \ldots, n_T$, define

$$G(k) := \left| \frac{1 + |\sigma \Delta W_k| + \sigma \Delta W_k}{1 + |\sigma \Delta W_k|} \right|. \tag{1.83}$$

Theorem 1.3.14 (A.s. asymptotic stability I of BIMs with variable h_n, [127]). *Assume that $\exists$ nonrandom sequence $b = (b_n)_{n \in \mathbb{N}}$ with $b_n \to +\infty$ as $n \to +\infty$ for a fixed choice of step sizes $h_n > 0$ such that*

$$\sum_{k=0}^{+\infty} \frac{Var(\ln(G(k)))}{b_k^2} < +\infty, \qquad \lim_{n \to +\infty} \frac{\sum_{k=0}^{n-1} \mathbb{E} \ln(G(k))}{b_n} < 0.$$

Then the BIMs (1.79) with scalar weights $c^0 = 0$ and $c^1 = |\sigma|$ applied to martingale test (1.78) with parameter $\sigma \in \mathbb{R}^1 \setminus \{0\}$ with the fixed sequence of variable step sizes h_n provide (globally) asymptotically a.s. stable sequences $Y = (Y_n)_{n \in \mathbb{N}}$.

Proof. We may apply Theorem 1.3.13 with $V(n) = |Y_n|$ since the assumptions are satisfied for the BIMs (1.26) with scalar weights $c^0 = 0$ and $c^1 = |\sigma|$ applied to martingale test (1.78). Hence, the proof is complete. ◇

Theorem 1.3.15 (A.s. asymptotic stability II of BIMs with variable h_n, S.[127]). *The BIMs (1.79) with scalar weights $c^0 = 0$ and $c^1 = |\sigma|$ applied to martingale test (1.78) with parameter $\sigma \in \mathbb{R}^1 \setminus \{0\}$ with any nonrandom variable step sizes h_k satisfying $0 < h_{min} \le h_k \le h_{max}$ provide (globally) asymptotically a.s. stable sequences $Y = (Y_n)_{n \in \mathbb{N}}$.*

Proof. We may again apply Theorem 1.3.13 with $V(n) = |Y_n|$. For this purpose, we check the assumptions. Define $b_n := n$. Note that the variance $Var(\ln(G(k)))$ is uniformly bounded since $\Delta W_n \in \mathcal{N}(0, h_n)$ and $0 < h_{min} \le h_k \le h_{max}$. More precisely, we have

$$\begin{aligned}
&Var(\ln(G(k))) \\
&\quad \le \mathbb{E}[\ln(G(k))]^2 = \mathbb{E}[I_{\{\Delta W_n > 0\}} \ln(G(k))]^2 + \mathbb{E}[I_{\{\Delta W_n < 0\}} \ln(G(k))]^2 \\
&\quad < p_n[\ln(2)]^2 + \mathbb{E}[\ln(1 + |\sigma \Delta W_n|)]^2 \le p_n[\ln(2)]^2 + \mathbb{E}[\ln(\exp(|\sigma \Delta W_n|))]^2 \\
&\quad \le p_n[\ln(2)]^2 + \mathbb{E}[\sigma \Delta W_n]^2 = p_n[\ln(2)]^2 + \sigma^2 h_n \le p_n[\ln(2)]^2 + \sigma^2 h_{max}
\end{aligned}$$

for $G(k)$ as defined in (1.83), where $I_{\{Q\}}$ denotes the indicator function of the inscribed set Q and $p_n = \sqrt{\mathbb{P}\{\Delta W_n > 0\}}$. Note that $0 < p_n = \sqrt{2}/2 < 1$ if ΔW_n is Gaussian distributed. Therefore, there is a finite real constant $K_2^G < (\ln(2))^2 + \sigma^2 h_{max}$ such that

$$\sum_{k=1}^{+\infty} \frac{Var(\ln(G(k)))}{k^2} \leq \sum_{k=1}^{+\infty} \frac{K_2^G}{k^2} = K_2^G \frac{\pi^2}{6} < +\infty.$$

It remains to check whether

$$\lim_{n \to +\infty} \frac{\sum_{k=1}^{n} \mathbb{E} \ln(G(k))}{n} < 0.$$

For this purpose, we only note that $\mathbb{E} \ln(G(k))$ is decreasing for increasing $\sqrt{h_k}$ for all $k \in \mathbb{N}$ (see the proof of Theorem 1.3.12). Therefore, we can estimate this expression by

$$\mathbb{E} \ln(G(k)) \leq \frac{1}{2} \mathbb{E} \left[\ln \left| 1 - \left(\frac{\sigma \sqrt{h_{min}} \xi}{1 + |\sigma \sqrt{h_{min}} \xi|} \right)^2 \right| \right] := K_1^G < 0$$

where $\xi \in \mathcal{N}(0,1)$ is a standard Gaussian distributed random variable and K_1^G the negative real constant as defined above. Thus,

$$\lim_{n \to +\infty} \frac{\sum_{k=1}^{n} \mathbb{E} \ln(G(k))}{n} \leq K_1^G < 0.$$

Hence, thanks to Theorem 1.3.13 (or Theorem 1.3.14), the proof is completed. ◇

Remark 1.3.13. One may also estimate stability exponents of numerical methods, i.e. estimates of the rate of exponential decay or growth. For more details, see [123, 127].

1.3.7 Nonlinear Weak V-Stability in General

So far we discussed stability along "stability-controlling" functions V (or Lyaponuv-type function) which are governed by linear inequalities (or linear difference inclusions). Now, in what follows, we shall investigate fully nonlinear relations. For this purpose, define $v_n := \mathbb{E} V(n)$ and $\Delta v_n = v_{n+1} - v_n$ for all $n \in \mathbb{N}$ and V as given below.

Theorem 1.3.16 (Discrete Bihari inequality for moments, S. [127]). *Let* $V : \mathbb{N} \longrightarrow \mathbb{R}_+^1$ *be a sequence of random variables* $V(n) : (\Omega, \mathcal{F}_n, \mathbb{P}) \to (\mathbb{R}_+^1, \mathcal{B}(\mathbb{R}_+^1))$ *satisfying*

$$v_{n+1} = \mathbb{E}[V(n+1)] \leq \mathbb{E}[V(n)] + c(n) g(\mathbb{E}[V(n)]) \tag{1.84}$$

with nonrandom $c(n) \in \mathbb{R}^1$ *for all* $n \in \mathbb{N}$*, where* $g : \mathbb{R}^1_+ \to \mathbb{R}^1_+$ *is a Borel-measurable, non-decreasing function satisfying the integrability-condition*

$$-\infty < G(u) := \int_{v_0}^{u} \frac{dz}{g(z)} < +\infty \tag{1.85}$$

for all $u > 0$ *and* $(v_n)_{n\in\mathbb{N}}$ *is non-decreasing. Assume that* $\mathbb{E}V(0) < +\infty$*. Then, for all* $n \in \mathbb{N}$*, we have*

$$v_{n+1} = \sup_{k=0,1,\ldots,n} \mathbb{E}V(k) \le G^{-1}\Big(G(\mathbb{E}[V(0)]) + \sum_{k=0}^{n} c(k)\Big) \tag{1.86}$$

where G^{-1} *is the inverse function belonging to* G*.*

Proof (S. [127]). Suppose that $v_0 > 0$ at first. Then, inequality (1.84) implies that

$$\frac{\Delta v_n}{g(v_n)} \le c(n).$$

Therefore, by simple integration under (1.85), we obtain

$$G(v_{k+1}) - G(v_k) = \int_{v_k}^{v_{k+1}} \frac{dv}{g(v)} \le \int_{v_k}^{v_{k+1}} \frac{dv}{g(v_k)} = \frac{\Delta v_k}{g(v_k)} \le c(k)$$

for all $k \in \mathbb{N}$. Summing up these inequalities leads to

$$G(v_{n+1}) - G(v_0) = \sum_{k=0}^{n} G(v_{k+1}) - G(v_k) \le \sum_{k=0}^{n} c(k)$$

for all $n \in \mathbb{N}$, which is equivalent to

$$G(v_{n+1}) \le G(v_0) + \sum_{k=0}^{n} c(k).$$

Note that the inverse G^{-1} of G exists and both G and G^{-1} are increasing since G satisfying (1.85) is increasing. Hence, we arrive at

$$v_{n+1} \le G^{-1}\left(G(v_0) + \sum_{k=0}^{n} c(k)\right).$$

Note also that $c(k) \ge 0$ due to the assumption v is non-decreasing. If $v_0 = 0$ then one can repeat the above calculations for all $v_0 = \varepsilon > 0$. It just remains to take the limit as ε tends to zero in the obtained estimates. Thus, the proof of (1.86) is complete. ◇

Remark 1.3.14. Theorem 1.3.16 can be understood as a discrete version of the continuous time Lemma of Bihari [12] and is due to S. [127].

1.3.8 Girsanov's Equation to Test Moments and Positivity

The simplest test equation for asymptotic stability, boundedness of moments, monotonicity and positivity of numerical methods is the one-dimensional Girsanov SDE (originating from Girsanov [34]) of Itô type

$$dX(t) = \sigma([X(t)]_+)^\alpha \, dW(t) \tag{1.87}$$

with $X(0) = X_0 \geq 0$ (a.s.) in view of its solutions and strong uniqueness. Without loss of generality, we may suppose that $\sigma \geq 0$. If $\alpha \in [0,1]$ and $\mathbb{E}X_0^2 < +\infty$ we obtain continuous time solutions which are martingales with respect to the σ-algebra generated by the driving Wiener process $W = (W(t))_{t \geq 0}$. If $\alpha = 0$ or $\alpha \in [1/2, 1]$ these (non-anticipating) martingale-solutions are unique with probability one (by the help of Osgood-Yamada-Watanabe results, cf. Karatzas and Shreve [70]). Due to the pathwise continuity, the non-negative cone $\mathbb{R}_+ = \{x \in \mathbb{R}^1 : x \geq 0\}$ is left invariant (a.s.). A nontrivial question is whether related numerical approximations are stable, converge to the underlying analytic solution and have the same invariance property. We are able to answer these problems. Here, we are only interested in stability and invariance of "appropriately balanced discretizations" (for convergence, see also in Sect. 1.4 below). For this purpose, consider the balanced implicit methods (BIMs)

$$Y_{n+1} = Y_n + \sigma([Y_n]_+)^\alpha \Delta W_n + \sigma([Y_n]_+)^{\alpha-1} |\Delta W_n| (Y_n - Y_{n+1}) \tag{1.88}$$

as a linear-implicit member of the more general class of BTMs (1.1).

Theorem 1.3.17 (Stability of 2nd Moments of BIMs, S.[127]). *The BIMs (1.88) applied to Girsanov's SDE (1.87) with $0 < \alpha < 1$ leave the non-negative cone $\mathbb{R}_+$ invariant (a.s.) and provide polynomially stable numerical sequences. More precisely, if $\sigma > 0$, $0 < \mathbb{E}Y_0^2 < +\infty$ and Y_0 is independent of the σ-algebra $\mathscr{F} = \sigma\{W(t) : t \geq 0\}$ then their second moments are strictly increasing as n increases and they are governed by*

$$\mathbb{E}[Y_n^2] \leq \left((\mathbb{E}[Y_0^2])^{1-\alpha} + (1-\alpha)\sigma^2(t_{n+1} - t_0) \right)^{1/(1-\alpha)} \tag{1.89}$$

Proof. Suppose that $0 < \alpha < 1$. At first we rewrite (1.88) as the explicit scheme

$$Y_{n+1} = Y_n \frac{1 + \sigma([Y_n]_+)^{\alpha-1}\Delta W_n + \sigma([Y_n]_+)^{\alpha-1}|\Delta W_n|}{1 + \sigma([Y_n]_+)^{\alpha-1}|\Delta W_n|} \tag{1.90}$$

which immediately gives the a.s. invariance with respect to the non-negative cone $\mathbb{R}_+$, provided that $Y_0 \geq 0$ (a.s.). Therefore, we may drop the taking of positive part by $[.]_+$ in the above form. Now, rewrite (1.90) as

$$Y_{n+1} = Y_n + \frac{\sigma(Y_n)^{\alpha}\Delta W_n}{1 + \sigma(Y_n)^{\alpha-1}|\Delta W_n|}. \tag{1.91}$$

Taking the square and expectation yields

$$v_{n+1} := \mathbb{E}[Y_{n+1}^2] = \mathbb{E}[Y_n]^2 + \mathbb{E}\left[\frac{\sigma(Y_n)^{\alpha}\Delta W_n}{1 + \sigma(Y_n)^{\alpha-1}|\Delta W_n|}\right]^2. \tag{1.92}$$

Thus, due to the positivity of all summands at the right hand side, we may conclude that the second moments $(v_n)_{n\in\mathbb{N}}$ are non-decreasing and, in fact if $\sigma > 0$, v_n is strictly increasing. It remains to apply Theorem 1.3.16. For this purpose, estimate (1.92) by Jensen's inequality for concave functions in order to obtain

$$v_{n+1} \leq v_n + \sigma^2\mathbb{E}[(Y_n)^{2\alpha}\Delta W_n] \;=\; v_n + \sigma^2\mathbb{E}[(Y_n)^{2\alpha}\mathbb{E}[\Delta W_n|\mathscr{F}_n]] \tag{1.93}$$

$$= v_n + \sigma^2\mathbb{E}[(Y_n)^{2\alpha}\mathbb{E}[\Delta W_n]] \;\leq\; v_n + \sigma^2(v_n)^{\alpha}h_n. \tag{1.94}$$

Therefore, we may take $V(x) = x^2$, $g(z) = z^{\alpha}$, $c(n) = \sigma^2 h_n$ and apply Theorem 1.3.16 with the conclusion (1.86) in order to get to (1.89). Note also that

$$G(u) = \int_{v_0}^{u}\frac{dz}{g(z)} = \int_{v_0}^{u}\frac{dz}{z^{\alpha}} = \frac{z^{1-\alpha}}{1-\alpha}\Big|_{z=v_0}^{z=u} = \frac{u^{1-\alpha} - (v_0)^{1-\alpha}}{1-\alpha},$$

$$G^{-1}(z) = \left(z(1-\alpha) + v_0^{1-\alpha}\right)^{1/(1-\alpha)}.$$

If $v_0 = 0$ then one can repeat the above calculations for $v_0 = \varepsilon > 0$. It just remains to take the limit as ε tends to zero in the obtained estimates. Thus, the proof is complete. ◇

One could also compare the moment evolutions of the explicit Euler method $Y^E = (Y_n^E)_{n\in\mathbb{N}}$ with that of BIMs $Y^B = (Y_n^B)_{n\in\mathbb{N}}$ governed by (1.26). Then, it is fairly easy to recognize that $\mathbb{E}(Y_n^B)^{2\kappa} \leq \mathbb{E}(Y_n^E)^{2\kappa}$ for all integers $\kappa \in \mathbb{N}$, provided that $\mathbb{E}(Y_0^B)^{2\kappa} \leq \mathbb{E}(Y_0^E)^{2\kappa}$. It is also interesting to note that the explicit Euler methods cannot preserve the a.s. invariance property with respect to the non-negative cone $\mathbb{R}_+$. In fact, they exit that cone with positive probability, independently of the choice of any nonrandom step sizes Δ_n. Summarizing, the underlying explicit solution to (1.87) has very similar analytic properties as BIMs (1.88). Numerical experiments for Girsanov SDEs are conducted in S. [127].

1.4 Finite Convergence and Orders (Rates of Global Consistency)

1.4.1 Mean Square Convergence

The concept of global mean square convergence is understood as follows (similar for p-th mean convergence).

Definition 1.4.1. A numerical method Y with one-step representation $Y_{s,y}(t)$ is said to be **(globally) mean square convergent to X with rate r_g** on $[0,T]$ iff $\exists$ Borel-measurable function $V:\mathbb{D}\to\mathbb{R}^1_+$ and $\exists$ real constants $K_g = K_g(T) \geq 0$, $K_S^Y = K_S^Y(b^j), 0 < h_{max} \leq \delta_0 \leq 1$ such that $\forall(\mathscr{F}_0, \mathscr{B}(\mathbb{D}))$-measurable random variables $Z(0)$ with $\mathbb{E}[\|Z(0)\|_d^2] < +\infty$ and $\forall t : 0 \leq t \leq T$

$$\left(\mathbb{E}[||X_{0,Z(0)}(t) - Y_{0,Z(0)}(t)||_d^2|\mathscr{F}_0]\right)^{1/2} \leq K_g \exp\left(K_S^Y t\right) V(Z(0)) h_{max}^{r_g} \quad (1.95)$$

along any nonrandom partitions $0 = t_0 \leq t_1 \leq \ldots \leq t_{n_T} = T$.

Using the results of the previous section, the following theorem is rather obvious in conjunction with standard L^2-convergence theorems following stochastic Lax-Richtmeyer principles as presented and proven in [125,129]. Recall the assumptions (A0)–(A5) from subsection 1.2.1.

Theorem 1.4.1 (Mean Square Convergence of BTMs (1.1)). *Assume that (A0)–(A5) with $V(x) = (1 + ||x||_d^2)^{1/2}$ on $\mathbb{D} \subseteq \mathbb{R}^d$ are satisfied.*
Then, the BTMs (1.1) with autonomous implicitness $\Theta \in \mathbb{R}^{d\times d}$ are mean square convergent with rate $r_g = 0.5$.

Proof. Recall that, under assumptions (A0)–(A5), we have verified the local rates $r_0 \geq 1.5$ of mean consistency and $r_2 \geq 1.0$ of mean square consistency with $V(x) = (1 + ||x||_d^2)^{1/2}$. Furthermore, the Hölder-continuity of (a, b^j) implies that the exact solution X of SDEs (1.2) is mean square contractive and Hölder-continuous with mean square Hölder exponent $r_{sm} = 0.5$. Moreover, the BTMs (1.1) are weakly V-stable along V. Hence, an application of the general Kantorovich-Lax-Richtmeyer approximation principle (cf. subsection 1.4.2) proved in [129] yields that

$$r_g = r_2 + r_{sm} - 1.0 \geq 1.0 + 0.5 - 1.0 = 0.5,$$

hence the mean square rate $r_g \geq 0.5$ is confirmed. ◇

Remark 1.4.1. In the case $\mathbb{D} = \mathbb{R}^d$ and uniformly Lipschitz-continuous coefficients (a, b^j), one can also apply convergence Theorem 1.1 from Milstein [95] (originating from [94]) in order to establish the rate $r_g = 0.5$ of mean square convergence for BTMs (1.1).

1.4.2 Mean Square Convergence Along Functionals (KLR-Principle [81])

Let $X_{s,Z}(t), Y_{s,Z}(t)$ be the one-step representations of stochastic processes X, Y evaluated at time $t \geq s$, started from $Z \in H_2([0,s], \mu, H)$ (i.e., more precisely, we have $X_{s,Z}(u) = Z_u = Y_{s,Z}(u)$ for $0 \leq u \leq s$ and $X_{s,Z}(t), Y_{s,Z}(t)$ are interpreted as the values of the stochastic processes X and Y in H at time $t \geq s$, respectively, with fixed history (memory) given by Z up to time $s \geq 0$). They are supposed to be constructable along any $(\mathscr{F}_t)$-adapted discretization of the given time-interval $[0, T]$ and could depend on a certain maximum mesh size h_{max}. Assume that there are deterministic real constants $r_0, r_{sm}, r_2 \geq 0, 0 < \delta_0 \leq 1$ such that we have:

(H1) **Strong $(\mathbb{D}_t)$-invariance of** X, Y, i.e. $\exists(\mathscr{F}_t)$-adapted, closed subsets

$$\mathbb{D}_t \subseteq H_2([0,t], \mu, H)$$

such that, for all $0 \leq s < T$, we have

$$\mathbb{P}\left\{(X_{s,X(s)}(u))_{s \leq u \leq t}, (Y_{s,Y(s)}(u))_{s \leq u \leq t} \in \mathbb{D}_t : s \leq t \leq T \,|\, X(s), Y(s) \in \mathbb{D}_s\right\} = 1,$$

(H2) V**-Stability of** Y, i.e. $\exists$ functional $V : H_2([0,t], \mu, H) \to \mathbb{R}_+$ for all $0 \leq t \leq T$ such that $\forall\, Y(t) \in \mathbb{D}_t : \mathbb{E}V(Y(t)) < +\infty$, $V(Y(t))$ is $(\mathscr{F}_t)$-adapted and $\exists$ real constant K_S^Y $\forall\, t, h : 0 \leq h \leq \delta_0, 0 \leq t, t+h \leq T$

$$\mathbb{E}[V(Y_{t,Y(t)}(t+h))|\mathscr{F}_t] \leq \exp(2K_S^Y h) \cdot V(Y(t)),$$

(H3) **Mean square contractivity of** X, i.e. $\exists$ real constant K_C^X such that $\forall\, X(t), Y(t) \in \mathbb{D}_t$ (where $X(t), Y(t)$ are $(\mathscr{F}_t)$-adapted) $\forall\, t, h : 0 \leq h \leq \delta_0, 0 \leq t, t+h \leq T$

$$\mathbb{E}[||X_{t,X(t)}(t+h) - X_{t,Y(t)}(t+h)||_H^2 | X(t), Y(t)] \leq \exp(2K_C^X h)||X_{t,X(t)}(t) - X_{t,Y(t)}(t)||_H^2,$$

(H4) **Mean consistency of** (X, Y) **with rate** $r_0 > 0$, i.e. $\exists$ real constant K_0^C such that $\forall\, Z(t) \in \mathbb{D}_t$ (where $(Z_u(t))_{0 \leq u \leq t}$ is $(\mathscr{F}_t, \mathscr{B}(H))$-measurable) $\forall\, t, h : 0 \leq h \leq \delta_0, 0 \leq t, t+h \leq T$

$$||\mathbb{E}[X_{t,Z(t)}(t+h)|Z(t)] - \mathbb{E}[Y_{t,Z(t)}(t+h)|Z(t)]||_H \leq K_0^C \cdot V(Z(t)) \cdot h^{r_0},$$

(H5) **Mean square consistency of** (X,Y) **with rate** $r_2 > 0$, i.e. $\exists$ real constant K_2^C such that $\forall\, Z(t) \in \mathbb{D}_t$ $\forall\, t,h: 0 \le h \le \delta_0, 0 \le t, t+h \le T$

$$\left(\mathbb{E}\Big[||X_{t,Z(t)}(t+h) - Y_{t,Z(t)}(t+h)||_H^2 | Z(t)\Big]\right)^{1/2} \le K_2^C \cdot V(Z(t)) \cdot h^{r_2}$$

(H6) **Mean square Hölder-type smoothness of diffusive (martingale) part of** X **with rate** $r_{sm} \in [0, \frac{1}{2}]$, i.e. $\exists$ real constant $K_{SM} \ge 0$ such that $\forall\, X(t), Y(t) \in \mathbb{D}_t$ (where $X(t), Y(t)$ are $(\mathscr{F}_t)$-adapted) $\forall t, h: 0 \le h \le \delta_0$, $0 \le t, t+h \le T$

$$\mathbb{E}||M_{t,X(t)}(t+h) - M_{t,Y(t)}(t+h)||_H^2 \le (K_{SM})^2 \cdot \mathbb{E}||X(t) - Y(t)||_H^2 \cdot h^{2r_{sm}}$$

where $M_{t,z}(t+h) = X_{t,z}(t+h) - \mathbb{E}[X_{t,z}(t+h)|\mathscr{F}_t]$ for $z = X(t), Y(t)$,

(H7) **Interplay between consistency rates** given by $r_0 \ge r_2 + r_{sm} \ge 1.0$,

(H8) **Initial moment** V**-boundedness** $\mathbb{E}[V(X_0)] + \mathbb{E}[V(Y_0)] < +\infty$.

Stochastic approximation problems satisfying the assumptions $(H1)$–$(H8)$ on H_2 are called **well-posed**. In the classical case of stochastic dynamics with Lipschitz-continuous vector coefficients like that of SDEs driven by Wiener processes one often takes the function $V((X(s))_{0 \le s \le t}) = (1 + ||X(t)||_H^2)^{p/2}$ or $||X(t)||_H^p$ as the required functional V. Then, V plays the role of a Lyapunov function controlling the stability behavior of considered stochastic process and the smoothness condition $(A6)$ of the martingale part with $r_{sm} = 0.5$ is obviously satisfied. Of course, if $V(X) = 0$ is chosen, then X and Y must be identical and any derived convergence assertions based on above assumptions are meaningless.

The following fairly general approximation principle [129] can be established. Define the **point-wise** L^2**-error**

$$\varepsilon_2(t) = \sqrt{\mathbb{E} < X(t) - Y(t), X(t) - Y(t) >_H}$$

for the processes $X, Y \in H_2$, and the deterministic bounds $h_{min} = \inf_{i \in \mathbb{N}} h_i \le h_n \le h_{max} = \sup_{i \in \mathbb{N}} h_i$ on the mesh sizes h_n on which (at least one of) X, Y are usually based.

Theorem 1.4.2 (Axiomatic Approach of S. [129]). *Assume that the conditions* $(H1) - -(H8)$ *are satisfied and that* $\mathbb{E}||X(0) - Y(0)||_H^2 < K_{init} h_{max}^{r_g}$. *Then the stochastic processes* $X, Y \in H_2([0,T], \mu, H)$ *converge to each another on* H_2 *with respect to the naturally induced metric* $m(X,Y) = (< X - Y, X - Y >_{H_2})^{1/2}$ *with convergence rate* $r_g = r_2 + r_{sm} - 1.0$. *More precisely, for any* $\rho \ne 0$ *and for any choice of deterministic step sizes* h_i *(variable or constant) with* $0 < h_i \le h_{max} \le \delta_0$, *we have the universal error estimates*

$$\varepsilon_2(t) \le \exp((K_C^X + \rho^2)(t-s))\varepsilon_2(s) + K_I(\rho)\exp(K_S^Y t)\sqrt{\frac{\exp(2(K_C^X + \rho^2 - K_S^Y)(t-s)) - 1}{2(K_C^X + \rho^2 - K_S^Y)}}\, h_{max}^{r_g} \quad (1.96)$$

for all $0 \leq s \leq t \leq T$, *where* s, t *are deterministic, and*

$$\sup_{0 \leq t \leq T} \varepsilon_2(t) \leq \exp([K_C^X + \rho^2]_+ T)\varepsilon_2(0) \tag{1.97}$$

$$+ K_I(\rho) \exp([K_S^Y]_+ T) \sqrt{\frac{\exp(2(K_C^X + \rho^2 - K_S^Y)T) - 1}{2(K_C^X + \rho^2 - K_S^Y)}} \, h_{max}^{r_g}$$

with appropriate constant

$$K_I(\rho) = K_{max} \cdot \frac{\sqrt{(K_0^C)^2 + (K_2^C)^2[\rho^2 + (K_{SM})^2]}}{\rho} \cdot \mathbb{E}[V(Y(0))], \tag{1.98}$$

$$K_{max} = \exp(([K_C^X]_- + [K_S^Y]_-)h_{max}).$$

Remark 1.4.2. The complete proof broken down in a series of auxiliary lemmas is found in S. [129]. This Theorem 1.4.2 forms the main fundament of the approximation principle which our entire paper is based on (cf. its table of contents) and is an extension of ideas due to Kantorovich, Lax and Richtmeyer as they are well-known in deterministic analysis.

1.4.3 Strong Mean Square Convergence

Definition 1.4.2. A numerical method Y with continuous one-step representation is said to be **(globally) strongly mean square convergent to** X **with rate** r_g^* on $[0, T]$ iff $\exists$ Borel-measurable function $V : \mathbb{D} \to \mathbb{R}_+^1$ and $\exists$ real constants $K_g = K_g(T) \geq 0$, $K_S^Y = K_S^Y(b^j), 0 < h_{max} \leq \delta_0 \leq 1$ such that $\forall(\mathscr{F}_0, \mathscr{B}(\mathbb{D}))$-measurable random variables $Z(0)$ with $\mathbb{E}[\|Z(0)\|_d^2] < +\infty$ we have

$$\left(\mathbb{E}[\max_{0 \leq t \leq T} ||X_{0,Z(0)}(t) - Y_{0,Z(0)}(t)||_d^2 | \mathscr{F}_0]\right)^{1/2} \leq K_g \exp\left(K_S^Y t\right) V(Z(0)) h_{max}^{r_g^*} \tag{1.99}$$

along any nonrandom partitions $0 = t_0 \leq t_1 \leq \ldots \leq t_{n_T} = T$.

Remark 1.4.3. Of course, strong mean square convergence represents a stronger requirement than just the "simple" mean square convergence given by definition 1.4.1. In fact, the rates of strong mean convergence carry over to that of "simple" mean square convergence. No proofs for strong mean square convergence are known to us other than that under standard uniform Lipschitz-continuity and presuming the existence and uniform boundedness of higher order moments.

The proof of following theorem is left to interested reader, but it can be carried out in a similar fashion as previous convergence proofs by applying a theorem due to Milstein [94, 95].

Theorem 1.4.3. *Assume that all coefficients a and b^j are uniformly Lipschitz continuous and assumptions (A0)–(A5) hold with $V(x) = (1 + ||x||_d^4)^{1/4}$. Then, the BTMs (1.1) with equidistant step sizes $h \leq \delta_0 < 1$ are strongly mean square convergent with rate $r_g^* = 0.5$.*

This theorem is proved by Milstein's Theorem stated as follows. We leave the work out of related details to the interest of our readership.

Theorem 1.4.4 (Milstein's strong mean square convergence theorem [95]). *Assume that all coefficients a and b^j are uniformly Lipschitz continuous on $[0, T] \times \mathbb{R}^d$ and the local rates of consistency r_0 and r_2 satisfy*

$$r_2 \geq \frac{3}{4}, \quad r_0 \geq r_2 + \frac{1}{2}.$$

Moreover, let $\exists K_4 \, \forall t \in [0, T] \, \forall h : 0 < h \leq \delta_0 \, \forall z \in \mathbb{R}^d$

$$\mathbb{E} \, \|X_{t,z}(t+h) - Y_{t,z}(t+h)\|_d^4 \;\leq\; (K_4)^4 (1 + \|z\|_d^4) h^{4r_2 - 1}.$$

Then, equidistant numerical methods Y with nonrandom initial values $X(0) = Y_0 = x_0 \in \mathbb{R}^d$ permit estimates

Fundamental Strong Mean Square Convergence Relation

$$\varepsilon_2^*(T) = \left(\mathbb{E} \sup_{0 \leq t \leq T} \|X_{0,x_0}(t) - Y_{0,x_0}(t)\|_d^2 \right)^{1/2} \;\leq\; K_2^{SC} (1 + \|x_0\|_d^4)^{1/4} h^{r_2 - \frac{1}{2}}$$

where K_2^{SC} is a real constant, and $h \leq 1$, i.e. $\gamma_g^*(2) = r_2 - \frac{1}{2}$.

1.4.4 Milstein's Weak Convergence Theorem

Sometimes one is only interested in moments of solutions of SDEs. This concept rather corresponds to the concept of weak convergence from functional analysis. A numerical version of that concept is introduced and studied independently by Milstein [93] and Talay [139]. Let F denote a nonempty class of sufficiently smooth test functions $f : \mathbb{R}^d \to \mathbb{R}^1$ (or functionals) and $\mathbb{D} \subseteq \mathbb{R}^d$.

Definition 1.4.3. A numerical method Y with one-step representation $Y_{s,y}(t)$ is said to be **(globally) weakly convergent to** X **with rate** $r_w \in (0, +\infty)$ on $[0, T] \times \mathbb{D}$

(with respect to the test class F) iff $\exists$ Borel-measurable function $V : \mathbb{D} \to \mathbb{R}^1_+$ and $\exists$ real constants $K_w = K_w(F, T, b^j) \geq 0$ and δ_0 with $0 < h_{max} \leq \delta_0 \leq 1$ such that

$$\forall x \in \mathbb{D} : \sup_{f \in F} \sup_{0 \leq t \leq T} \left| \mathbb{E}\, f\,(X_{0,x}(t)) - \mathbb{E}\, f(Y^h_{0,x}(t)) \right| \leq K_w V(x) \cdot h^{r_w} \quad (1.100)$$

where Y^h is constructed along any nonrandom partitions $0 = t_0 \leq t_1 \leq \ldots \leq t_{n_T} = T$ with maximum step size h based on scheme-values $(Y_n)_{n \in \mathbb{N}}$ at discrete instants t_n.

Usually one investigates weak convergence with respect to continuous functions $f \in F$. However, step functions (elementary functions) such as indicator functions I_S of certain sets S or convex functions are also common to guarantee convergence of related probabilities. However, the main result on weak convergence is borrowed from Milstein [95] and reads as follows. Let $\mathscr{P}G(r)$ be the set of functions $f : \mathbb{R}^d \to \mathbb{R}^1$ which does not possess polynomial growth more than power $r \in \mathbb{N}$.

Theorem 1.4.5 (Milstein's weak convergence theorem [95]). *Assume that a, b^j are uniformly Lipschitz continuous on $[0, T] \times \mathbb{R}^d$, $f : \mathbb{R}^d \to \mathbb{R}$ and a, b^j together with their derivatives up to degree $2r + 2$ are polynomially bounded of degree $2p$, $r \in \mathbb{N}$, X is governed by SDE (1.2) and:*

(i) $X_{0,x_0}(0) = x_0 \in \mathbb{R}^d$ *independent of* $\mathscr{F}^j = \sigma(W^j(s), s \geq 0)$ *with* $\mathbb{E}[\|x_0\|_d^{2p}] < +\infty$.
(ii) $\sup_{(h_n)_{n\in\mathbb{N}}: h_n \leq \delta_0 \leq 1} \sup_{n \in \mathbb{N}} \mathbb{E} \| Y^{(h_n)}_{0,x_0}(t_n)\|_d^{2p} \leq K_b < +\infty$.
(iii) $\exists K_0\ \forall t \in [0, T]\ \forall x \in \mathbb{R} : \|a(t, x)\|_d^2 + \sum_{j=1}^m \|b^j(t, x)\|_d^2 \leq (K_B)^2 (1 + \|x\|_d^2)$.
(iv) $\exists K_c\ \forall t \in [0, T]\ \forall x, y \in \mathbb{R} : \|a(t, x) - a(t, y)\|_d^2 + \sum_{j=1}^m \|b^j(t, x) - b^j(t, y)\|_d^2 \leq (K_L)^2 \|x - y\|_d^2$.
(v) $\exists g = g(x) \in F \subseteq \mathscr{P}G(2p)\ \forall h \leq h_{max}\ \forall [t, t + h] \subset [0, T]\ \forall l = 1, 2, \ldots, 2r + 1$

$$\left| \mathbb{E} \left(\prod_{k=1}^{l} (X_{t,x}(t + h) - x)^{i_k} - \prod_{k=1}^{l} (Y_{t,x}(t + h) - x)^{i_k} \right) \right| \leq g(x) h^{r+1}$$

$$\mathbb{E} \prod_{k=1}^{2r+2} |Y_{t,x}(t + h) - x|^{i_k} \leq g(x) h^{r+1}$$

Then, weak convergence of Y to X with rate $r_w = r$ with respect to the test class $f \in F \subseteq \mathscr{P}G(2p) \cap C^2(\mathbb{R}^d)$ with $V(x) = (1 + ||x||_d^{2p}$ on $\mathbb{D} = \mathbb{R}^d$ is established, i.e. more precisely

$$\varepsilon_0^w(T) = \sup_{f \in F} \sup_{0 \leq t \leq T} \|\mathbb{E}[f(X_{0,x_0}(t))] - \mathbb{E}[f(Y_{0,x_0}(t))]\|_d$$

$$\leq K(T,a,b^j)(1+\mathbb{E}\|x_0\|_d^{2p})h_{max}^r$$

where all K represent real constants (only depending on T, a, b^j), and $h_{max} \leq 1$.

The proof is found in Milstein [95]. Forward and backward Euler, stochastic Theta methods with vanishing weight matrices c^j $(j = 1, 2, \ldots, m)$ as well as families of implicit Milstein and Theta-Milstein methods have weak convergence order $r_w = 1.0$. There are other methods such as Talay-Tubaro's extrapolation method which can achieve weak order 2.0 or even higher order of weak convergence for moments of functions of solutions of SDEs at fixed deterministic terminal time T. The situation is more complicated for approximation of functionals depending on the past or even whole memory of trajectories. Then it may happen that the orders (rates) shrink to much lower orders. A similar reduction of orders can happen if functions are not very smooth or a control at all times T up to infinity is necessary. For some analysis of convex functionals, see S. [133]. In this case and when non-smoothness is given, one can exploit the more robust L^p-convergence analysis to justify a minimum rate of convergence (an at least guaranteed order).

1.4.5 Weak Convergence of BTMs (1.1) with Nonrandom Weights

For simplicity, set $F = C^2_{b(\kappa)}(\mathbb{R}^d, \mathbb{R}^1)$ and consider the subclass of linear-implicit BTMs (by putting $\Theta = 0$ in (1.1))

$$Y_{k+1} = Y_k + \sum_{j=0}^{m} b^j(t_k, Y_k)\Delta W_k^j + \sum_{j=0}^{m} c^j(t_k, Y_k)|\Delta W_k^j|(Y_k - Y_{k+1}) \quad (1.101)$$

which we also abbreviate by **BIMs**.

For approximations in the weak sense w.r.t. test class F, one should rather take the weights $c^j \equiv \mathcal{O}$ for $j = 1, 2, \ldots, m$ to guarantee the maximum rate of weak convergence. More degree of freedom is in the choice of c^0. A preferable choice is $c^0(t, x) = 0.5\nabla a(t, x)$ due to a reasonable replication of the p-th moment stability behavior of such BIMs compared to the underlying SDEs. This choice would also coincide with linearly drift-implicit midpoint and trapezoidal methods for bilinear SDEs. Let $C^l_{b(\kappa)}(\mathbb{R}^d, \mathbb{R}^1)$ denote the set of all l-times $(l \in \mathbb{N})$ continuously differentiable functions $f : \mathbb{R}^d \to \mathbb{R}^1$ with uniformly bounded derivatives up to l-th order such that

$$\max\{|f(x)|, ||\nabla f(x)||_d, \|\nabla^2 f(x)\|_{d\times d}, \ldots\} \leq K_f \cdot (1 + \|x\|_d^\kappa)$$

for all $x \in \mathbb{R}^d$, where K_f and κ are appropriate real constants.

Theorem 1.4.6 (Weak Convergence of BIMs, S.[128]). *Assume that* $(A1)$–$(A4)$ *with* $V \in C^2_{b(\kappa)}(\mathbb{R}^d, \mathbb{R}^1_+)$ *with* $V(x) \le (1 + ||x||^2_d)^{1/2}$ *hold,* $\mathbb{E}\|Y_0\|^{4\kappa} < +\infty$ *for an integer* $\kappa \ge 1$*, all coefficients* $a, b^j \in C^2_{b(\kappa)}([0,T] \times \mathbb{R}^d, \mathbb{R}^d)$ *of SDE (1.2) are uniformly Lipschitz-continuous with Lipschitz constants* K^j_L *with respect to x and*

$$\forall t \in [0,T]\, \forall x \in \mathbb{R}^d \quad \sum_{j=0}^{m} \|c^0(t,x) b^j(t,x)\|^4_d \le (K_{4c})^4 (1 + \|x\|^{4\kappa}_d). \quad (1.102)$$

Then the subclass of BIMs (1.101) with weights $c^j(t,x) \equiv \mathcal{O}$ *for* $j = 1, 2, \ldots, m$ *(i.e. BIMs with nonrandom weights) is weakly converging with rate* $r_w = 1.0$ *with respect to the test class* $f \in C^2_{b(\kappa)}(\mathbb{R}^d, \mathbb{R}^1)$*. More precisely, for all test functions* $f \in C^2_{b(\kappa)}(\mathbb{R}^d, \mathbb{R}^1)$ *for which the standard Euler method weakly converges with rate* $r^E_w = 1.0$*, there is a real constant* $K_w = K_w(T, K_f, b^j)$ *(for its estimate, see at the end of following proof) such that*

$$|\mathbb{E} f(X(T)) - \mathbb{E} f(Y_{n_T})| \le K_w \cdot \Big(\max_{k=0,1,\ldots,n_T} \mathbb{E}(1 + \|Y_k\|^{4\kappa}_d) \Big) \cdot h_{max} \quad (1.103)$$

where the maximum step size h_{max} *satisfies the condition*

$$2\kappa(4\kappa - 1) m K^2_M (K^j_{b(2)})^2 h_{max} < 1 \quad (1.104)$$

with constants $K^j_{b(2)}$ *chosen as in (1.107) for all* $j = 1, 2, \ldots, m$ *(i.e. for* b^j *instead of* f*).*

Proof. Recall that the forward Euler methods weakly converge with worst case global rate $r^E_w = 1.0$ and error-constants $K^E_w = K^E_w(T) \ge 0$ under the given assumptions (see Milstein [95] and Talay [140]). Let $f \in C^2_{b(\kappa)}(\mathbb{R}^d, \mathbb{R}^1)$ have uniformly bounded derivatives satisfying

$$\max \Big(|f(x)|, \|\nabla f(x)\|_d, \|\nabla^2 f(x)\|_{d\times d} \Big) \le K_f (1 + \|x\|^{4\kappa})^{1/4} \le K_f (1 + \|x\|^{\kappa})$$

with constant K_f. Moreover, for such functions f, one can find an appropriate real constant $K^E_w = K^E_w(T, f, b^j)$ such that it satisfies the conditional estimates of the local weak error

$$\Big| \mathbb{E}[f(X_{s,x}(t)) - f(Y^E_{s,x}(t))] \Big| \le K^E_w \cdot (1 + \|x\|^{4\kappa}_d) \cdot (t-s)^2$$

for sufficiently small $0 \le t - s \le h_{max} \le \delta_0$ and $x \in \mathbb{R}^d$, and the global weak error

$$\Big| \mathbb{E}[f(X_{0,x}(T)) - f(Y^E_{0,x}(T))] \Big| \le K^E_w \cdot (1 + \|x\|^{4\kappa}_d) \cdot T \cdot h_{max},$$

for sufficiently small $h_{max} \leq \delta_0 \leq \min(1, T)$. Now, define the auxiliary functions $u : [0, T] \times \mathbb{R}^d \to \mathbb{R}^1$ by

$$u(s, x) \ = \ \mathbb{E} f(X_{s,x}(t_{k+1}))$$

for $0 \leq s \leq t_{k+1}$. Suppose that $0 \leq h_k \leq \delta_0 \leq \min(1, T)$. For simplicity, assume that X and Y are constructed on one and the same complete probability space (which does not exhibit a real restriction due to Kolmogorov's extension theorem). Then, by following similar ideas as in Milstein [95] extended to the variable step size case, we arrive at

$$\begin{aligned}
\varepsilon_0(t_{k+1}) :&= \Big|\mathbb{E}[f(X_{0,x_0}(t)) - f(Y_{0,y_0}(t))]\Big| \\
&= \Big|\sum_{i=0}^{k-1}\Big(\mathbb{E}[u(t_{i+1}, X_{t_i,Y_i}(t_{i+1}))] - \mathbb{E}[u(t_{i+1}, Y_{t_i,Y_i}(t_{i+1}))]\Big) \\
&\quad + \mathbb{E}[f(X_{t_k,Y_k}(t_{k+1}))] - \mathbb{E}[f(Y_{t_k,Y_k}(t_{k+1}))]\Big| \\
&\leq \sum_{i=0}^{k-1}\mathbb{E}\Big|\mathbb{E}[u(t_{i+1}, X_{t_i,Y_i}(t_{i+1})) - u(t_{i+1}, Y_{t_i,Y_i}(t_{i+1}))|\mathscr{F}_{t_i}]\Big| \\
&\quad + \mathbb{E}\Big|\mathbb{E}[f(X_{t_k,Y_k}(t_{k+1})) - f(Y_{t_k,Y_k}(t_{k+1}))|\mathscr{F}_{t_k}]\Big| \\
&\leq \sum_{i=0}^{k-1}\mathbb{E}\Big|\mathbb{E}[u(t_{i+1}, X_{t_i,Y_i}(t_{i+1})) - u(t_{i+1}, Y^E_{t_i,Y_i}(t_{i+1}))|\mathscr{F}_{t_i}]\Big| \\
&\quad + \sum_{i=0}^{k-1}\mathbb{E}\Big|\mathbb{E}[u(t_{i+1}, Y^E_{t_i,Y_i}(t_{i+1})) - u(t_{i+1}, Y_{t_i,Y_i}(t_{i+1}))|\mathscr{F}_{t_i}]\Big| \\
&\quad + \mathbb{E}\Big|\mathbb{E}[f(X_{t_k,Y_k}(t_{k+1})) - f(Y^E_{t_k,Y_k}(t_{k+1}))|\mathscr{F}_{t_k}]\Big| \\
&\quad + \mathbb{E}\Big|\mathbb{E}[f(Y^E_{t_k,Y_k}(t_{k+1})) - f(Y_{t_k,Y_k}(t_{k+1}))|\mathscr{F}_{t_k}]\Big| \\
&\leq K^E_w \cdot \max_{i=0,1,\ldots,k+1}(1 + \mathbb{E}\|Y_i\|_d^{4\kappa}) \cdot \sum_{i=0}^{k} h_i^2 \\
&\quad + \sum_{i=0}^{k-1}\mathbb{E}\Big|\mathbb{E}[u(t_{i+1}, Y^E_{t_i,Y_i}(t_{i+1})) - u(t_{i+1}, Y_{t_i,Y_i}(t_{i+1}))|\mathscr{F}_{t_i}]\Big| \\
&\quad + \mathbb{E}\Big|\mathbb{E}[f(Y^E_{t_k,Y_k}(t_{k+1})) - f(Y_{t_k,Y_k}(t_{k+1}))|\mathscr{F}_{t_k}]\Big| \\
&= K^E_w \cdot \max_{i=0,1,\ldots,k+1}(1 + \mathbb{E}\|Y_i\|_d^{4\kappa}) \cdot t_{k+1} \cdot h_{max} + m_1(k) + m_2(k).
\end{aligned}$$

$$\text{where} \quad m_1(k) = \sum_{i=0}^{k-1} \mathbb{E}\Big|\mathbb{E}[u(t_{i+1}, Y^E_{t_i,Y_i}(t_{i+1})) - u(t_{i+1}, Y_{t_i,Y_i}(t_{i+1}))|\mathscr{F}_{t_i}]\Big|,$$

$$m_2(k) = \mathbb{E}\Big|\mathbb{E}[f(Y^E_{t_k,Y_k}(t_{k+1})) - f(Y_{t_k,Y_k}(t_{k+1}))|\mathscr{F}_{t_k}]\Big|.$$

Next, we analyze the remaining terms m_1 and m_2. For this purpose, suppose that $g \in C^2_{b(\kappa)}(\mathbb{R}^d, \mathbb{R}^1)$. Then, the expressions m_1 and m_2 have only terms of the form $\mathbb{E}|\mathbb{E}[g(Y^E_{t_k,Y_k}(t_{k+1})) - g(Y_{t_k,Y_k}(t_{k+1}))|\mathscr{F}_{t_k}]|$. Thus, it remains to estimate them by $K_k h_k^2$ with constants K_k. Note also that $M_{s,x}(t) = I_d + c^0(s,x)(t-s)$ is nonrandom and invertible, $Y_{t_k,Y_k}(t_{k+1}) = Y_{k+1}$ by definition, and

$$d_{s,x}(t) := Y^E_{s,x}(t) - Y_{s,x}(t) \;=\; M^{-1}_{s,x}(t) \sum_{j=0}^{m} c^0(s,x) b^j(s,x)(W^j_t - W^j(s))(t-s).$$

Now, we obtain

$$\begin{aligned}
m(k) :&= \mathbb{E}\Big|\mathbb{E}[g(Y^E_{t_k,Y_k}(t_{k+1})) - g(Y_{t_k,Y_k}(t_{k+1}))|\mathscr{F}_{t_k}]\Big| \\
&= \mathbb{E}\Big|\mathbb{E}[< \nabla g(Y_k), Y^E_{t_k,Y_k}(t_{k+1}) - Y_{t_k,Y_k}(t_{k+1}) >_d |\mathscr{F}_{t_k}] \\
&\quad + \mathbb{E}[< \nabla g(\eta_1(t_{k+1})) - \nabla g(Y_k)), Y^E_{t_k,Y_k}(t_{k+1}) - Y_{t_k,Y_k}(t_{k+1}) >_d |\mathscr{F}_{t_k}]\Big| \\
&= \mathbb{E}\Big| < \nabla g(Y_k), \mathbb{E}[d_{t_k,Y_k}(t_{k+1})|\mathscr{F}_{t_k}] >_d \\
&\quad + \mathbb{E}[< \nabla^2 g(\eta_2(t_{k+1}))(\eta_1(t_{k+1}) - Y_k), d_{t_k,Y_k}(t_{k+1}) >_d |\mathscr{F}_{t_k}]\Big| \\
&= \mathbb{E}\Big| < \nabla g(Y_k), \mathbb{E}[d_{t_k,Y_k}(t_{k+1})|\mathscr{F}_{t_k}] >_d \\
&\quad + \mathbb{E}[\theta^1_k < \nabla^2 g(\eta_2(t_{k+1})) d_{t_k,Y_k}(t_{k+1}), d_{t_k,Y_k}(t_{k+1}) >_d |\mathscr{F}_{t_k}]\Big| \\
&= \mathbb{E}\Big| < \nabla g(Y_k), M^{-1}_{t_k,Y_k}(t_{k+1}) c^0(t_k,Y_k) a(t_k,Y_k) >_d h_k^2 \\
&\quad + \mathbb{E}[\theta^1_k < \nabla^2 g(\eta_2(t_{k+1})) d_{t_k,Y_k}(t_{k+1}), d_{t_k,Y_k}(t_{k+1}) >_d |\mathscr{F}_{t_k}]\Big| \\
&\leq K_M \Big(\mathbb{E}[\|\nabla g(Y_k)\|^2_{d\times d}]\Big)^{1/2} \Big(\mathbb{E}[\|c^0(t_k,Y_k) a(t_k,Y_k)\|^2_d\Big)^{1/2} h_k^2 \\
&\quad + \Big(\mathbb{E}[\|\nabla^2 g(\eta_2(t_{k+1}))\|^2_{d\times d}]\Big)^{1/2} \Big(\mathbb{E}[\|d_{t_k,Y_k}(t_{k+1})\|^4_d]\Big)^{1/2} \\
&\leq K_M \Big(\mathbb{E}[\|\nabla g(Y_k)\|^4_{d\times d}]\Big)^{1/4} \Big(\mathbb{E}[\|c^0(t_k,Y_k) a(t_k,Y_k)\|^4_d\Big)^{1/4} h_k^2 \\
&\quad + \Big(\mathbb{E}[\|\nabla^2 g(\eta_2(t_{k+1}))\|^4_{d\times d}]\Big)^{1/4} \Big(\mathbb{E}[\|d_{t_k,Y_k}(t_{k+1})\|^4_d]\Big)^{1/2}
\end{aligned}$$

$$\leq 2^{3/2} K_g K_M K_{4c} \Big(\mathbb{E}[1 + \|Y_k\|_d^{4\kappa}] \Big)^{1/2} h_k^2 +$$

$$+\sqrt{3}\, 2^{3/2} (m+1)^{3/2} K_g K_M^2 K_{4c}^2 \Big(\mathbb{E}[1 + \|Y_k\|_d^{4\kappa}] \Big)^{3/4} h_k^3$$

$$\leq 2^{3/2} K_g K_M K_{4c} (1 + \sqrt{3}\,(m+1)^{3/2} K_M K_{4c}) \cdot \Big(\max_{i=0,1,\ldots,k} \mathbb{E}[1 + \|Y_{i+1}\|_d^{4\kappa}] \Big) \cdot h_k^2$$

where $\eta(t)$ is an intermediate value between $Y^E_{t_k,Y_k}(t)$ and $Y_{t_k,Y_k}(t)$, i.e. $\eta(t) = Y_k + \theta_k (Y^E_{t_k,Y_k}(t) - Y_{t_k,Y_k}(t))$ with scalar $\theta_k \in [0,1]$. Therefore, we may conclude that

$$m_1(k)$$

$$\leq 2^{3/2} K_f K_M K_{4c} (1 + \sqrt{3}\,(m+1)^{3/2} K_M K_{4c}) \Big(\max_{i=0,1,\ldots,k} \mathbb{E}[1 + \|Y_{i+1}\|_d^{4\kappa}] \Big) \sum_{i=0}^{k-1} h_i^2$$

$$\leq 2^{3/2} K_f K_M K_{4c} \Big(1 + \sqrt{3}(m+1)^{3/2} K_M K_{4c} \Big) \Big(\max_{i=0,1,\ldots,k} \mathbb{E}[1 + \|Y_{i+1}\|_d^{4\kappa}] \Big) t_k h_{max},$$

and

$$m_2(k)$$

$$\leq 2^{3/2} K_f K_M K_{4c} (1 + \sqrt{3}\,(m+1)^{3/2} K_M K_{4c}) \Big(\max_{i=0,1,\ldots,k} \mathbb{E}[1 + \|Y_{i+1}\|_d^{4\kappa}] \Big) h_k h_{max}.$$

Consequently, for all $k = 0, 1, \ldots, n_T - 1$, the weak error ε_0 of $BIMs$ (1.101) with nonrandom weights c^0 must satisfy

$$\begin{aligned} \varepsilon_0(t_{k+1}) &\leq K_w(t_{k+1}) \cdot \max_{i=0,1,\ldots,k} \mathbb{E}[1 + \|Y_{i+1}\|_d^{4\kappa}] \cdot h_{max} \\ &\leq K_w(T) \cdot \max_{i=0,1,\ldots,n_T-1} \mathbb{E}[1 + \|Y_{i+1}\|_d^{4\kappa}] \cdot h_{max}. \end{aligned}$$

where

$$K_w(t) \leq (K_w^E + 2^{3/2} K_f K_M K_{4c} (1 + \sqrt{3}\,(m+1)^{3/2} K_M K_{4c}))t.$$

The $p = 4\kappa$-moments of the BIMs (1.101) with vanishing weights c^j ($j = 1, 2, \ldots, m$) and sufficiently small step sizes $h_k \leq h_{max}$ are uniformly bounded. Thus, weak convergence with worst case rate $r_w \geq 1.0$ can be established under the given assumptions of Theorem 1.4.6, hence the proof is complete. ◇

Remark 1.4.4. Theorem 1.4.6 says that the BIMs with nonrandom weights and BTMs with $\Theta = 0$ have the same rate of weak convergence as the forward Euler methods have. For further details and more general classes of functionals F, see Talay [140]. One can also find estimates of K_w which are monotonically increasing in K_f, thanks to Theorem 1.4.6. Therefore, we obtain uniform weak convergence with respect to all test functions $f \in C^2_{b(\kappa)}(\mathbb{R}^d, \mathbb{R}^1)$ which have boundedness constants bounded by $K_f \le c < +\infty$.

BIMs are implementable very easily while gaining numerical stability compared to explicit methods (as that of Euler-Maruyama) and maintaining the same convergence rates as their explicit counterparts. Thus, we can justify them as a useful and remarkable alternative to the most used numerical methods for SDEs.

Remark 1.4.5. An expansion of the global error process for BIMs or BTMs as in [141] is not known so far. Such an expansion could give the possibility to exploit extrapolation procedures to increase the order of weak accuracy of those methods. This is still an open problem.

1.4.6 Supplement: Linear-Polynomial Boundedness of $f \in C^0_{Lip}$

In the proof above, we used the fact that linear-polynomial boundedness of Lipschitz continuous functions can be established too. To see this, let $C^0_{b(\kappa)}([0,T] \times \mathbb{R}^d, \mathbb{R}^l)$ denote the set of all continuous functions $f : [0,T] \times \mathbb{R}^d \to \mathbb{R}^l$ which are uniformly polynomially bounded such that

$$\|f(t,x)\|_l \;\le\; K_f \cdot (1 + \|x\|_l^\kappa)$$

for all $x \in \mathbb{R}^d$, where $K_f \ge 0$ and $\kappa \ge 0$ are appropriate real constants.

Lemma 1.4.1. *Assume that $f \in C^0_{b(\kappa)}([0,T] \times \mathbb{R}^d, \mathbb{R}^l)$ with constants $\kappa \ge 0$ and K_f is uniformly Lipschitz continuous with constant K_L, i.e.*

$$\forall t \in [0,T]\ \forall x, y \in \mathbb{R}^d \quad \|f(t,x) - f(t,y)\|_l \;\le\; K_L\, \|x - y\|_l. \quad (1.105)$$

Then, there exist constants $K_{b(p)} = K_{b(p)}(p, T, K_f, K_L)$ such that $\forall t \in [0,T]\ \forall x \in \mathbb{R}^d$

$$\|f(t,x)\|_l \le 2^{-(p-1)/p} K_{b(p)} \cdot (1 + \|x\|_l) \le K_{b(p)} \cdot (1 + \|x\|_l^p)^{1/p} \quad (1.106)$$

for all $p \ge 1$, where the real constants $K_{b(p)}$ can be estimated by

$$0 \;\le\; K_{b(p)} \;\le\; 2^{(p-1)/p} \cdot \max\{K_f, K_L\}. \quad (1.107)$$

Proof of Lemma 1.4.1. Estimate

$$\begin{aligned} 0 &\le \|f(t,x)\|_l \le \|f(t,0)\|_l + \|f(t,x) - f(t,0)\|_l \le K_f + K_L \|x\|_l \\ &\le \max\{K_f, K_L\}(1 + \|x\|_l) \;\le\; 2^{(p-1)/2} \max\{K_f, K_L\}(1 + \|x\|_l^p)^{1/p}. \end{aligned}$$

Therefore, constant $K_{b(p)}$ can be chosen as in (1.107). Thus, the proof is complete. ◇

Remark 1.4.6. In fact, it suffices that $\sup_{0\leq t\leq T} \|f(t,x_*)\|_l < +\infty$ for some $x_* \in \mathbb{R}^d$ and f is Lipschitz continuous in $x \in \mathbb{R}^d$ with Lipschitz constant $K_L(t)$ which is uniformly bounded with respect to $t \in [0,T]$. However, $K_{b(p)}$ may depend on κ too. Now, we can apply Lemma 1.4.1 in order to recognize that the requirement (A2) of Hölder continuity of system (a, b^j) yields linear-polynomial boundedness of both a and all b^j coefficients of diffusion processes X.

Remark 1.4.7. Weak approximations and weak rates of convergence of Euler-type methods are also studied in Kohatsu-Higa et al. [75,76], in Kannan et al. [67,68], in Kushner et al. [84, 85], in Ogawa [103–105], or in Talay [140].

Remark 1.4.8. Strong convergence under nonclassical conditions has been investigated in Deelstra & Delbaen [24], Gyöngy [42], Higham et al. [49], Hu [55], Ogawa [104], Tudor [142]. Moreover, asymptotic efficiency of classes of Runge-Kutta methods is considered in Newton [101].

1.5 Positivity

It is well-known that the **geometric Brownian motion** X satisfying the Itô SDE

$$dX(t) = \lambda\, X(t)\, dt \;+\; \gamma\, X(t)\, dW(t) \tag{1.108}$$

driven by the Wiener process W possesses the exact solution (with constants λ, $\gamma \in \mathbb{R}^1$)

$$X(t) = \exp((\lambda - \gamma^2/2)t + \gamma W(t)) \cdot X(0) \tag{1.109}$$

for $\mathscr{F}_0$-adapted initial data $X(0) \in \mathbb{R}^1$. Remarkable facts are that X remains non-negative for non-negative initial values $X(0) = x_0$, X remains strictly positive for strictly positive initial values, and X with $X(0) < 0$ or $X(0) > 0$ does not change its sign as time t advances. Let us investigate the behaviour of representatives of Theta methods (1.1) with respect to this observation of sign-preservation. For this purpose, we suppose $\gamma > 0$ for the further consideration.

1.5.1 Non-Positivity of Standard Euler Methods for Linear ODEs

In deterministic numerical analysis a very simple example is well–known. Consider the equation (ODE)

$$\dot{x} = \lambda x \quad \text{with} \quad x(0) = x_0 \geq 0$$

and its non-negative exact solution $x(t) = \exp(\lambda t) \cdot x_0$. Then the Euler scheme (1.1) gives

$$y_{n+1} = y_n + \lambda y_n h_n = (1 + \lambda h_n) y_n = y_0 \prod_{i=0}^{n} (1 + \lambda h_i). \qquad (1.110)$$

Obviously, started at $y_0 > 0$, this solution is always positive if $\lambda \geq 0$ or $|\lambda| h_i < 1$ for all $i = 0, 1, \ldots, n$. Thus negative values may occur under the assumption $y_0 > 0$, $\lambda < 0$ and h_i large enough (indeed infinitely often). This step size restriction even gets worse for large values of λ and a uniform restriction of h is not possible for all parameters $\lambda < 0$ in order to guarantee (strict) positivity (or sign-preservation).

1.5.2 Positivity of Balanced Theta Methods for Linear ODEs

In contrast to the standard Euler scheme, in the case $y_0 > 0$ and $\lambda < 0$, we can always prevent negative outcomes or even "explosions" in numerical methods with arbitrary step sizes h_i for linear differential equations. For this purpose we introduce the family of implicit Euler-type schemes (1.1) with

$$\begin{aligned} y_{n+1} &= y_n + (\theta y_{n+1} + (1-\theta) y_n) \lambda h_n \\ &= \frac{1 + (1-\theta)\lambda h_n}{1 - \theta \lambda h_n} y_n = y_0 \prod_{i=0}^{n} \frac{1 + (1-\theta)\lambda h_i}{1 - \theta \lambda h_i}, \end{aligned} \qquad (1.111)$$

hence it gives exclusively positive values under $y_0 > 0$ if $1 + (1-\theta)\lambda h_i > 0$ for all $i \in \mathbb{N}$. A generalization of these schemes is presented by the deterministic balanced methods

$$\begin{aligned} y_{n+1} &= y_n + \lambda y_n h_n + c h_n (y_n - y_{n+1}) \\ &= \frac{1 + (\lambda + c) h_n}{1 + c h_n} y_n = y_0 \prod_{i=0}^{n} \frac{1 + (\lambda + c) h_i}{1 + c h_i} \end{aligned} \qquad (1.112)$$

for an appropriate constant $c > 0$. Consequently, numerical solutions generated by (1.112) with $c \geq |\lambda|$ or by (1.111) with $\theta = 1$ are positive and monotonically decreasing for all $y_0 > 0$, $\lambda < 0$ and arbitrary step sizes $h_i \geq 0$. They do not have any explosions, and do not vanish for positive start values as well. Indeed they provide sign-preserving approximations. All these properties are features of the underlying SDE. Hence, BIMs possess adequate realizations.

1.5.3 Negativity of Drift-Implicit Theta Methods for Linear SDEs

After previous elementary illustrations with ODEs, we return to the stochastic case. For the sake of simplicity, we confine ourselves to equidistant partitions with uniform step size h (a similar result can be formulated for partitions with variable, nonrandom step sizes h_i, but with "some care").

Theorem 1.5.1 (Non-adequate Negativity of BTMs without Balanced Terms). *Suppose X satisfies (1.108) with $X(0) > 0$ (a.s.), $\gamma \neq 0$ and $(1-\theta\lambda h) > 0$. Then the drift-implicit Theta approximation (1.1) applied to the linear SDE (1.108) and started in $Y_0^\theta = X(0)$ with nonrandom equidistant step size $h > 0$, non-autonomous nonrandom implicitness $\theta \in \mathbb{R}^1$ and without balanced terms $c^j \equiv 0$ has negative values with positive probability, i.e. there is a stopping time $\tau = \tau(\omega) : \Omega \to \mathbb{N}$ such that*

$$\mathbb{P}\left(\{\omega \in \Omega : Y_\tau^\theta(\omega) < 0\}\right) > 0.$$

Proof. The family of drift-implicit Theta schemes (1.1) applied to (1.108) with $c^j \equiv 0$ and autonomous scalar implicitness $\theta = \theta_n$ is governed by

$$\begin{aligned} Y_{n+1}^\theta &= Y_n^\theta + \theta\lambda Y_{n+1}^\theta h + (1-\theta)\lambda Y_n^\theta h + \gamma Y_n^\theta \Delta W_n \\ &= \frac{1+(1-\theta)\lambda h + \gamma \Delta W_n}{1-\theta\lambda h} Y_n^\theta = Y_0^\theta \prod_{i=0}^n \left(\frac{1+(1-\theta)\lambda h + \gamma\sqrt{h}\xi_i}{1-\theta\lambda h}\right). \end{aligned}$$

Define the events $E_i \subseteq \Omega$ on $(\Omega, \mathscr{F}, \mathbb{P})$ (for $i \in \mathbb{N}$) by

$$E_i := \{\omega \in \Omega : 1 + (1-\theta)\lambda h + \gamma\sqrt{h}\xi_i(\omega) < 0\}$$

for i.i.d. $\xi_i \in \mathscr{N}(0,1)$ (standard Gaussian distributed). Then the event

$$E := \{\omega \in \Omega : \exists \tau(\omega) < +\infty, \tau(\omega) \in \mathbb{N} : Y_{\tau(\omega)}^\theta < 0\}$$

can be substituted by the events E_i.

$$\mathbb{P}(E_0) = \mathbb{P}\left(\{\omega \in \Omega : \xi_0(\omega) < -\frac{1+(1-\theta)\lambda h}{\gamma\sqrt{h}}\}\right) =: p$$

Notice that

$$\mathbb{P}(E_0) = \mathbb{P}\left(\{\omega \in \Omega : \xi_0(\omega) < -\frac{1+(1-\theta)\lambda h}{|\gamma|\sqrt{h}}\}\right) = p$$

as ξ_0 is symmetric about 0 (recall the assumption of Gaussian distribution of ξ_0). Thus, one obtains

$$\begin{aligned} \mathbb{P}(E) &= \mathbb{P}(E \cap (E_0 \cup \overline{E_0})) = \mathbb{P}(E|E_0)\mathbb{P}(E_0) + \mathbb{P}(E|\overline{E}_0)\mathbb{P}(\overline{E}_0) \\ &= p + (1-p)\mathbb{P}(E|\overline{E}_0) = p + (1-p)\mathbb{P}(E \cap (E_1 \cup \overline{E}_1)|\overline{E}_0) \end{aligned}$$

$$= p + (1-p)\left(\mathbb{P}\,(E|\overline{E}_0, \overline{E}_1)\mathbb{P}\,(E_1) + \mathbb{P}\,(E|\overline{E}_0, \overline{E}_1)\mathbb{P}\,(\overline{E}_1)\right)$$
$$= p + p(1-p) + (1-p)^2\mathbb{P}\,(E|\overline{E}_0, \overline{E}_1)$$

$$= p + p(1-p) + p(1-p)^2 + (1-p)^3\mathbb{P}\,(E|\overline{E}_0, \overline{E}_1, \overline{E}_2)$$
$$\ldots$$
$$= p\sum_{i=0}^{\infty}(1-p)^i = p\left(\frac{1}{1-(1-p)}\right) = \frac{p}{p} = 1.$$

Note that $0 < p < 1$. Thus it must exist (a.s.) a finite stopping time $\tau_n = n(w)h$ such that $Y^\theta(\tau_n) = Y_n^\theta < 0$.

Alternatively, with the help of the well–known Lemma of Borel–Cantelli (or Kolmogorov's 0–1–law, see Shiryaev [134]) one also finds a short proof of Theorem 1.5.1 (see original idea with $\theta = 0$ in S. [119]). For this purpose, we define

$$A_n = \{\omega \in \Omega : \exists i \le n : Y_i^\theta < 0\}$$

for $n \in \mathbb{N}^+$. Then it follows that

$$\overline{E} = \cap_{n=1}^{\infty} \cup_{k=n}^{\infty} \overline{A}_k = \overline{\cup_{n=1}^{\infty} \cap_{k=n}^{\infty} A_k}\,.$$

Because of $\mathbb{P}\,(\overline{A}_n) = (1 - \mathbb{P}\,(E_0))^n \quad (n = 1, 2, \ldots)$, we obtain

$$\sum_{n=0}^{\infty}\mathbb{P}\,(\overline{A}_n) = \frac{1}{\mathbb{P}\,(E_0)} < +\infty$$

where

$$1 > \mathbb{P}\,(E_0) = \mathbb{P}\,(\{\omega \in \Omega : 1 + (1-\theta)\lambda h + |\gamma|\sqrt{h}\xi_0(\omega) < 0\})$$
$$= \frac{1}{\sqrt{2\pi}}\int_{-\infty}^{\left(-\frac{1+(1-\theta)\lambda h}{|\gamma|\sqrt{h}}\right)} \exp(-x^2/2)dx$$
$$= \Phi\left(-\frac{1+(1-\theta)\lambda h}{|\gamma|\sqrt{h}}\right) > 0 \quad \forall h \in (0,\infty)$$

and Φ denotes the probability distribution function of the standard Gaussian distribution $\mathcal{N}(0,1)$, hence the assertion $\mathbb{P}\,(\overline{E}) = 0$ is true. Thus, drift-implicit Theta schemes with any $\theta \in \mathbb{R}^1$ and without balanced terms always possess a trajectory with negative outcome under the assumptions of Theorem 1.5.1.

1.5.4 Positivity of Balanced Theta Methods for Linear SDEs

In contrast to the Euler methods, we find numerical methods which only have non-negative values and are sign-preserving. A corresponding assertion is formulated by Theorem 1.5.2.

Theorem 1.5.2 (A.s. Positivity of Balanced Methods, S.[119]). *Assume that process X satisfies (1.108) and $X(0) \geq 0$ (a.s.).*
Then the balanced methods with non-negative constants c^0 and c^1 exclusively have positive outcomes at all instants $n \in \mathbb{N}$, provided that

$$1 + (c^0 + \lambda)h \geq 0 \quad \textit{and} \quad c^1 \geq |\gamma| \,. \tag{1.113}$$

Proof. This claim follows immediately from the structure of the balanced methods (1.1) applied to the (1.108) with $\Theta = 0$, appropriate weights c^0 and c^1 satisfying (1.113). One encounters

$$\begin{aligned} Y_{n+1}^B &= Y_n^B + \lambda Y_n^B h + \gamma Y_n^B \Delta W_n + (c^0 h + c^1 |\Delta W_n|)(Y_n^B - Y_{n+1}^B) \\ &= \frac{1 + (c^0 + \lambda)h + \gamma \Delta W_n + c^1 |\Delta W_n|}{1 + c^0 h + c^1 |\Delta W_n|} Y_n^B \,. \end{aligned}$$

Thereby, $Y_{i+1}^B \geq 0$ iff

$$1 + (c^0 + \lambda)h + \gamma \Delta W_i + c^1 |\Delta W_i| \geq 0$$

for all $i \in \mathbb{N}$. Obviously, this is true under (1.113). ◇

Remark 1.5.1. An optimal choice of weights c^j to maintain positivity under $\lambda < 0$ is given by $c^1 = |\gamma|$ and $c^0 = [-\lambda]_+$ where $[z]_+$ is the positive part of inscribed expression z (for optimality, see also exact contraction-monotone and energy-exact methods below). Numerical methods with variable step sizes could also prevent negativity, however this may lead to inefficient implementations in random settings and throws out the nontrivial question of convergence of random instants t_n to T as $n \to +\infty$. Moreover, the predictable measurability of instants t_n is given up (i.e. one deals with another category of stochastic approximations!).

1.5.5 Positivity of Balanced Theta Methods for Nonlinear SDEs

Consider nonlinear autonomous SDEs (1.2) of the form

$$dX(t) = \left[f(X(t))dt + \sum_{j=1}^{m} g_j(X(t))dW^j(t) \right] X(t) \tag{1.114}$$

with $a(t,x) = xf(x)$ and $b^j(t,x) = xg_j(x)$ where f and g_j are nonrandom matrices in $\mathbb{R}^{d\times d}$. Such (1.114) are met in biology and ecology (see Lotka-Volterra equations or logistic Pearl-Verhulst models). There X approximates the number of populations of certain species such as fish, flies, rabbits, fox, mice and owls or the dynamic evolution of offspring-to-parents ratios. Naturally, X must have only positive outcomes for its practical meaningful modeling character.

Let us discretize SDEs (1.114) by the balanced methods (1.1) of the form

$$X_{n+1} = \left[I + f(X_n)h_n + \sum_{j=1}^{m} g_j(X_n)\Delta W_n^j \right] X_n$$
$$+ \left[||f(X_n)||_{max} h_n + \sum_{j=1}^{m} ||g_j(X_n)||_{max} |\Delta W_n^j| \right] (X_n - X_{n+1}) \quad (1.115)$$

where $||.||_{max}$ is the maximum norm of all entries of inscribed matrices and I is the unit matrix in $\mathbb{R}^{d\times d}$.

Theorem 1.5.3 (Positivity of BIMs for Nonlinear SDE). *The balanced implicit methods (1.115) applied to SDEs (1.114) with same initial value $X(0) > 0$ and any step sizes h_n maintain the positivity at all instants $n \in \mathbb{N}$.*

Proof. Rewrite scheme (1.115) to equivalent expression

$$C_n X_{n+1} = \Big[I + (f(X_n) + ||f(X_n)||_{max} I)h_n$$
$$+ \sum_{j=1}^{m} (g_j(X_n)\Delta W_n^j + ||g_j(X_n)||_{max} |\Delta W_n^j|) \Big] X_n \quad (1.116)$$

where C_n represents the invertible $d \times d$ diagonal matrix

$$C_n = \left[1 + ||f(X_n)||_{max} h_n + \sum_{j=1}^{m} ||g_j(X_n)||_{max} |\Delta W_n^j| \right] I.$$

Now, apply complete induction on $n \in \mathbb{N}$ and exploit the positivity of both the right hand side of (1.116) and the matrix C_n to conclude the assertion of Theorem 1.5.3. It is worth noting that the choice of step sizes $h_n > 0$ does not play a role in keeping positivity as n advances in our previous argumentation. ◇

Remark 1.5.2. A slightly more general theorem on positivity of convergent multi-dimensional BIMs is found in S. [119, 120].

1.5.6 Positivity of Balanced Milstein Methods

The lack of preservation of positivity can not be removed by the use of "higher order methods" such as standard Milstein or Taylor-type methods (while noting that the probabilities of negative outcomes can be reduced by using them). However, by appropriate introduction of further balancing terms in them, one may achieve the goal of positivity (and sign-preservation) and higher order of convergence (and consistency as well). This fact is demonstrated by the following discussion which is related to higher order numerical methods applied to SDEs driven exclusively by diagonal noise (for the sake of simplicity). Recall that systems with diagonal noise are characterized by $\mathscr{L}^i b^j(t,x) = 0$ for all $i \neq j$ where $i,j = 1,\ldots,m$. In this case the **balanced Milstein methods (BMMs)** with diagonal noise follow the scheme

$$Y_{n+1} = Y_n + \sum_{j=0}^{m} b^j(t_n, Y_n) I_{(j),t_n,t_{n+1}} + \sum_{j=1}^{m} \mathscr{L}^j b^j(t_n, Y_n) I_{(j,j),t_n,t_{n+1}} + \left(d^0(t_n, Y_n) I_{(0),t_n,t_{n+1}} + \sum_{j=1}^{m} d^j(t_n, Y_n) I_{(j,j),t_n,t_{n+1}} \right) (Y_n - Y_{n+1}) . \tag{1.117}$$

This class has been introduced and studied in Kahl and S. [60] in a more general form referring to multiple Itô integrals $I_{\alpha,t_n,t_{n+1}}$.

Theorem 1.5.4 (Positivity of Balanced Milstein Methods [60]). *The one-dimensional BMMs (1.117) applied to one-dimensional SDEs (1.2) with diagonal noise along partitions*

$$t_0 < t_1 < \ldots < t_n < t_{n+1} < \ldots$$

and applied to SDEs (1.2) with diagonal noise maintain positivity of initial conditions $X(0) \in \mathbb{R}_+$ if the following conditions hold

(A6) $\forall t \in [0,T]\ \forall x \in \mathbb{R}_+ : d^0(t,x) - \frac{1}{2}\sum_{j=1}^{m} d^j(t,x)$ *is positive semi-definite.*
(A7) $\forall j = 1,2,\ldots,m\ \forall t \in [0,T]\ \forall x \in \mathbb{R}_+ : d^j(t,x)$ *is positive semi-definite.*
(A8) $\forall j = 1,\ldots,m\ \forall t_n \in [0,T]$ *and* $\forall x \in \mathbb{R}_+$.

$$b^j(t_n,x)\frac{\partial}{\partial x} b^j(t_n,x) + d^j(t_n,x)x > 0, \tag{1.118}$$

(A9) $\forall t_n \in [0,T]$ *and* $\forall x \in \mathbb{R}_+$.

$$x - \sum_{j=1}^{m} \frac{(b^j(t_n,x))^2}{2b^j(t_n,x)\frac{\partial}{\partial x} b^j(t_n,x) + 2d^j(t_n,x)x} > 0, \tag{1.119}$$

(A10) If $D(t_n,x) = a(t_n,x) - \frac{1}{2}\sum_{j=1}^{m} b^j(t_n,x)\frac{\partial}{\partial x}b^j(t_n,x) + d^0(t_n,x)x - \frac{1}{2}\sum_{j=1}^{m} d^j(t_n,x)x < 0$ *for a value* $x \in \mathbb{R}_+$ *at time-instant* $t_n \in [0,T]$ *then the current step size* h_n *is chosen such that* $\forall t_n \in [0,T]$, $\forall x \in \mathbb{R}_+$.

$$h_n < \frac{x + N(t_n,x)}{-D(t_n,x)} \tag{1.120}$$

where

$$N(t_n,x) = -\sum_{j=1}^{m} \frac{(b^j(t_n,x))^2}{2b^j(t_n,x)\frac{\partial}{\partial x}b^j(t_n,x) + 2d^j(t_n,x)x}.$$

Remark 1.5.3. The assumption (A8) guarantees that the BMMs inherit the positivity preserving structure of the underlying Milstein method. Condition (A9) is more technical, but in many applications this is valid without the use of the weight function d^1. (A10) is only necessary if $D(t_n,x) < 0$, otherwise we can drop this restriction for positivity. So, we obtain a first idea to apply BMMs as advanced Milstein-type methods to preserve positivity by choosing d^0 and d^1 in such a way that $D(t_n,x)$ is greater than zero and we do not have to restrict the step size through (1.120) in this case. Furthermore, for global mean square convergence of BMMs with worst case rate 1.0 and positivity at the same time, we need to require that $D(t,x) \geq 0$. Note that the adapted, but random step size selection depending on current random outcomes Y_n by condition (A10) in the case of $D(t,x) < 0$ would contradict to the exclusive use of nonrandom step sizes as exploited in standard convergence proofs. Moreover, a restricted step size selection as given by (A10) throws out the problem of proving that any terminal time T can be reached in a finite time with probability one. So it is advantageous to require $D(t,x) \geq 0$ for all $x \geq 0$ and $0 \leq t \leq T$ for meaningful and practically relevant approximations.

Proof. Set $x = Y_n$. Using the one-step representation of the BMM (1.117) we obtain

$$\begin{aligned}
&\left(1 + d^0(t_n,x) + \frac{1}{2}\sum_{j=1}^{m} d^j(t_n,x)\left((\Delta W_n^j)^2 - h_n\right)\right) Y_{n+1} \\
&= \Bigg(x + a(t_n,x)h_n + \sum_{j=1}^{m} b^j(t_n,x)\Delta W_n^j \\
&\quad + \frac{1}{2}\sum_{j=1}^{m} b^j(t_n,x)\frac{\partial}{\partial x}b^j(t_n,x)\left((\Delta W_n^j)^2 - h_n\right) \\
&\quad + d^0(t_n,x)xh_n + \frac{1}{2}\sum_{j=1}^{m} d^j(t_n,x)x\left((\Delta W_n^j)^2 - h_n\right)\Bigg) \\
&= R(t_n,Y_n).
\end{aligned}$$

The expression $(\ldots)$ in-front of Y_{n+1} at the left hand side of this equation is positive due to (A6) and (A7). Rewriting the right hand side leads to

$$R(t_n,x) = x + \left(a(t_n,x) - \frac{1}{2}\sum_{j=1}^{m} b^j(t_n,x)\frac{\partial}{\partial x}b^j(t_n,x) + d^0(t_n,x)x \right.$$
$$\left. - \frac{1}{2}\sum_{j=1}^{m} d^j(t_n,x)x \right) h_n + g(\Delta W_n^1,\ldots,\Delta W_n^m)$$

with

$$g(\Delta W_n^1,\ldots,\Delta W_n^m) = \sum_{j=1}^{m} b^j(t_n,x)\Delta W_n^j$$
$$+ \frac{1}{2}\sum_{j=1}^{m}\left(b^j(t_n,x)\frac{\partial}{\partial x}b^j(t_n,x) + d^j(t_n,x)x \right)(\Delta W_n^j)^2.$$

The function $g : \mathbb{R}^m \to \mathbb{R}^1$ possesses a global minimum due to (A8). More precisely, an obvious calculation shows that

$$\min_{z\in\mathbb{R}^m} g(z) = -\sum_{j=1}^{m} \frac{(b^j(t_n,x))^2}{2\left(b^j(t_n,x)\frac{\partial}{\partial x}b^j(t_n,x) + d^j(t_n,x)x\right)}. \tag{1.121}$$

This enables us to estimate R from below by replacing the value of $g(\Delta W_n^1,\ldots,$ $\Delta W_n^m)$ by its minimum. So we arrive at

$$R(t_n,x) \geq x + \left(a(t_n,x) - \frac{1}{2}\sum_{j=1}^{m} b^j(t_n,x)\frac{\partial}{\partial x}b^j(t_n,x) + d^0(t_n,x)x \right.$$
$$\left. - \frac{1}{2}\sum_{j=1}^{m} d^j(t_n,x)x \right)\Delta_n - \sum_{j=1}^{m} \frac{(b^j(t_n,x))^2}{2\left(b^j(t_n,x)\frac{\partial}{\partial x}b^j(t_n,x) + d^j(t_n,x)x\right)}$$
$$= x + N(t_n,x) + D(t_n,x)h_n$$

We can clearly see that (A9)–(A10) under (A8) are needed to get positive values $Y_{n+1} > 0$ whenever $Y_n > 0$ for all $n \in \mathbb{N}$. More precisely, if

$$D(t_n,x) = \left(a(t_n,x) - \frac{1}{2}\sum_{j=1}^{m} b^j(t_n,x)\frac{\partial}{\partial x}b^j(t_n,x) \right.$$
$$\left. + d^0(t_n,x)x - \frac{1}{2}\sum_{j=1}^{m} d^j(t_n,x)x \right) \geq 0$$

then $R(t_n, x) > 0$ and we do not need any restriction of the step size h_n by (A10) at all. If $D(t_n, x) < 0$ then $x + N(t_n, x) + D(t_n, x)h_n \geq 0$ guarantees that $R(t_n, x) > 0$, hence condition (A10) is needed in this case. Therefore, assumptions (A8)-(A10) imply the property of positivity of BMMs (1.117). ◇

Remark 1.5.4. The proof of Theorem 1.5.4 shows that the condition (A9) can be relaxed to

$$x - \sum_{j=1}^{m} \frac{(b^j(t_n, x))^2}{2b^j(t_n, x)\frac{\partial}{\partial x}b^j(t_n, x) + 2d^j(t_n, x)x} \geq 0 \tag{1.122}$$

under the assumption $D(t_n, x) > 0$.

Moreover, in some cases it is more efficient to verify the following conditions instead of restrictions (A9) and (A10) known from Theorem 1.5.4.

Corollary 1.5.1 (Kahl and S. [60]). *The one-dimensional BMM (1.117) satisfying (A6)–(A8) along partitions*

$$t_0 < t_1 < \cdots < t_n < t_{n+1} < \cdots$$

maintains positivity if additionally

(A11) $\forall j = 1, \ldots, m\ \forall t_n \in [0, T]$ *and* $\forall x \in I\!R_+ : x + N(t_n, x) + D(t_n, x)h_n > 0$, *where the functions N and D are defined in Theorem 1.5.4.*

Example 6.1. Consider the one-dimensional geometric Brownian motion

$$dX(t) = \sum_{j=1}^{m} \sigma_j X(t)\, dW^j(t), \qquad X(0) = x_0 > 0 \tag{1.123}$$

without any drift which is a standard example for stability analysis for numerical integration schemes for SDEs (e.g. see [120, 127]) where the standard numerical methods possess serious step size restrictions or even fail to preserve stability and positivity. Using any BMM with

$$d^0(t_n, x) \geq \frac{m}{2}\sum_{j=1}^{m}\sigma_j^2,$$

$$\forall l = 1, \ldots, m \qquad d^l(t_n, x) = (m-1)\sigma_l^2$$

can solve this problem with higher order of accuracy very easily since

$$D(t_n, x) = -\frac{1}{2}\sum_{j=1}^{m}\sigma_j^2 x + d^0(t_n, x)x - \frac{m-1}{2}\sum_{j=1}^{m}\sigma_j^2 x$$

$$\geq -\frac{m}{2}\sum_{j=1}^{m}\sigma_j^2 x + \frac{m}{2}\sum_{j=1}^{m}\sigma_j^2 x \;=\; 0 \quad \text{and}$$

$$N(t_n,x) = -\sum_{j=1}^{m}\frac{\sigma_j^2 x^2}{2\sigma_j^2 x + 2(m-1)\sigma_j^2 x} \;=\; -\frac{x}{2} < 0$$

for $x \in \mathbb{R}_+ = (0,+\infty)$. Hence, the restriction (A10) on the step size h_n is not relevant here. However, notice that a restriction of the form $D(t_n,x) \geq -K = constant$ is important for the finiteness of related numerical algorithm (i.e. in particular in order to reach any desired terminal time $T > 0$ with probability one). Moreover, all assumptions (A6)–(A9) are satisfied. Consequently, the related BMMs provide positive-invariant numerical approximations to test SDE (1.123) with any choice of step sizes h_n and any $\mathcal{F}_0$-adapted initial data $X(0)$. Moreover, one can show that they are consistent, convergent and stable (i.e. dynamically consistent).

Example 6.2. Consider the mean-reverting process (ECIRM)

$$dR(t) = \kappa(\theta - R(t))dt + \sigma[R(t)]^p dW(t) \tag{1.124}$$

with $\theta, \kappa, \sigma > 0$ which is of great importance in financial mathematics as well as in other areas of applied science. Focusing on the financial meaning of this equation we obtain the well known Cox-Ingersoll-Ross model with exponent $p = 0.5$, describing the short-rate R in the interest rate market. Furthermore this SDE can be used to model stochastic volatility. As commonly known, these dynamics must be positive for practically meaningful models. Indeed one can show that for the above process X with exponents $p \in [0.5, 1)$ and positive initial data $R(0)$ (e.g. by using Fellers classification of boundary values). Then Kahl and S. [60] have shown that the BMM (1.117) satisfying (A6)-(A7) with $\sigma^2 > 0$ along partitions

$$t_0 < t_1 < \cdots < t_n < t_{n+1} < \cdots$$

has exclusively positive outcomes when applied to the mean-reverting process (1.124) with diffusion exponent $p \in (0.5, 1]$ with the following choice of the weight functions

$$d^0(x) = \alpha\kappa + \frac{1}{2}\sigma^2 p|x|^{(2p-2)}, \qquad d^1(x) \;=\; 0. \tag{1.125}$$

with relaxation parameter $\alpha \in [0,1]$ such that

$$h_n < \frac{2p-1}{2p\kappa(1-\alpha)}. \tag{1.126}$$

Here, the relaxation parameter α is similar to the implicitness parameter θ in the class of Theta methods (1.1) and gives some flexibility to adjust the BMM to specific problem issues. The fully implicit case $\alpha = 1$ is a safe choice as we do not have to restrict the step size in that case. On the other hand, numerical tests have indicated that a reduced level of implicitness leads to better approximation results. Therefore, we would recommend to use $\alpha = 0.5$ whenever the parameter configuration allows this choice, also supported by results from [120]. Similarly, but with some care, one may circumvent the problem of negative outcomes by BMMs in the critical case $p = 0.5$ of model (1.124). For more details, see [60].

Remark 1.5.5. An alternative numerical method to prevent inadequate negativity is given by splitting techniques of Moro and Schurz [99].

1.5.7 Non-positivity of Standard Euler Methods for Nonlinear SDE

Again, for simplicity, we confine ourselves to equidistant partitions of Euler approximations with step size h (e.g. $h = T/N$) and possibly state-dependent noise term

$$X_{n+1} = X_n + ha(X_n) + \sqrt{h}b(X_n)\xi_{n+1}, \quad n = 1, 2, \ldots, \quad X(0) = x_0 > 0, \tag{1.127}$$

where $\{\xi_n\}_{n\in\mathbb{N}}$ are independent identically distributed (i.i.d.) random variables on $(\Omega, \mathscr{F}_{t_n}, \mathbb{P})$. Then one is able to prove the following Theorem (see [2]) which extends the results from [119, 120] to nonlinear equations (1.127). As usual, let

$$(\Omega, \mathscr{F}, (\mathscr{F}_{t_n})_{n\in\mathbb{N}}, \mathbb{P})$$

be a complete filtered probability space and the filtration $(\mathscr{F}_{t_n})_{n\in\mathbb{N}}$ be naturally generated, namely that $\mathscr{F}_{t_n} = \sigma\{\xi_0, \xi_1, \ldots \xi_n\}$.

Theorem 1.5.5 (Non-positivity of Euler Methods for Nonlinear SDEs in 1D). *Assume that:*

(o) $h >, x_0 > 0$ *are nonrandom and fixed*
(i) $a : \mathbb{R} \to \mathbb{R}$ *and* $b : \mathbb{R} \to \mathbb{R}$ *are continuous, nonrandom functions with*

$$a(0) = b(0) = 0, \tag{1.128}$$

(ii) G *is well-defined by*

$$G(u) = \frac{u + ha(u)}{\sqrt{h}|b(u)|} \tag{1.129}$$

for all $u \neq 0$ *and satisfies*

$$\inf_{u>0} \{-G(u)\} > -\infty, \tag{1.130}$$

(iii) diffusion function b *is positive-definite on* $\mathbb{R}^1_+$, *i.e.*

$$\forall x > 0 \; : \; b(x) > 0, \tag{1.131}$$

(iv) $(\xi_n)_{n \in I\!N}$ *are i.i.d. Gaussian* $\mathcal{N}(0,1)$*-distributed random variables on the filtered probability space* $(\Omega, \mathcal{F}, (\mathcal{F}_{t_n})_{n \in I\!N}, \mathbb{P})$.

Then, for the standard Euler method (1.127) with equidistant step size h and positive initial value $X(0) = x_0 > 0$*, there exists an a.s. finite stopping time* $\tau : \Omega \to I\!N$ *such that*

$$X_\tau < 0. \quad (a.s.)$$

Remark 1.5.6. In fact, our main results are also valid when all ξ_n are i.i.d. random variables with symmetric continuous probability distribution F satisfying

$$\forall n \in \mathbb{N} \; : \; F_{\xi_n} = F_{-\xi_n}, supp(F_{\xi_n}) = \mathbb{R}^1. \tag{1.132}$$

Proof. In the proof we use a technique which was suggested by [119] and which is based on the Borel-Cantelli Lemma (for latter, see, e.g. [134]). For this purpose and fixed step size h, introduce the nonrandom real constant M by setting

$$\inf_{u>0} \{-G(u)\} = \inf_{u>0} \left\{ -\frac{u + ha(u)}{\sqrt{h}|b(u)|} \right\} =: M > -\infty. \tag{1.133}$$

Furthermore, define

$$\bar{E}_n = \{\omega \in \Omega : X_i(\omega) > 0, \forall i \leq n\}, \tag{1.134}$$

then

$$E_n = \{\omega \in \Omega : \exists i \leq n : X_i(\omega) \leq 0\}. \tag{1.135}$$

From the Borel-Cantelli Lemma we conclude that the solution X to equation (1.127) becomes non-positive with probability one if

$$\sum_{n=0}^{\infty} \mathbb{P}\,(\bar{E}_n) < \infty. \tag{1.136}$$

To prove (1.136) we estimate $\mathbb{P}\,(\bar{E}_n)$ from above. Set $X(0) = X_0$. Since

$$\begin{aligned}\mathbb{P}\,(\bar{E}_n) &= \mathbb{P}\,\{\omega \in \Omega : X_i(\omega) > 0, \forall i \leq n\} \\ &= \prod_{j=1}^{n} \mathbb{P}\,\{\omega \in \Omega : X_j(\omega) > 0 | X_{j-1} > 0, \ldots, X_0 > 0\},\end{aligned}$$

we need to estimate $\mathbb{P}\,\{\omega \in \Omega : X_j(\omega) > 0 | X_{j-1} > 0, \ldots, X_0 > 0\}$. We have for $j > 1$

$$\begin{aligned}
&\mathbb{P}\left\{\omega \in \Omega : X_j(\omega) > 0 \middle| X_{j-1} > 0, \ldots, X_0 > 0\right\} \\
&= \mathbb{P}\left\{\omega \in \Omega : X_{j-1} + ha(X_{j-1}) + \sqrt{h}b(X_{j-1})\xi_j(\omega) > 0 \middle| \bar{E}_{j-1}\right\} \\
&= \mathbb{P}\left\{\omega \in \Omega : \sqrt{h}b(X_{j-1})\xi_j(\omega) > -X_{j-1} - ha(X_{j-1}) \middle| \bar{E}_{j-1}\right\} \\
&= \mathbb{P}\left\{\omega \in \Omega : \xi_j(\omega) > -\frac{X_{j-1} + ha(X_{j-1})}{\sqrt{h}b(X_{j-1})} \middle| X_{j-1} > 0\right\} \\
&\leq \mathbb{P}\left\{\omega \in \Omega : \xi_j(\omega) > M \middle| X_{j-1} > 0\right\} = \mathbb{P}\left\{\omega \in \Omega : \xi_j(\omega) > M\right\}
\end{aligned}$$

under $b(u) > 0$ for $u > 0$. The last equality in the above estimation holds since X_{j-1} is defined by $\xi_0, \xi_1, \ldots, \xi_{j-1}$ and ξ_j is independent of $\xi_0, \xi_1, \ldots, \xi_{j-1}$, and therefore ξ_j is independent of X_{j-1}.

Now, put

$$\mathbb{P}\left\{\omega \in \Omega : \xi_j(\omega) > M\right\} = q_M,$$

and note that q_M does not depend on j and $q_M < 1$. The former statement is true since $M > -\infty$ and ξ_n are all $\mathcal{N}(0,1)$–distributed, and therefore all ξ_n are continuously distributed with support $(-\infty, +\infty)$. Therefore,

$$\mathbb{P}(\bar{E}_n) \leq (q_M)^n, \quad \sum_{n=0}^{\infty} \mathbb{P}(\bar{E}_n) \leq \frac{1}{1 - q_M} < \infty.$$

Furthermore, define

$$\tau(\omega) = \inf\{i \in \mathbb{N} : X_i(\omega) \leq 0\}, \tag{1.137}$$

and note that, by the above argumentation, the random variable τ_0 is a.s. finite and all events $\{\tau = n\}$ are $(\mathcal{F}_{t_n}, \mathcal{P}(\mathbb{N}))$-measurable, where $\mathcal{P}(\mathbb{N})$ is the power set of all natural numbers $\mathbb{N}$ (i.e. $\{\tau = n\} \in \mathcal{F}_{t_n}$ for all $n \in \mathbb{N}$ and τ is a finite Markov time). We also note that we actually have $X_\tau < 0$ (a.s.). This last statement is correct since X_n has a continuous probability distribution and, therefore, it cannot take on the value 0 with positive probability. ◇

Remark 1.5.7. Condition (1.131) is not so essential for the validity of the above proof. Indeed, one can proceed without imposing condition (1.131), but this is not in the scope of this survey.

Remark 1.5.8. Condition (1.130) is fulfilled in particular if, for some $K_a, K_b > 0$ and all $u \in \mathbb{R}$, we have

$$a(u) \geq K_a u \text{ if } u \leq 0, \quad a(u) \leq K_b u \text{ if } u \geq 0 \quad \text{and} \quad |b(u)| \geq K_b|u| \text{ if } u \in \mathbb{R}. \tag{1.138}$$

Moreover, for example thanks to Theorem 1.5.5, if there are constants $p>0$, $K_a>0$ and $K_b > 0$ such that

$$a(u) = -K_a|u|^p u \quad \text{and} \quad b(u) = K_b u, \quad \text{for all} \quad u \in \mathbb{R}, \tag{1.139}$$

then every solution X of (1.127) started at $X(0) = x_0 > 0$ will eventually possess negative values as the integration time t_n advances.

Remark 1.5.9. In passing, we note that the fact of non-positivity of numerical methods is not so a bad property for some classes of SDEs. For example, systems with additive noise (see e.g. discretizations of Ornstein-Uhlenbeck processes) must have that feature of changing signs infinitely. This stems from the inherent property of underlying Wiener processes to cross zero levels after some finite time always (cf. the law of iterated logarithm of Wiener processes). All in all, we recommend some care with requiring positivity in only adequate situations.

1.6 Boundedness (Finite Stability)

In this section, we shall study the topic of uniform boundedness of numerical approximations which is mostly omitted in the literature. The property of boundedness can be interpreted as stability on finite intervals or the absence of "inadequate explosions". This is also an important requirement in view of meaningful modeling under the absence of blow-ups. We can only indicate some results on the fairly complex property of boundedness.

1.6.1 Almost Sure Boundedness for Logistic Equations

Meaningful stochastic generalizations of logistic equations lead to the nonlinear SDEs of Itô-type

$$dX(t) = [(\rho+\lambda X(t))(K-X(t))-\mu X(t)]\,dt+\sigma X(t)^{\alpha}|K-X(t)|^{\beta} dW(t) \tag{1.140}$$

driven by a standard Wiener process $(W(t) : t \geq 0)$, started at $X_0 \in \mathbb{D} = [0, K] \subset \mathbb{R}^1$, where $\rho, \lambda, K, \mu, \sigma$ are positive and α, β non-negative real parameters. There ρ can be understood as coefficient of transition (self-innovation), λ as coefficient of imitation depending on the contact intensity with its environment, K as a somewhat "optimal" environmental carrying capacity and μ as natural death rate. However, in view of issues of practical meaningfulness, model (1.140) makes only sense within deterministic algebraic constraints, either given by extra boundary conditions or self-inherent properties resulting into natural barriers at 0 at least. This fact is supported by the limited availability of natural resources as known from the evolution of species in population ecology. In what follows we study **almost sure regularity** (boundedness on $\mathbb{D}$) of both exact and numerical solutions of (1.140).

Theorem 1.6.1 ([120, 130]). *Let $X(0) \in \mathbb{D} = [0, K]$ be independent of σ-algebra $\sigma(W(t), t \geq 0)$. Then, under the conditions that $\alpha \geq 1$, $\beta \geq 1$, $K \geq 1$, $\rho \geq 0$, $\lambda \geq 0$, $\mu \geq 0$, the stochastic process $\{X(t), t \geq 0\}$ governed by equation (1.140) is regular on $\mathbb{D} = [0, K]$, i.e. we have $\mathbb{P}\ (X(t) \in [0, K]) = 1$ for all $t \geq 0$. Moreover, regularity on $\mathbb{D}$ implies boundedness, uniqueness, continuity and Markov property of the strong solution process $\{X(t), t \geq 0\}$ of SDE (1.140) whenever $X(0) = 0$ (a.s.), $X(0) = K$ (a.s.) or*

$$\mathbb{E}[\ln(X(0)(K - X(0)))] > -\infty.$$

Remark 1.6.1. The proof of Theorem 1.6.1 is found in S. [130]. To avoid technical complications, define the diffusion coefficient $b(x)$ to be zero outside $[0, K]$. Note that the requirement $\alpha \geq 1$ is a reasonable one in ecology, marketing and finance. This can be seen from the fact that modeling in population models is motivated by modeling per-capita-growth rates (cf. S. [130]). Similar argumentation applies to models in finance (asset pricing) and marketing (innovation diffusion).

Proof. Define the drift function

$$a(x) = (\rho + \lambda x)(K - x) - \mu x$$

and diffusion function

$$b(x) = \sigma x^{\alpha}(K - x)^{\beta}$$

for $x \in [0, K]$. Take the sequence of open domains

$$\mathbb{D}_n := (\exp(-n), K - \exp(-n)), n \in \mathbb{N}.$$

Then, equation (1.140) is well–defined, has unique, bounded and Markovian solution up to random time $\tau^{s,x}(\mathbb{D}_n)$, due to Lipschitz continuity and (linear) boundedness of drift $a(x)$ and diffusion $b(x)$ on $\mathbb{D}_n$. Now, use Lyapunov function $V \in C^2(\mathbb{D})$ defined on $\mathbb{D} = (0, K)$ via

$$V(x) \;=\; K - \ln(x(K - x))\,.$$

Note that $V(x) = K - \ln(x(K - x)) = x - \ln(x) + K - x - \ln(K - x) \geq 2$ for $x \in \mathbb{D} = (0, K)$. Now, fix initial time $s \geq 0$, introduce a new Lyapunov function $W \in C^{1,2}([s, +\infty)\text{x}\mathbb{D})$ by $W(t, x) = \exp(-c(t-s))V(x)$ for all $(t, x) \in [s, +\infty)$ x $\mathbb{D}$, where

$$c = \frac{\rho + \lambda K + \sigma^2 K^{2\alpha+2\beta-2} + \mu}{2}.$$

Then $V \in C^2(\mathbb{D})$ and $W \in C^{1,2}([s,+\infty) \times \mathbb{D}_n)$. Define $\mathscr{L} = \mathscr{L}^0$ as infinitesimal generator as in (1.6). Calculate

$$\mathscr{L}V(x) = \Big((\rho+\lambda x)(K-x) - \mu x\Big)\left[\frac{-1}{x} + \frac{1}{K-x}\right] + \frac{\sigma^2}{2} x^{2\alpha}(K-x)^{2\beta}\left[\frac{1}{x^2} + \frac{1}{(K-x)^2}\right]$$

for $x \in \mathbb{D} = (0,K)$. An elementary calculus-based estimate leads to $\mathscr{L}V(x) \le c \cdot V(x)$ on $\mathbb{D}$. Consequently, we have

$$V(x) \ge 2, \qquad \inf_{y \in \mathbb{D}\setminus\mathbb{D}_n} V(y) > 1+n, \quad \mathscr{L}V(x) \le c \cdot V(x) \qquad \forall x \in \mathbb{D}.$$

Therefore one may conclude that $\mathscr{L}W(t,x) \le 0$, since $\mathscr{L}V(x) \le c \cdot V(x)$. Introduce $\tau_n := \min(\tau^{s,x}(\mathbb{D}_n), t)$. After applying Dynkin's formula (averaged Itô formula), one finds that $\mathbb{E}W(\tau_n, X_{\tau_n}) \le V(x)$ ($X(s) = x$ is deterministic!), hence

$$\mathbb{E}\Big[\exp(c(t-\tau_n))V(X_{\tau_n})\Big] \le \exp(c(t-s))V(x).$$

Using this fact, $x \in \mathbb{D}_n$ (n large enough), one estimates

$$\begin{aligned} 0 &\le \mathbb{P}\,(\tau^{s,x}((0,K)) < t) \le \mathbb{P}\,(\tau^{s,x}(\mathbb{D}_n) < t) = \mathbb{P}\,(\tau_n < t) = \mathbb{E}I\!I_{\tau_n < t} \\ &\le \mathbb{E}\left[\exp(c(t-\tau_n)) \cdot \frac{V(X_{\tau^{s,x}(\mathbb{D}_n)})}{\inf_{y\in\mathbb{D}\setminus\mathbb{D}_n} V(y)} \cdot I\!I_{\tau_n<t}\right] \\ &\le \exp(c(t-s)) \cdot \frac{V(x)}{\inf_{y\in\mathbb{D}\setminus\mathbb{D}_n} V(y)} \le \exp(c(t-s)) \cdot \frac{V(x)}{1+n} \quad \underset{n\to+\infty}{\longrightarrow} 0, \end{aligned}$$

for all fixed $t \in [s,+\infty)$, where $I\!I_{(.)}$ represents the indicator function of subscribed random set. Consequently

$$\mathbb{P}\,(\tau^{s,x}(\mathbb{D}) < t) \;=\; \lim_{n\to+\infty} \mathbb{P}\,(\tau^{s,x}(\mathbb{D}_n) < t) \;=\; 0,$$

for $x \in (0,K)$. After discussion of the trivial invariance behavior of $X(t)$ when $X_0 = 0$ or $X_0 = K$, (almost sure) regularity of $X(t)$ on $[0,K]$ follows immediately. Eventually, uniqueness, continuity and Markov property is obtained by a result from Khas'minskiĭ [71] (see Theorem 4.1, p. 84). ◇

Numerical regularization (the preservation of invariance of certain subsets under discretization while keeping convergence orders of related standard methods) is generally aiming at the construction of convergent and appropriately bounded numerical approximations for SDEs. First, we introduce the notion of regular discrete time processes.

Definition 1.6.1. A random sequence $(Z_i)_{i\in\mathbb{N}}$ is called **regular on** (or invariant with respect to given domain) $\mathbb{D} \subset \mathbb{R}^d$ iff $\mathbb{P}\,(Z_i \in \mathbb{D}) = 1$ for all $i \in \mathbb{N}$, otherwise **non-regular** (not invariant with respect to $\mathbb{D}$).

The following BIM solves the problem of numerical regularization on bounded domain $\mathbb{D} = [0, K]$, at least in the case of $\alpha \geq 1, \beta \geq 1$. Take

$$Y_{n+1} = \begin{cases} Y_n + \Big((\rho + \lambda Y_n)(K - Y_n) - \mu Y_n\Big)h_n + \sigma Y_n^{\alpha}(K - Y_n)^{\beta}\Delta W_n \\ +\Big(\mu h_n + C(K)Y_n^{\alpha-1}(K - Y_n)^{\beta-1}|\sigma\Delta W_n|\Big)(Y_n - Y_{n+1}), \end{cases} \tag{1.141}$$

where $C = C(K)$ is an appropriate positive constant and $Y_0 \in \mathbb{D} = [0, K]$ (a.s.). Then one finds the following assertion.

Theorem 1.6.2. *Assume that the initial value $Y_0 \in [0, K]$ (a.s.) is independent of σ-algebra $\sigma(W(t), t \geq 0)$ and $K > 0$, $\rho \geq 0$, $\lambda \geq 0$, $\mu \geq 0$. The numerical solution $(Y_n)_{n\in \mathbb{N}}$ governed by (1.141) is regular on $\mathbb{D} = [0, K]$ if additionally*

$$+\infty > C(K) \geq K > 0,\ \alpha \geq 1,\ \beta \geq 1,\ 0 < h_n \leq \frac{1}{\rho + \lambda K} \quad (\forall n \in \mathbb{N}).$$

Remark 1.6.2. It is rather obvious that standard Euler methods or Theta methods without using balancing terms can not provide almost surely bounded numerical approximations on $\mathbb{D} = [0, K]$ for the logistic (1.140). (One can even estimate their positive local probability of exiting $\mathbb{D}$ – an exercise we leave to the interested reader).

Proof. Use induction on $n \in \mathbb{N}$. Then, after explicit rewriting of (1.141), one finds

$$Y_{n+1} = Y_n + \frac{\Big[(\rho + \lambda Y_n)h_n + \sigma Y_n^{\alpha}(K - Y_n)^{\beta-1}\Delta W_n\Big](K - Y_n) - \mu Y_n h_n}{1 + \mu h_n + C(K)Y_n^{\alpha-1}(K - Y_n)^{\beta-1}|\sigma\Delta W_n|}$$

$$\leq Y_n + \delta_n \cdot (K - Y_n) \leq K$$

where

$$\delta_n = \frac{(\rho + \lambda Y_n)h_n + \sigma Y_n^{\alpha}(K - Y_n)^{\beta-1}\Delta W_n}{1 + \mu h_n + C(K)Y_n^{\alpha-1}(K - Y_n)^{\beta-1}|\sigma\Delta W_n|},$$

since $\delta_n \leq 1$ if $Y_n \in [0, K]$, $C(K) \geq K$ and $h_n \leq 1/(\rho + \lambda K)$. Otherwise, non-negativity of Y_{n+1} follows from the identity

$$Y_{n+1} = \frac{Y_n + (\rho+\lambda Y_n)(K-Y_n)h_n + Y_n^{\alpha}(K-Y_n)^{\beta-1}((K-Y_n)\sigma\Delta W_n + C(K)|\sigma\Delta W_n|)}{1 + \mu h_n + \sigma C(K)Y_n^{\alpha-1}(K-Y_n)^{\beta-1}|\Delta W_n|}$$

if $C = C(K) \geq K$. Consequently, we have $\mathbb{P}\ (0 \leq Y_n \leq K) = 1$ for all $n \in \mathbb{N}$. ◇

Note, a stochastic adaptation of step sizes would form an alternative to deterministic step size selection as above. For example, for regularity, it suffices to require $h_n < 1/[\rho + \lambda Y_n - \mu]_+$ for all $n \in \mathbb{N}$. However, then one has to find a truncation procedure to guarantee finiteness of corresponding algorithms to reach given terminal times T! This is particularly important for adequate long term simulations on computers.

The sequence $Y = (Y_n)_{n\in\mathbb{N}}$ following (1.141) is also regular on $\mathbb{D}$ under other conditions than those of Theorem 1.6.2. For example, if the condition $\alpha \geq 1$ is replaced by $\alpha \in [0, 1)$, the weights $c(x) = |\sigma|C(K)x^{\alpha-1}(K - x)^{\beta-1}$ of BIMs (1.141) guarantee the a.s. invariance of $\mathbb{D} = [0, K]$. However, they are unbounded in this case. One even obtains regularity and boundedness of all numerical increments $Y_{n+1} - Y_n$ here, but we may suspect to loose convergence speed with such methods. So the open question arises how to maintain standard convergence rates and almost sure regularity of numerical methods on $\mathbb{D}$ when $\alpha \in [0, 1)$. Who knows the right answer? (At least, the case $0.5 \leq \alpha < 1$ would be physically relevant.)

Remark 1.6.3. The rates of mean square consistency and convergence of BIMs (1.141) to the strong solutions of logistic SDE (1.140) is studied in S. [120, 130].

1.6.2 *P-th Mean Boundedness of Theta Methods for Monotone Systems*

Consider nonlinear SDEs (1.2) discretized by drift-implicit Theta methods (1.1) with slightly modified class of schemes

$$X_{n+1} = X_n + [\theta_n a(t^*_{n+1}, X_{n+1}) + (1 - \theta_n)a(t^*_n, X_n)]h_n + \sum_{j=1}^{m} b^j(t^*_n, X_n)\Delta W^j_n \tag{1.142}$$

where $\theta_n \in \mathbb{R}^1$ and $t^*_{n+1} \in [t_n, t_{n+1}]$ are nonrandom. Let Π be a time-scale (i.e. discrete ($\Pi = \mathbb{N}$) or continuous ($\Pi = [0, T)$)) and $p \in \mathbb{R}^1 \setminus \{0\}$.

Definition 1.6.2. A stochastic process $X = (X(t))_{t\in\Pi}$ is said to be (unconditionally) **uniformly** p**-th mean bounded** along Π iff

$$\sup_{t\in\Pi} \mathbb{E}||X(t)||^p < +\infty.$$

If $p = 2$ we also speak of **(uniform) mean square boundedness** along Π.

SDEs (1.2) with p-th monotone coefficient systems (a, b^j) have uniformly bounded solutions X (see S. [120]). Let us investigate this aspect for numerical approximations. For technical reasons, set $\theta_{-1} = \theta_0$, $h_{-1} = h_0$, $t^*_{-1} = 0$. $||.||$ denotes the Euclidean norm in $\mathbb{R}^d$ and $< .,. >$ the Euclidean scalar product.

Theorem 1.6.3 (Uniform M.S. Boundedness of Theta Methods with $\theta_n \geq 0.5$). *Assume that $p = 2$ and $\exists K_a \leq 0, K_b$ (constants) $\forall t \geq 0\ \forall x \in \mathbb{R}^d$*

$$< a(t,x), x > \leq K_a ||x||^2, \tag{1.143}$$

$$\sum_{j=1}^{m} ||b^j(t,x)||^2 \leq K_b, \tag{1.144}$$

$$2K_a + K_b \leq 0, \tag{1.145}$$

$$\forall n \in \mathbb{N} : \theta_n h_n \leq \theta_{n-1} h_{n-1}. \tag{1.146}$$

Then the drift-implicit Theta methods with non-increasing, nonrandom sequence $(\theta_n h_n)_{n \in \mathbb{N}}$ and non-increasing, nonrandom step sizes h_n are uniformly mean square bounded along $\Pi = \mathbb{N}$ and

$$\begin{aligned} V_{n+1} &:= (1 - 2\theta_n h_n K_a)\mathbb{E}||X_{n+1}||^2 + \theta_n^2 h_n^2 \mathbb{E}[||a(t_{n+1}^*, X_{n+1})||^2] \\ &\leq \mathbb{E}||X_0||^2 + h_0^2(1 - \theta_0^2)\mathbb{E}[||a(t_0^*, X_0)||^2] = V_0 < +\infty \end{aligned} \tag{1.147}$$

for all $n \in \mathbb{N}$, whenever $\mathbb{E}[||a(0, X_0)||^2] < +\infty$ and started at $X_0 \in L^2(\Omega, \mathscr{F}_0, \mathbb{P})$.

Proof. First, separate the terms with indices $n + 1$ and n to obtain

$$X_{n+1} - \theta_n a(t_{n+1}^*, X_{n+1})h_n = X_n + (1 - \theta_n)a(t_n^*, X_n)h_n + \sum_{j=1}^{m} b^j(t_n^*, X_n)\Delta W_n^j.$$

Second, one takes the square of the Euclidean norm on both sides, leading to

$$||X_{n+1} - \theta_n a(t_{n+1}^*, X_{n+1})h_n||^2 = ||X_n + (1-\theta_n)a(t_n^*, X_n)h_n + \sum_{j=1}^{m} b^j(t_n^*, X_n)\Delta W_n^j||^2.$$

Third, taking the expectation on both sides and using the independence of all ΔW_n^j from X_n yield that

$$\begin{aligned} &\mathbb{E}||X_{n+1}||^2 - 2\theta_n h_n \mathbb{E} < X_{n+1}, a(t_{n+1}^*, X_{n+1}) > + \theta_n^2 h_n^2 \mathbb{E}||a(t_{n+1}^*, X_{n+1})||^2 \\ &= \mathbb{E}||X_n||^2 + \left[2(1 - \theta_n)\mathbb{E} < X_n, a(t_n^*, X_n) > + \mathbb{E}\sum_{j=1}^{m} ||b^j(t_n^*, X_n)||^2 \right] h_n \\ &\quad + h_n^2 \Big[(1 - 2\theta_n)\mathbb{E}||a(t_n^*, X_n)||^2 + \theta_n^2 \mathbb{E}||a(t_n^*, X_n)||^2 \Big] \\ &\leq (1 + h_n[2(1 - \theta_n)K_a + K_b])\mathbb{E}||X_n||^2 + \theta_n^2 h_n^2 \mathbb{E}||a(t_n^*, X_n)||^2 \end{aligned}$$

for all $n \in \mathbb{N}$, provided that $\theta_n \geq 0.5$. Notice that both θ_n and h_n are supposed to be nonrandom. Fourth, the assumption of non-increasing products $\theta_n h_n$ is equivalent to

$$-2\theta_n h_n K_a \leq -2\theta_{n-1} h_{n-1} K_a$$

for all $n \in \mathbb{N}$, since $K_a \leq 0$. This gives the estimation

$$1 - 2\theta_n h_n K_a + K_b h_n \leq 1 - 2\theta_{n-1} h_{n-1} K_a + K_b h_{n-1},$$

hence (recall the assumptions $K_a \leq 0$, $h_n > 0$)

$$1 + [2(1-\theta_n)K_a + K_b]h_n \leq 1 - 2\theta_{n-1} h_{n-1} K_a + K_b h_{n-1}.$$

Fifth, applying this latter estimation to the final estimation in step "Third" implies that

$$\begin{aligned}(1 - 2\theta_n K_a h_n)\mathbb{E}||X_{n+1}||^2 + \theta_n^2 h_n^2 \mathbb{E}||a(t_{n+1}^*, X_{n+1})||^2 \\ \leq (1 - \theta_{n-1} K_a h_{n-1})\mathbb{E}||X_n||^2 + \theta_{n-1}^2 h_{n-1}^2 \mathbb{E}||a(t_n^*, X_n)||^2.\end{aligned}$$

Sixth, define the functional V by

$$V_n := (1 - \theta_{n-1} K_a h_{n-1})\mathbb{E}||X_n||^2 + \theta_{n-1}^2 h_{n-1}^2 \mathbb{E}||a(t_n^*, X_n)||^2$$

for all $n \in \mathbb{N}$. By complete induction on $n \in \mathbb{N}$ while using the estimate from "Fifth" step, we find that

$$\begin{aligned}V_n \leq V_0 &= (1 - \theta_{-1} K_a h_{-1})\mathbb{E}||X_0||^2 + \theta_{-1}^2 h_{-1}^2 \mathbb{E}||a(t_{-1}^*, X_0)||^2 \\ &= (1 - \theta_0 K_a h_0)\mathbb{E}||X_0||^2 + \theta_0^2 h_0^2 \mathbb{E}||a(0, X_0)||^2 < +\infty\end{aligned}$$

for all $n \in \mathbb{N}$. Recall the conventions $\theta_{-1} = \theta_0$ and $h_{-1} := h_0$. This concludes the proof of Theorem 1.6.3 for $\theta_n \in [0.5, 1]$. ◇

A similar theorem for the estimation from below can be found for $\theta_n \in [0, 0.5]$. So Theta methods have uniformly bounded moments. This justifies also to study the longterm convergence of all moments $||X_n||^p$ for $0 < p \leq 2$ as n advances to $+\infty$ (e.g. as a consequence, $\liminf_{n\to+\infty}$ and $\limsup_{n\to+\infty}$ of those expressions must exist). Moreover, one has some reason for the preference of Theta methods with all parameters $\theta_n = 0.5$ and midpoints $t_n^* = (t_{n+1} - t_n)/2$ (i.e. trapezoidal in space, midpoint in time, cf. also exact norm-monotone methods below). For linear systems and constant θ_n, some of the above estimates are exact. For example, take $a(t,x) = -||x||^{p-2}x$ with $K_a = -1$ and $b(t,x) = \sigma||x||^{(p-2)/2}x$ with $K_b = \sigma^2$ for $p = 2$, and consider the drift-implicit Theta methods with all $\theta_n = 0.5$ and equidistant step sizes $h_n = h > 0$.

1.6.3 *Preservation of Boundary Conditions Through Implicit Methods*

For simple illustration, consider **Brownian Bridges** (pinned Brownian motion). They can be generated by the one-dimensional SDE

$$dX(t) = \frac{b - X(t)}{T - t} dt + dW(t) \tag{1.148}$$

started at $X_0 = a$, pinned to $X_T = b$ and defined on $t \in [0, T]$, where a and b are some fixed real numbers. According to the Corollary 6.10 of Karatzas and Shreve (1991), the process

$$X(t) = \begin{cases} a(1 - \frac{t}{T}) + b\frac{t}{T} + (T - t)\int_0^t \frac{dW(s)}{T - s} & \text{if} \quad 0 \le t < T \\ b & \text{if} \quad t = T \end{cases} \tag{1.149}$$

is the pathwise unique solution of (1.148) with the properties of having Gaussian distribution, continuous paths (a.s.) and expectation function

$$m(t) = \mathbb{E}[X(t)] = a\left(1 - \frac{t}{T}\right) + b\frac{t}{T} \quad \text{on}[0, T] \tag{1.150}$$

Here problems are caused by unboundedness of drift

$$a(t, x) = \frac{b - x}{T - t}.$$

What happens now with approximations when we are taking the limit toward terminal time T? Can we achieve a preservation of the boundary condition $X(T) = b$ in approximations Y under non-boundedness of the drift part of the underlying SDE at all?

A partial answer is given as follows. Consider the behavior of numerical solutions by the **family of drift-implicit Theta methods**

$$Y_{n+1}^{\theta} = Y_n^{\theta} + \left[\theta\frac{b - Y_{n+1}^{\theta}}{T - t_{n+1}} + (1 - \theta)\frac{b - Y_n^{\theta}}{T - t_n}\right] h_n + \Delta W_n \tag{1.151}$$

where $\theta \in \mathbb{R}_+ = [0, +\infty)$, $Y_0 = a$ and $n = 0, 1, \ldots, n_T - 1$. Obviously, in the case $\theta = 0$, it holds that

$$Y^0(T) := Y_{n_T}^0 = \lim_{n \to n_T} Y_n^0 = b + \Delta W_{n_T - 1}. \tag{1.152}$$

Thus, the *explicit Euler method ends in random terminal values*, which is a contradiction to the behavior of exact solution (1.149)! Otherwise, in the case $\theta > 0$, rewrite (1.151) as $Y_{n+1}^{\theta} =$

$$\frac{T - t_{n+1}}{T - t_{n+1} + \theta h_n} Y_n^{\theta} - \frac{(1-\theta)(T - t_{n+1})h_n}{(T - t_n)(T - t_{n+1} + \theta h_n)} Y_n^{\theta} + \frac{T - t_{n+1}}{T - t_{n+1} + \theta h_n} \Delta W_n$$

$$+ \frac{(1-\theta)(T - t_{n+1})h_n}{(T - t_n)(T - t_{n+1} + \theta h_n)} b + \frac{\theta h_n}{T - t_{n+1} + \theta h_n} b \qquad (1.153)$$

which implies

$$Y^{\theta}(T) := Y_{n_T}^{\theta} = \lim_{n \to n_T} Y_n^{\theta} = b. \qquad (1.154)$$

Thus, the drift-implicit Theta methods with positive implicitness $\theta > 0$ can preserve (a.s.) the boundary conditions!

Theorem 1.6.4 (S. [119]). *For any choice of step sizes* $h_n > 0, n = 0, 1, \ldots, n_{T-1}$, *we have*

[1]. $\mathbb{E} Y_{n_T}^{\theta} = b$ *if* $\theta \geq 0$
[2]. $\mathbb{E}(Y_{n_T}^{\theta} - b)^2 = h_{n_T - 1}$ *if* $\theta = 0$
[3]. $\mathbb{P}(Y_{n_T}^{\theta} = b) = 0$ *if* $\theta = 0$
[4]. $\mathbb{P}(Y_{n_T}^{\theta} = b) = 1$ *if* $\theta > 0$

where the random sequence $(Y_n^{\theta})_{n=0,1,\ldots,n_T}$ *is generated by drift-implicit Theta method (1.151) with step size* $\Delta W_n \in \mathcal{N}(0, h_n)$ *where* $\mathcal{N}(0, h_n)$ *denotes the Gaussian distribution with mean* 0 *and variance* h_n *(supposing deterministic step size).*

Remark 1.6.4. The proof of Theorem 1.6.4 is carried out in [119].

Remark 1.6.5. As we clearly recognize, discontinuities in the drift part of SDEs may destroy rates of convergence. A guarantee of algebraic constraints at the ends of time-intervals through implicit stochastic numerical methods can be observed. The example of Brownian Bridges supports the preference of implicit techniques, not only in so-called stiff problems as often argued with in literature.

Remark 1.6.6. A practical and efficient alternative to preserve boundary conditions is given by the splitting-step algorithm of Moro and Schurz [99].

1.7 Oscillations

Unfortunately, there is not much known on oscillatory behaviour of stochastic numerical methods so far. So this section gives a first excursion into this important, but very complex field. For this purpose, consider the following definitions.

Definition 1.7.1. A numerical method Z with values $(Z_n)_{n\in\mathbb{N}}$ at instants $(t_n)_{n\in\mathbb{N}}$ of partitions of $[0,T]$ is called **strictly oscillatory** about c iff

$$\forall n \in \mathbb{N} : \frac{Z_{n+1} - c}{Z_n - c} < 0. \tag{1.155}$$

A numerical method Z with values $(Z_n)_{n\in\mathbb{N}}$ at instants $(t_n)_{n\in\mathbb{N}}$ of partitions of $[0,T]$ is called **quasi-oscillatory** about c iff

$$\frac{Z_{n+1} - c}{Z_n - c} < 0 \tag{1.156}$$

for infinitely many $n \in \mathbb{N}$.
Moreover, a numerical method Z with values $(Z_n)_{n\in\mathbb{N}}$ at instants $(t_n)_{n\in\mathbb{N}}$ of partitions of $[0,T]$ is called (asymptotically) **oscillatory about** c **in the wide sense** iff

$$\forall n \in \mathbb{N} : \liminf_{n\to+\infty} \frac{Z_{n+1} - c}{Z_n - c} < 0 \quad \text{and} \quad \limsup_{n\to+\infty} \frac{Z_{n+1} - c}{Z_n - c} > 0. \tag{1.157}$$

If none of above conditions is true then we say Z is **non-oscillatory**.

We shall study the most important case of oscillation about $c = 0$ throughout this section. Oscillations in the wide sense and quasi-oscillations mean that the numerical method has outcomes infinitely often above and below the level c. Oscillations in the wide sense can not belong to asymptotically stable approximations at equilibrium 0, whereas quasi-oscillations can be a feature of asymptotically stable ones.

Definition 1.7.2. A method Z is said to have **spurious oscillations** iff it has (quasi-) oscillations of any kind from above and the underlying SDE (1.2) does not show any oscillatory behaviour at all.

It is well-known that the **geometric Brownian motion** X satisfying the Itô SDE (1.108) does not show any oscillations for all $\mathscr{F}_0$-adapted initial data $X(0) \in \mathbb{R}^1$. In fact, this process preserves the sign of $\mathscr{F}_0$-adapted initial data $X(0)$ (a.s) as integration time advances. Now, let us study the oscillatory behaviour of related numerical methods applied to this simple test (1.108) and some further nonlinear equations. For this purpose, we suppose that $\gamma > 0$ for the further consideration.

1.7.1 *Spurious Oscillations of Standard Euler Methods for Linear ODE*

In deterministic numerical analysis a very simple example is well–known. Consider the equations (ODEs)

$$\dot{x} = \lambda x \quad \text{with} \quad x(0) = x_0 \in \mathbb{R}^1 \tag{1.158}$$

and its sign-preserving exact solution $x(t) = \exp(\lambda t) \cdot x_0$. Then the standard Euler scheme (1.1) gives

$$y_{n+1} = y_n + \lambda y_n h_n = (1 + \lambda h_n) y_n = y_0 \prod_{i=0}^{n} (1 + \lambda h_i) . \tag{1.159}$$

Obviously, started at $y_0 \neq 0$, we have that $y_n \neq 0$ for all $n \in \mathbb{N}$ provided that $\lambda h_i \neq -1$ for all $i \in \mathbb{N}$ and

$$\frac{y_{n+1}}{y_n} = 1 + \lambda h_n .$$

This expression is always positive if $\lambda h_i > -1$ for all $i = 0, 1, \ldots, n$. However, negative values may occur under the assumption that $\lambda < 0$ and h_i large enough (indeed always), i.e. we observe no sign changes whenever $\lambda h_i < -1$. This step size restriction for oscillations about 0 even gets less restrictive for large negative values of $\lambda < 0$ and a uniform restriction of h_i is not possible for all possible parameters $\lambda < 0$ in order to guarantee (strict) positivity of ratios y_{n+1}/y_n and hence no strict oscillations at all. In contrast to that, the underlying exact solution X (as exponentials do not either) does not show any oscillatory behaviour at all. This argumentation proves the following theorem on existence of spurious oscillations through Euler-type discretizations.

Theorem 1.7.1 (Spurious Strict Oscillations of Euler Methods in 1D). *Assume that*

$$\forall n \in \mathbb{N} : \lambda h_n < -1. \tag{1.160}$$

Then the forward Euler method applied to linear ODEs (1.158) with step sizes h_n possesses strict spurious oscillations for any initial data $x(0) = y_0$.
Assume that, for infinitely many $n \in \mathbb{N}$, we have

$$\lambda h_n < -1. \tag{1.161}$$

Then the forward Euler method applied to linear ODEs (1.158) with step sizes h_n possesses spurious quasi-oscillations for any initial data $x(0) = y_0$.

Remark 1.7.1. If $\lambda h_n \geq -1$ for all $n \in \mathbb{N}$ then the Euler method has no oscillations at all. However, this might be very restrictive for long-term simulations with negative values of $\lambda << -1$. At least, for nonlinear equations, the situation becomes even worse (here thresholds imposed on h_n for the absence of spurious oscillations are not known in general).

1.7.2 Non-oscillations of (Balanced) Theta Methods for Linear ODEs

In contrast to the standard Euler scheme, in the case $\lambda < 0$, we can always prevent oscillations in numerical methods with arbitrary step sizes h_i for linear differential equations in 1D. For this purpose we return to the deterministic family of Euler-type Theta schemes (1.1) with implicitness $\theta_n \in \mathbb{R}^1$

$$y_{n+1} = y_n + (\theta_n y_{n+1} + (1-\theta_n)y_n)\lambda h_n$$
$$= \frac{1+(1-\theta_n)\lambda h_n}{1-\theta_n \lambda h_n} y_n = y_0 \prod_{i=0}^{n} \frac{1+(1-\theta_i)\lambda h_i}{1-\theta_i \lambda h_i}. \quad (1.162)$$

Thus, we find that

$$\frac{y_{n+1}}{y_n} = \frac{1+(1-\theta_n)\lambda h_n}{1-\theta_n \lambda h_n}$$

whenever $\theta_n \lambda h_n \neq 1$. Hence, when $\lambda < 0$ and all $\theta_i \lambda h_i < 1$, these schemes possess strict oscillations if $1+(1-\theta_i)\lambda h_i < 0$ for all $i \in \mathbb{N}$ and quasi-oscillations if $1+(1-\theta_n)\lambda h_n < 0$ for infinitely many $n \in \mathbb{N}$.

A generalization of these schemes is presented by the deterministic balanced methods (as a subclass of (1.1))

$$y_{n+1} = y_n + \lambda y_n h_n + c_0 h_n (y_n - y_{n+1})$$
$$= \frac{1+(\lambda+c_0)h_n}{1+c_0 h_n} y_n = y_0 \prod_{i=0}^{n} \frac{1+(\lambda+c_0)h_i}{1+c_0 h_i} \quad (1.163)$$

for an appropriate constant $c_0 > 0$. This leads to the identity

$$\frac{y_{n+1}}{y_n} = \frac{1+(\lambda+c_0)h_n}{1+c_0 h_n}$$

for all $n \in \mathbb{N}$. Consequently, as a summary, we gain the following theorem.

Theorem 1.7.2. *Numerical approximations generated by balanced methods (1.163) with $c_0 \geq [-\lambda]_+$ or by drift-implicit Theta methods (1.162) with parameters satisfying $\theta_n = 1$ or $(1-\theta_n)\lambda h_n > 0$ are non-oscillatory for all test (1.158) with any $\lambda \leq 0$ along any choice of arbitrary step sizes $h_n \geq 0$. Therefore, they possess adequate realizations without spurious oscillations along any partition.*

Remark 1.7.2. Suppose that $\lambda \leq 0$. Then, the balanced methods (1.163) with the weight $c_0 = [-\lambda]_+$ coincide with drift-implicit backward Euler methods (1.1) with $\theta_n = 1$ for all $n \in \mathbb{N}$ while applying to ODEs. Hence, there are drift-implicit

Theta methods (1.1) without any spurious oscillations for any choice of step sizes h_n. However, the situation changes in stochastic settings where nontrivial weights c^j are needed in order to limit the possibility of inadequate oscillations (see below).

1.7.3 Spurious Oscillations of Theta Methods for Linear SDEs

After previous elementary illustrations with ODEs, we return to the stochastic case. For the sake of simplicity, we confine ourselves to partitions with nonrandom step sizes h_n. Note that

$$\forall t \geq s \geq 0 : \mathbb{P}\left(\left\{\omega \in \Omega : \frac{X(t)(\omega)}{X(s)(\omega)} < 0\right\}\right) = 0$$

for the stochastic process X satisfying (1.110).

Theorem 1.7.3. *Assume that X satisfies (1.108) with $X(0) \neq 0$ (a.s.), $\gamma \neq 0$ and*

$$\forall n \in \mathbb{N} \;:\; 1 \neq \theta_n \lambda h_n, \quad \xi_n \in \mathcal{N}(0,1).$$

Then the drift-implicit Theta methods (1.1) applied to the linear SDE (1.108) and started in $Y_0^\theta = X(0)$ with nonrandom step sizes h_n, nonrandom parameters $\theta_n \in \mathbb{R}^1$ and values $(Y_n^\theta)_{n \in \mathbb{N}}$ have oscillations about 0 with positive probability, i.e.

$$\forall n \in \mathbb{N} : \mathbb{P}\left(\left\{\omega \in \Omega : \frac{Y_{n+1}^\theta(\omega)}{Y_n^\theta(\omega)} < 0\right\}\right) > 0. \tag{1.164}$$

Proof. The family of drift-implicit Euler-Theta schemes (1.1) applied to one-dimensional linear SDE (1.108) with $c^j \equiv 0$ and implicitness θ_n is governed by

$$\begin{aligned} Y_{n+1}^\theta &= Y_n^\theta + \theta_n \lambda Y_{n+1}^\theta h_n + (1-\theta_n)\lambda Y_n^\theta h_n + \gamma Y_n^\theta \Delta W_n \\ &= \frac{1 + (1-\theta_n)\lambda h_n + \gamma \Delta W_n}{1 - \theta_n \lambda h_n} Y_n^\theta = Y_0^\theta \prod_{i=0}^{n} \left(\frac{1 + (1-\theta_i)\lambda h_i + \gamma \sqrt{h_i}\xi_i}{1 - \theta_i \lambda h_i} \right). \end{aligned}$$

Therefore we have

$$\frac{Y_{n+1}^\theta}{Y_n^\theta} = \frac{1 + (1-\theta_n)\lambda h_n + \gamma \sqrt{h_n}\xi_n}{1 - \theta_n \lambda h_n} = \frac{1 + (1-\theta_n)\lambda h_n}{1 - \theta_n \lambda h_n} + \frac{\gamma \sqrt{h_n}}{1 - \theta_n \lambda h_n}\xi_n$$

for all $N \in \mathbb{N}$. Recall that $1 - \theta_n \lambda h_n \neq 0$, $\gamma \neq 0$ and ξ_n are Gaussian $\mathcal{N}(0,1)$-distributed. That means that

$$\frac{Y_{n+1}^\theta}{Y_n^\theta} \in \mathcal{N}\left(\frac{1 + (1-\theta_n)\lambda h_n}{1 - \theta_n \lambda h_n}, \frac{\gamma^2}{(1 - \theta_n \lambda h_n)^2} h_n \right).$$

Consequently, under $\gamma \neq 0$ and $1 - \theta_n \lambda h_n \neq 0$, this fact implies that (1.164). ◇

Remark 1.7.3. After, returning to the proof above, one may extract the probability of local sign changes of Theta methods Y^θ and obtains

$$\mathbb{P}\left(\left\{\frac{Y^\theta_{n+1}}{Y^\theta_n} < 0\right\}\right) = \mathbb{P}\left(\left\{\xi_n < -\frac{1+(1-\theta_n)h_n}{|\gamma|\sqrt{h_n}}\right\}\right)$$
$$= \frac{1}{2\pi}\int_{-\infty}^{-\frac{1+(1-\theta_n)h_n}{|\gamma|\sqrt{h_n}}} \exp\left(-\frac{x^2}{2}\right) dx$$

which tends to 0 as $h_n \to 0$ since ξ_n is Gaussian distributed, provided that

$$\inf_{n \in \mathbb{N}} (1-\theta_n)\sqrt{h_n} > -\infty.$$

1.7.4 Non-oscillatory Behaviour by Balanced Methods for Linear SDE

Consider the balanced methods

$$Y^B_{n+1} = Y^B_n + \lambda Y^B_n h_n + \gamma Y^B_n \Delta W_n + (c_0 h_n + c_1|\Delta W_n|)(Y^B_n - Y^B_{n+1}). \quad (1.165)$$

Theorem 1.7.4. *Suppose X satisfies (1.108) with adapted initial values $X(0) \geq 0$. Then the balanced methods (1.165) with constants c^0, $c^1 \geq 0$ and step sizes h_n, started at $Y^B_0 = X(0)$ do not allow any sign changes as n advances, provided that*

$$1 + (c^0 + \lambda)h_n \geq 0 \quad \textit{and} \quad c^1 \geq |\gamma|\,. \quad (1.166)$$

for all $n \in \mathbb{N}$. Therefore, they do not possess any spurious oscillations in contrast to all solutions X of linear SDEs (1.108).

Proof. This claim follows immediately from the construction of the balanced methods (1.165) applied to the linear equation (1.108). One receives then

$$Y^B_{n+1} = Y^B_n + \lambda Y^B_n h_n + \gamma Y^B_n \Delta W_n + (c^0 h_n + c^1|\Delta W_n|)(Y^B_n - Y^B_{n+1})$$
$$= \frac{1 + (c^0+\lambda)h_n + \gamma \Delta W_n + c^1|\Delta W_n|}{1 + c^0 h_n + c^1|\Delta W_n|} Y^B_n\,.$$

Thereby, $Y^B_{i+1} \geq 0$ for all $i \in \mathbb{N}$ iff

$$1 + (c^0 + \lambda)h_n + \gamma \Delta W_i + c^1|\Delta W_i| \geq 0$$

for all $i \in \mathbb{N}$. Obviously, this is the case under (1.166). ◇

Remark 1.7.4. The standard Euler methods and drift-implicit Theta methods without any balancing terms c^j have spurious oscillations. This fact can be concluded from the following two subsections.

1.7.5 Non-oscillatory Behaviour by Balanced Methods for Nonlinear SDE

Spurious oscillations about 0 can be also excluded for balanced methods applied to nonlinear SDEs of the form (1.114).

Theorem 1.7.5 (Non-oscillations of BIMs for Nonlinear SDE in $\mathbb{R}^d$). *The balanced implicit methods (1.115) applied to SDEs (1.114) with same adapted initial value $X(0)$ and any step sizes h_n do not possess any spurious oscillations about* 0.

Proof. This fact follows immediately from Theorem 1.5.3 on the positivity of BIMs (1.115) applied to SDEs (1.114). The related proof shows the sign-preservation of them during the entire course of numerical integration. Hence, the assertion of Theorem 1.7.5 is verified. ◇

1.7.6 Oscillatory Behaviour of Euler Methods for Nonlinear SDE

Almost surely oscillatory behaviour of Euler methods applied to certain nonlinear SDE can be established. For this purpose, let

$$\sup_{u<0}\{-G(u)\} = \sup_{u<0}\left\{-\frac{u+ha(u)}{\sqrt{h}|b(u)|}\right\} =: L < \infty \tag{1.167}$$

and

$$b(x) = 0 \Rightarrow x = 0. \tag{1.168}$$

First, we will prove an auxiliary lemma that X has never the value 0 (a.s.).

Lemma 1.7.1 ([2]). *Assume that conditions (1.128), (1.133), (1.167) and (1.168) hold. Then the solution X to (1.127) with arbitrary nonrandom initial value $X(0) = x_0 \neq 0$ obeys*

$$\mathbb{P}\{X_n \neq 0 \text{ for all } n \in \mathbb{N}\} = 1.$$

Proof of Lemma 1.7.1. We shall use complete induction on $n \in \mathbb{N}$. First, note that $X_1 = x_0 + a(x_0) + b(x_0)\xi_1$, so as $x_0 \neq 0$, and *supp* $\xi_1 = \mathbb{R}$, X_1 is a continuous random variable and takes the value 0 with probability zero. Second, as the induction assumption at level n, suppose that X_n is a continuous random variable. This is true at level $n = 1$. To proceed in general, we note that because X_n is a continuous random variable, and a and b are continuous, $a(X_n)$ and $b(X_n)$ are continuous random variables. Moreover, as $b(x) = 0$ only when $x = 0$, and X_n being continuous is non–zero (a.s.), the random variable $b(X_n) \neq 0$ a.s. Since ξ_{n+1} is normal, and therefore also a continuous random variable, $X_{n+1} = X_n + ha(X_n) + \sqrt{h}b(X_n)\xi_{n+1}$ is a continuous random variable. Hence, by induction on n, X_n is a continuous random variable for all $n \in \mathbb{N}$.

This fact implies that $\mathbb{P}\{X_n = 0\} = 0$ for each $n \in \mathbb{N}$. Thus $\mathbb{P}\left(\cup_{n=1}^{\infty}\{X_n = 0\}\right) = 0$. This implies that

$$\mathbb{P}\left(\cap_{n=1}^{\infty}\{X_n \neq 0\}\right) = \mathbb{P}\left(\overline{\cup_{n=1}^{\infty}\{X_n = 0\}}\right) = 1,$$

as stated in conclusion of Lemma 1.7.1. ◇

Now, one is able to prove the following Theorem (cf. [2]). As usual, let

$$(\Omega, \mathscr{F}, (\mathscr{F}_{t_n})_{n \in \mathbb{N}}, \mathbb{P})$$

be a complete filtered probability space and the filtration $(\mathscr{F}_{t_n})_{n \in \mathbb{N}}$ be naturally generated, namely that $\mathscr{F}_{t_n} = \sigma\{\xi_0, \xi_1, \ldots \xi_n\}$.

Theorem 1.7.6 (Oscillations of Euler Methods for Nonlinear SDEs in 1D). *Assume that*

- *(o)* $h >, x_0 > 0$ *are nonrandom and fixed.*
- *(i)* $a : \mathbb{R} \to \mathbb{R}$ *and* $b : \mathbb{R} \to \mathbb{R}$ *are continuous, nonrandom functions vanishing at* 0 *(i.e. (1.128)).*
- *(ii)* G *is well-defined by (1.129) for all* $u \neq 0$ *and satisfies both (1.167) and (1.133).*
- *(iii) diffusion function* b *vanishes only at* 0 *on* $\mathbb{R}^1_+$, *i.e. (1.168) holds.*
- *(iv)* $(\xi_n)_{n \in \mathbb{N}}$ *are i.i.d. Gaussian* $\mathscr{N}(0, 1)$*-distributed random variables on the filtered probability space* $(\Omega, \mathscr{F}, (\mathscr{F}_{t_n})_{n \in \mathbb{N}}, \mathbb{P})$.

Then, the standard Euler method (1.127) with equidistant step size h *and any adapted initial value* $X(0) = x_0 \neq 0$ *is oscillating (a.s.) about zero.*

Remark 1.7.5. In fact, Theorem 1.7.6 remains valid when all ξ_n are i.i.d. random variables with symmetric continuous probability distribution F satisfying (1.132).

Proof. We note at the outset that by (1.168) and the continuity of b, we either have $b(x) < 0$ for all $x < 0$ or $b(x) > 0$ for all $x < 0$.

Suppose that $x_0 > 0$. Theorem 1.5.5 proves that $X_\tau < 0$ for some a.s. finite stopping time $\tau \in \mathbb{N}$. We define

$$p_L = \mathbb{P}\{\omega \in \Omega : \zeta(\omega) < L\}, \tag{1.169}$$

where ζ is a $\mathscr{N}(0, 1)$–distributed random variable. For each $n \in \mathbb{N}$, define

$$\bar{E}_n = \{\omega \in \Omega : X_i(\omega) < 0, \forall i = \tau(\omega), \tau(\omega) + 1, \ldots, \tau(\omega) + n\}. \tag{1.170}$$

Since

$$E_n = \{\omega \in \Omega : \exists i \text{ among the integers } \tau(\omega)+1,\dots,\tau(\omega)+n : X_i(\omega) \geq 0\}, \tag{1.171}$$

we can conclude from the Borel-Cantelli Lemma that solutions X to equation (1.127) become non–negative with probability 1 on the set $\{\tau, \tau+1,\dots,\}$ if

$$\sum_{n=0}^{\infty} \mathbb{P}\,(\bar{E}_n) < \infty. \tag{1.172}$$

To see this, with $p_0 = p_0(N) := \mathbb{P}\,\{\tau = N\}$, we estimate $\mathbb{P}\,(\bar{E}_n)$ from above by

$$\begin{aligned}
\mathbb{P}\,(\bar{E}_n) &= \mathbb{P}\,\{\omega \in \Omega : X_i(\omega) < 0, \forall i \in \{\tau(\omega)+1,\dots,\tau(\omega)+n\}\} \\
&= \sum_{N=1}^{\infty} \mathbb{P}\big\{\omega \in \Omega : X_i(\omega) < 0, \forall i \in \{\tau(\omega)+1,\dots,\tau(\omega)+n\}\big|\tau=N\big\}\, p_0 \\
&= \sum_{N=1}^{\infty} \left(\prod_{j=N+1}^{N+n} \mathbb{P}\big\{\omega\in\Omega : X_j(\omega)<0\big|X_{j-1}<0,\dots,X_N<0, \tau = N\big\} \right) p_0 \\
&= \sum_{N=1}^{\infty} \left(\prod_{j=N+1}^{n+N} \mathbb{P}\Big\{X_{j-1}+ha(X_{j-1})+\sqrt{h}b(X_{j-1})\xi_j<0\Big|X_{j-1}<0, \tau=N\Big\} \right) p_0.
\end{aligned}$$

To analyze this probability, we write (a.s.)

$$\begin{aligned}
&\Big\{X_{j-1}+hf(X_{j-1})+\sqrt{h}b(X_{j-1})\xi_j < 0\Big\} \\
&= \left\{ \xi_j < -\frac{X_{j-1}+ha(X_{j-1})}{\sqrt{h}b(X_{j-1})}, b(X_{j-1})>0 \right\} \\
&\cup \left\{ \xi_j < -\frac{X_{j-1}+ha(X_{j-1})}{-\sqrt{h}b(X_{j-1})}, b(X_{j-1})<0 \right\}.
\end{aligned}$$

It is not necessary to condition on $b(X_{j-1}) = 0$, because by (1.168) this implies $X_{j-1} = 0$, in which case $X_{j-1} + ha(X_{j-1}) + \sqrt{h}b(X_{j-1})\xi_j = 0$. Therefore

$$\mathbb{P}\,\Big\{X_{j-1} + ha(X_{j-1}) + \sqrt{h}b(X_{j-1})\xi_j < 0\Big|X_{j-1} < 0, \tau = N\Big\}$$

$$= \mathbb{P}\left\{ \xi_j < -\frac{X_{j-1}+ha(X_{j-1})}{\sqrt{h}b(X_{j-1})}, b(X_{j-1}) > 0 \,\middle|\, X_{j-1} < 0, \tau = N \right\}$$

$$+\mathbb{P}\left\{\xi_j < -\frac{X_{j-1}+ha(X_{j-1})}{-\sqrt{h}b(X(j-1))}, b(X_{j-1}) < 0 \Big| X_{j-1} < 0, \tau = N\right\}.$$

We now consider the sign of b on $(-\infty, 0)$. If $b(x) > 0$ for $x < 0$, then

$$\begin{aligned}
&\mathbb{P}\left\{X_{j-1}+ha(X_{j-1})+\sqrt{h}b(X_{j-1})\xi_j < 0 \big| X_{j-1} < 0, \tau = N\right\}\\
&= \mathbb{P}\left\{\xi_j < -\frac{X_{j-1}+ha(X_{j-1})}{\sqrt{h}b(X_{j-1})}\Big| X_{j-1} < 0, \tau = N\right\}\\
&= \mathbb{P}\left\{\xi_j < -\frac{X_{j-1}+ha(X_{j-1})}{\sqrt{h}|b(X_{j-1})|}\Big| X_{j-1} < 0, \tau = N\right\}.
\end{aligned}$$

If, on the other hand when $b(x) < 0$ for $x < 0$, then

$$\begin{aligned}
&\mathbb{P}\left\{X_{j-1}+ha(X_{j-1})+\sqrt{h}b(X_{j-1})\xi_j < 0 \big| X_{j-1} < 0, \tau = N\right\}\\
&= \mathbb{P}\left\{\xi_j < -\frac{X_{j-1}+ha(X_{j-1})}{-\sqrt{h}b(X_{j-1})}\Big| X_{j-1} < 0, \tau = N\right\}\\
&= \mathbb{P}\left\{\omega \in \Omega : \xi_j(\omega) < -\frac{X_{j-1}(\omega)+ha(X_{j-1}(\omega))}{\sqrt{h}|b(X_{j-1}(\omega))|}\Big| X_{j-1} < 0, \tau = N\right\}.
\end{aligned}$$

Therefore, irrespective of the sign of b on $(-\infty, 0)$, we have

$$\begin{aligned}
&\mathbb{P}\left\{X_{j-1}+ha(X_{j-1})+\sqrt{h}b(X_{j-1})\xi_j < 0 \big| X_{j-1} < 0, \tau = N\right\}\\
&= \mathbb{P}\left\{\xi_j < -\frac{X_{j-1}+ha(X_{j-1})}{\sqrt{h}|b(X_{j-1})|}\Big| X_{j-1} < 0, \tau = N\right\}.
\end{aligned}$$

Now, as ξ_j is independent of X_{j-1} (since ξ_j is independent of $\xi_0, \xi_1, \ldots, \xi_{j-1}$), and ξ_j is also independent of events $\{\tau = N\}$ which belong to the σ-algebra $\mathscr{F}_N$ with $N < j$, we have

$$\begin{aligned}
&\mathbb{P}\left\{X_{j-1}+ha(X_{j-1})+\sqrt{h}b(X_{j-1})\xi_j < 0 \big| X_{j-1} < 0, \tau = N\right\}\\
&= \mathbb{P}\left\{\xi_j < -\frac{X_{j-1}+ha(X_{j-1})}{\sqrt{h}|b(X_{j-1})|}\Big| X_{j-1} < 0, \tau = N\right\}\\
&= \mathbb{P}\left\{\xi_j < -\frac{X_{j-1}+ha(X_{j-1})}{\sqrt{h}|b(X_{j-1})|}\right\} = \mathbb{P}\ \{\zeta < -G(X_{j-1})\},
\end{aligned}$$

where ζ is a standard normal random variable. Hence

$$\mathbb{P}\left\{X_{j-1}+ha(X_{j-1})+\sqrt{h}b(X_{j-1})\xi_j<0 \middle| X_{j-1}<0,\tau=N\right\}\leq \mathbb{P}\ \{\zeta<L\}=p_L.$$

Thus

$$\prod_{j=N+1}^{n+N}\mathbb{P}\left\{X_{j-1}+ha(X_{j-1})+\sqrt{h}b(X_{j-1})\xi_j<0 \middle| X_{j-1}<0,\tau=N\right\}\leq p_L^n.$$

Therefore, we have

$$\mathbb{P}\,(\bar{E}_n)\leq\sum_{N=1}^{\infty}\left(\prod_{j=N+1}^{n+N}\mathbb{P}\ \{\zeta<L\}\right)\mathbb{P}\,\{\tau=N\}\leq p_L^n\sum_{N=1}^{\infty}\mathbb{P}\,\{\tau=N\}=p_L^n.$$

Thus,

$$\sum_{n=0}^{\infty}\mathbb{P}\,(\bar{E}_n)\leq\frac{1}{1-p_L}<\infty,$$

and we conclude that there exists an a.s. finite stopping time $\tau_1>\tau_0:=\tau$ such that $X_{\tau_1}\geq 0$ a.s. But by Lemma 1.7.1, it follows that we must have $X_{\tau_1}>0$ (a.s.). By repeating the same approach while requiring (1.133) and (1.167), and using mathematical induction, we obtain that X_n changes its sign infinitely often and with probability 1. Similarly, we verify the assertion for the case $X(0)<0$ (just start with the event $\{\tau_0=0\}$ of negative initial values and proceed as above). Thus, the proof of Theorem 1.7.6 is complete. ◇

Example 8.1. Consider Itô SDE

$$dX(t)=-[X(t)]^3dt+X(t)dW(t)$$

with initial value $X(0)=x_0>0$ and its Euler discretization

$$Y_{n+1}^E=Y_n^E(1-h[Y_n^E]^2+\Delta W_n)$$

started at the same value $Y_0^E=X(0)=x_0>0$, where ΔW_n are independent and identically distributed increments. It can be shown that the exact solution X is a.s. positive for all times $t\geq 0$ and never oscillates. However, an application of Theorem 1.7.6 gives that the related standard Euler method is quasi-oscillatory (a.s.), hence it shows spurious oscillations. However, an application of balanced methods with $c^0(x)=x^2$ and $c^1(x)=1$ governed by

$$Y_{n+1}^B=Y_n^B(1-h[Y_n^B]^2+\Delta W_n)+([Y_n^B]^2h+|\Delta W_n|)(Y_n^B-Y_{n+1}^B)$$

$$= Y_n^B \left(1 + \frac{-h[Y_n^B]^2 + \Delta W_n}{1 + [Y_n^B]^2 h + |\Delta W_n|}\right)$$

$$= Y_n^B \frac{1 + \Delta W_n + |\Delta W_n|}{1 + [Y_n^B]^2 h + |\Delta W_n|}$$

leads to an adequate absence of oscillations since their values stay positive (a.s.) for all positive initial values (apply Theorem 1.5.3 for the verification of the latter fact or analyze the chain of equations above).

1.8 Energy-Exact Methods

Consider stochastic differential equations (stochastic oscillator with additive Gaussian noise)

$$\ddot{x} + \omega^2 x = \sigma \xi(t) \tag{1.173}$$

driven by white noise ξ, with real eigenfrequency $\omega \geq 0$ and real noise intensity $\sigma \neq 0$ (we suppose that the initial data $x(0)$ and $\dot{x}(0)$ are $(\mathscr{F}_0, \mathscr{B}^1)$-measurable and have finite second moments). This equation for a stochastic oscillator can be rewritten to the equivalent two-dimensional test system of SDEs

$$dX(t) = Y(t)dt \tag{1.174}$$

$$dY(t) = -\omega^2 X(t)dt + \sigma dW(t) \tag{1.175}$$

driven by the standard Wiener process W (i.e. $W(t) = \int_0^t \xi(s)ds$) and started at $(\mathscr{F}_0, \mathscr{B}(\mathbb{R}^2))$-measurable initial data $X(0) = X_0 = x_0$, $Y(0) = Y_0 = y_0$.
The energy $E(t)$ of this system (1.173) is well-known and given by the following lemma. Let

$$\mathscr{E}(t) = \frac{\omega^2[X(t)]^2 + [Y(t)]^2}{2} = \frac{\omega^2[x(t)]^2 + [v(t)]^2}{2} \tag{1.176}$$

where $v(t) = \dot{x}(t)$ is the velocity of displacement $x(t)$ at time $t \geq 0$.

Lemma 1.8.1 (Trace Formula of Expected Energy). *The total energy (as the sum of potential and kinetic energy) $E(t)$ of system (1.173) defined by (1.176) follows the linear relation (i.e., a perturbed conservation law)*

$$e(t) := \mathbb{E}[\mathscr{E}(t)] = e(0) + \frac{\sigma^2}{2}t \tag{1.177}$$

for all $t \geq 0$.

Let us test numerical methods whether they can follow this **perturbed conservation law** for its discretized energy functional. In particular, for the sake of simplicity, consider the drift-implicit θ-methods applied to the system (1.174)–(1.175) and governed by the scheme

$$X_{n+1} = X_n + (\theta_n Y_{n+1} + (1-\theta_n)Y_n)h_n$$
$$Y_{n+1} = Y_n - \omega^2(\theta_n X_{n+1} + (1-\theta_n)X_n)h_n + \sigma \Delta W_n \qquad (1.178)$$

with nonrandom parameter-sequence $(\theta_n)_{n\in\mathbb{N}}$, where

$$\theta_n \in \mathbb{R}^1, h_n = t_{n+1} - t_n, \Delta W_n = W(t_{n+1}) - W(t_n) \in \mathcal{N}(0, h_n)$$

along nonrandom partitions (or adapted partitions) of $[0, T]$ of the form

$$0 = t_0 \leq t_1 \leq \ldots \leq t_n \leq \ldots \leq t_{n_T} = T.$$

Definition 1.8.1. A numerical method Z based on instants $(Z_n)_{n\in\mathbb{N}}$ is called **energy-exact** for the SDE (1.173) iff its expected energy $e = e(t)$ satisfies the perturbed conservation law (1.177) at all instants $(t_n)_{n\in\mathbb{N}}$ along any nonrandom partition of non-random time-intervals $[0, T]$. Otherwise, Z is said to be **non-energy-exact**.

1.8.1 Non-energy-Exact Methods

The following theorem shows that the simplest choice of stochastic Runge-Kutta methods such as forward and backward Euler methods as well as trapezoidal methods fail to preserve simple energy laws (despite of their asymptotically consistent behaviour).

Theorem 1.8.1. *Assume that initial data $X(0) = x_0$, $Y(0) = \dot{x}_0$ are $L^2(\Omega, \mathcal{F}_0, \mathbb{P})$-integrable. Then the drift-implicit θ-methods (1.178) applied to (1.173) are not energy-exact for any sequence of nonrandom implicitness parameters $(\theta_n)_{n\in\mathbb{N}}$ and any nonrandom step-sizes $(h_n)_{n\in\mathbb{N}}$. More precisely, their expected energy $e(t) = \mathbb{E}[\mathcal{E}(t_n)]$ is finite for all $n \in \mathbb{N}$, e can be uniformly estimated by linear functions in t from below and above, and it satisfies $\forall \omega, \sigma \in \mathbb{R}^1$ $\forall X_0, Y_0 \in L^2(\Omega, \mathcal{F}_0, \mathbb{P})$*

$$0 \leq \mathbb{E}[\mathcal{E}(0)] + \frac{\sigma^2}{2} t_n \min_{k=0,1,\ldots,n-1} \frac{1}{1+\omega^2 h_k^2/4}$$
$$\leq \mathbb{E}[\mathcal{E}(t_n)] + \frac{\omega^2}{2} \sum_{k=0}^{n-1} (1-2\theta_k)\mathbb{E}[Y_k X_{k+1} - Y_{k+1} X_k] h_k$$

$$= \mathbb{E}[\mathscr{E}(0)] + \frac{\sigma^2}{2}\sum_{k=0}^{n-1}\frac{h_k}{1+\omega^2 h_k^2/4}$$

$$\leq \mathbb{E}[\mathscr{E}(0)] + \frac{\sigma^2}{2}t_n \max_{k=0,1,\ldots,n-1}\frac{1}{1+\omega^2 h_k^2/4} \leq \mathbb{E}[\mathscr{E}(0)] + \frac{\sigma^2}{2}t_n < +\infty$$

which renders to be a chain of equations for equidistant partitions (i.e. we may replace $\leq$ by $=$ signs above).

Proof. Define the statistical average values

$$\bar{X}_n = \frac{X_{n+1}+X_n}{2}, \quad \bar{Y}_n = \frac{Y_{n+1}+Y_n}{2}. \tag{1.179}$$

First, rewrite the system (1.178) of equations for (X,Y) to as

$$X_{n+1} = X_n + (2\theta_n\bar{Y}_n + (1-2\theta_n)Y_n)h_n$$
$$Y_{n+1} = Y_n - \omega^2(2\theta_n\bar{X}_n + (1-2\theta_n)X_n)h_n + \sigma\Delta W_n.$$

Second, multiply the components of these equations by $\omega^2\bar{X}_n$ and $\bar{Y}_n$ (resp.) to arrive at

$$\frac{\omega^2}{2}(X_{n+1}^2 - X_n^2) = (2\theta_n\omega^2\bar{Y}_n\bar{X}_n + (1-2\theta_n)\omega^2 Y_n\bar{X}_n)h_n$$
$$\frac{1}{2}\left(Y_{n+1}^2 - Y_n^2\right) = -\omega^2\left(2\theta_n\bar{Y}_n\bar{X}_n + (1-2\theta_n)\bar{Y}_n X_n\right)h_n + \sigma\bar{Y}_n\Delta W_n.$$

Third, adding both equations leads to

$$\mathscr{E}_{n+1} = \mathscr{E}_n - \omega^2(1-2\theta_n)[Y_n\bar{X}_n - \bar{Y}_n X_n]h_n + \sigma\bar{Y}_n\Delta W_n. \tag{1.180}$$

Fourth, note that

$$[Y_n\bar{X}_n - \bar{Y}_n X_n] = \frac{1}{2}[Y_n X_{n+1} - Y_{n+1}X_n],$$
$$\bar{Y}_n = \frac{2Y_n - \omega^2(X_n + \theta_n(1-2\theta_n)Y_n h_n)h_n + \sigma\Delta W_n}{2(1+\omega^2\theta_n^2 h_n^2)},$$
$$\mathbb{E}[\sigma\bar{Y}_n\Delta W_n] = \frac{\sigma^2}{2(1+\omega^2\theta_n^2 h_n^2)}h_n.$$

Fifth, pulling over expectations and summing over n in equation (1.180) for the pathwise evolution of related energy yield that

$$\mathbb{E}[\mathscr{E}(t_n)] = \mathbb{E}[\mathscr{E}(0)] + \frac{\sigma^2}{2}\sum_{k=0}^{n-1}\frac{h_k}{1+\omega^2\theta_k^2h_k^2}$$
$$-\frac{\omega^2}{2}\sum_{k=0}^{n-1}(1-2\theta_k)\mathbb{E}[Y_kX_{k+1}-Y_{k+1}X_k]h_k < +\infty.$$

Now, it remains to set $\theta_n = 0.5$ for all $n \in \mathbb{N}$. Thus, one obtains

$$\mathbb{E}[\mathscr{E}(t_n)] = \mathbb{E}[\mathscr{E}(0)] + \frac{\sigma^2}{2}\sum_{k=0}^{n-1}\frac{h_k}{1+\omega^2h_k^2/4} < +\infty$$

since the expected initial energy is finite under (ii). Consequently, by estimating the series with minimum and maximum in a standard fashion, the conclusion of Theorem 1.8.1 is confirmed. ◇

Remark 1.8.1. The energy behaviour of midpoint methods (1.5) has been studied in Hong, Scherer and Wang [52] along equidistant partitions and in S. [131] along any variable, but nonrandom partitions. The previous Theorem 1.8.1 is an extension of their results to the more general class of stochastic Theta methods.

Remark 1.8.2. Theorem 1.8.1 remains true if all ΔW_n are independent quantities with $\mathbb{E}[\Delta W_n] = 0$ and $\mathbb{E}[(\Delta W_n)^2] = h_n$ (So Gaussian property is not essential for its validity). Theorem 1.8.1 says also that midpoint methods with their expected energy e^M underestimate the exact mean energy

$$\forall t \geq 0 \quad \forall \omega, \sigma \in \mathbb{R}^1 :$$
$$e^M(t_n) = \mathbb{E}[\mathscr{E}(0)] + \frac{\sigma^2}{2}\sum_{k=0}^{n-1}\frac{h_k}{1+\omega^2\theta_k^2h_k^2} < \mathbb{E}[\mathscr{E}(t)]$$
$$= \mathbb{E}[\mathscr{E}(0)] + \frac{\sigma^2}{2}t \tag{1.181}$$

of underlying continuous SDE (1.173) (however they are consistent as maximum step size tends to zero). The proof of Theorem 1.8.1 also shows that the situation of inadequate replication of expected energy is not improving with the use of more general drift-implicit θ-methods (including forward Euler and backward Euler methods as well).

Extracting results from the previous proof of Theorem 1.8.1 gives the following immediate consequence.

Corollary 1.8.1 (Expected Energy Identity for Midpoint Methods). *Under the same assumptions as in Theorem 1.8.1, we have the expected energy identity*

$$\mathbb{E}[\mathscr{E}(t_n)] = \mathbb{E}[\mathscr{E}(0)] + \frac{\sigma^2}{2}\sum_{k=0}^{n-1}\frac{h_k}{1+\omega^2\theta_k^2h_k^2} \tag{1.182}$$

along any nonrandom partition $(t_n)_{n\in\mathbb{N}}$ for the drift-implicit θ-methods (1.178) with all implicitness parameters $\theta_k = 0.5$, any nonrandom constants $\omega, \sigma \in \mathbb{R}^1$ and any random initial data $X_0, Y_0 \in L^2(\Omega, \mathscr{F}_0, \mathbb{P})$.

Remark 1.8.3. It is obvious that the standard midpoint methods (1.178) with $\theta_n = 0.5$ do not possess an exact evolution of the energy. In fact, they underestimate the exact mean energy functional for any choice of parameters and any adapted initial conditions. This is in strong contrast to the deterministic situation with $\sigma = 0$ where midpoint methods are known to be exact-energy integrators (see simplectic methods) obeying the law of conservation of energy.

1.8.2 Existence of Energy-Exact Methods

Indeed, there exist numerical methods which can exactly preserve energy and conservation laws. These are nonstandard numerical methods. In stochastics, this was firstly noted and proved by S. [131]. The observed bias in the energy-evolution under discretization can be even removed by the energy-exact **improved midpoint methods**

$$X_{n+1} = X_n + \bar{Y}_n h_n, \quad Y_{n+1} = Y_n - \omega^2 \bar{X}_n h_n + \sigma\sqrt{1+\omega^2 h_n^2/4}\Delta W_n \quad (1.183)$$

where the involved quantities are defined as in (1.179). In passing, we note that this numerical method is fairly new to the best of our knowledge. Moreover, it is a consistent one with an exact replication of the temporal evolution of underlying continuous time energy and represents a stochastic correction of widely known standard midpoint methods.

Theorem 1.8.2 (Exact Energy Identity for Improved Midpoint Methods). *Under the same assumptions as in Theorem 1.8.1, we have the exact energy identity (also called* **trace formula** *in a more general context, see S. [132])*

$$\mathbb{E}[\mathscr{E}(t_n)] = \mathbb{E}[\mathscr{E}(0)] + \frac{1}{2}\sigma^2 t_n \quad (1.184)$$

along any nonrandom partition $(t_n)_{n\in\mathbb{N}}$ for the methods (1.183) with any nonrandom constants $\omega, \sigma \in \mathbb{R}^1$ and any random initial data $X_0, Y_0 \in L^2(\Omega, \mathscr{F}_0, \mathbb{P})$.

Proof. More general, consider the **improved implicit θ-methods**

$$\begin{aligned} X_{n+1} &= X_n + (\theta_n Y_{n+1} + (1-\theta_n)Y_n)h_n \\ Y_{n+1} &= Y_n - \omega^2(\theta_n X_{n+1} + (1-\theta_n)X_n)h_n + \sigma_n \Delta W_n \end{aligned} \quad (1.185)$$

with nonrandom implicitness-parameters θ_n, where

$$\sigma_n=\sigma\sqrt{1+\omega^2\theta_n^2 h_n^2},\theta_n \in \mathbb{R}^1, h_n=t_{n+1}-t_n, \Delta W_n = W(t_{n+1})-W(t_n) \in \mathcal{N}(0,h_n).$$

First, rewrite this system of equations for (X,Y) to as

$$\begin{aligned} X_{n+1} &= X_n + (2\theta_n \bar{Y}_n + (1-2\theta_n)Y_n)h_n \\ Y_{n+1} &= Y_n - \omega^2(2\theta_n \bar{X}_n + (1-2\theta_n)X_n)h_n + \sigma_n \Delta W_n. \end{aligned}$$

Second, multiply the components of these equations by $\omega^2 \bar{X}_n$ and $\bar{Y}_n$ (resp.) to get

$$\begin{aligned} \frac{\omega^2}{2}\left(X_{n+1}^2 - X_n^2\right) &= (2\theta_n\omega^2\bar{Y}_n\bar{X}_n + (1-2\theta_n)\omega^2 Y_n\bar{X}_n)h_n \\ \frac{1}{2}\left(Y_{n+1}^2 - Y_n^2\right) &= -\omega^2(2\theta_n\bar{Y}_n\bar{X}_n + (1-2\theta_n)\bar{Y}_n X_n)h_n + \sigma_n\bar{Y}_n\Delta W_n. \end{aligned}$$

Third, adding both equations leads to

$$\mathcal{E}_{n+1} = \mathcal{E}_n - \omega^2(1-2\theta_n)[Y_n\bar{X}_n - \bar{Y}_n X_n]h_n + \sigma_n\bar{Y}_n\Delta W_n.$$

Fourth, note that

$$\begin{aligned} [Y_n\bar{X}_n - \bar{Y}_n X_n] &= \frac{1}{2}[Y_n X_{n+1} - Y_{n+1}X_n], \\ \bar{Y}_n &= \frac{2Y_n - \omega^2(X_n + \theta_n(1-2\theta_n)Y_n h_n)h_n + \sigma_n\Delta W_n}{2(1+\omega^2\theta_n^2 h_n^2)}, \\ \mathbb{E}[\sigma_n\bar{Y}_n\Delta W_n] &= \frac{\sigma^2}{2}h_n. \end{aligned}$$

Fifth, pulling over expectations and summing over n yield that

$$\mathbb{E}[\mathcal{E}(t_n)] = \mathbb{E}[\mathcal{E}(0)] + \frac{\sigma^2}{2}t_n - \frac{\omega^2}{2}\sum_{k=0}^{n-1}(1-2\theta_k)\mathbb{E}[Y_k X_{k+1} - Y_{k+1}X_k]h_k < +\infty.$$

Recall that all step sizes h_n, parameters ω, σ and θ_n are supposed to be nonrandom. It remains to set $\theta_n = 0.5$ to verify the energy-identity (1.184). ◇

Extracting results from the previous proof of Theorem 1.8.2 yields the following.

Corollary 1.8.2 (Expected Energy Identity for θ-Methods (1.185)). *Under the same assumptions as in Theorem 1.8.1, we have the expected energy identity*

$$\mathbb{E}[\mathcal{E}(t_n)]=\mathbb{E}[\mathcal{E}(0)]+\frac{\sigma^2}{2}t_n - \frac{\omega^2}{2}\sum_{k=0}^{n-1}(1-2\theta_k)\mathbb{E}[Y_k X_{k+1} - Y_{k+1}X_k]h_k \quad (1.186)$$

along any nonrandom partition $(t_n)_{n\in\mathbb{N}}$ *for the improved implicit* θ*-methods (1.185) with any nonrandom parameters* $\theta_k \in \mathbb{R}^1$, *any nonrandom constants* $\omega, \sigma \in \mathbb{R}^1$ *and any random initial data* $X_0, Y_0 \in L^2(\Omega, \mathscr{F}_0, \mathbb{P})$.

Remark. Notice that relations (1.181) for continuous energy of SDE (1.173) and (1.184) for discrete energy of numerical methods (1.183) are indeed identical at the partition-instants t_n for all parameters ω, σ and initial values $X_0, Y_0 \in L^2(\Omega, \mathscr{F}_0, \mathbb{P})$! Thus we answer the question that such numerical methods indeed exist (which are consistent too). The conclusions of Theorems 1.8.1, 1.8.2 and Corollaries 1.8.1, 1.8.2 are still valid if all ΔW_n are independent random variables with $\mathbb{E}[\Delta W_n] = 0$ and $\mathbb{E}[(\Delta W_n)^2] = h_n$.

1.8.3 General Energy Identity of Numerical Methods and Monotonicity

One is able to establish a general energy balance identity for numerical methods. This will show that midpoint-type methods are designed to adequately replicate the increasing or decreasing evolution of energy balance for random initial data. Recall that any difference method for the approximation of any differential equation is constructed from general scheme-structure

$$X_{n+1} = X_n + \Phi_n(X) \tag{1.187}$$

where $\Phi_n(X)$ represents the **increment functional** of related numerical method. Recall $\bar{X}_n = (X_{n+1} + X_n)/2$ and let $||.||$ be the Euclidean norm in $\mathbb{R}^d$, $< .,. >$ the Euclidean scalar product.

Theorem 1.8.3 (General Energy Identity). *For all numerical methods in* $\mathbb{R}^d$ *satisfying (1.187) with increment functional* Φ_n, *we have*

$$||X_{n+1}||^2 = ||X_n||^2 + 2 < \bar{X}_n, \Phi_n(X) > \tag{1.188}$$

where $||.||$ *is the Euclidean norm in* $\mathbb{R}^d$ *and* $< .,. >$ *the Euclidean scalar product.*

Proof. First, for the Euclidean norm $||.||$ and scalar product $< .,. >$, recall the identity

$$||u+v||^2 = ||u||^2 + 2 < u, v > + ||v||^2$$

for all vectors $u, v \in \mathbb{R}^d$. Now, set

$$u := \bar{X}_n, \quad v := \Phi_n(X).$$

Note that, for numerical methods (1.187) with increment functional Φ_n, we have

$$v = X_{n+1} - X_n.$$

Second, after returning to the norm-identity (1.189), we get to

$$
\begin{aligned}
2 < \bar{X}_n, \Phi_n(X) > &= ||\bar{X}_n + \Phi_n(X)||^2 - ||\bar{X}_n||^2 - ||\Phi_n(X)||^2 \\
&= ||\bar{X}_n + X_{n+1} - X_n||^2 - ||\bar{X}_n||^2 - ||X_{n+1} - X_n||^2 \\
&= ||\bar{X}_n||^2 + 2 < \bar{X}_n, X_{n+1} - X_n > + ||X_{n+1} - X_n||^2 - ||\bar{X}_n||^2 \\
&\quad - ||X_{n+1} - X_n||^2 \\
&= < X_{n+1} + X_n, X_{n+1} - X_n > = \; ||X_{n+1}||^2 - ||X_n||^2.
\end{aligned}
$$

This is equivalent to the energy identity (1.188). ◇

Remark 1.8.4. For stochastic numerical methods, the energy identity (1.188) holds almost surely too since their increment functional Φ_n is random. This identity (1.188) also explains why midpoint-type numerical integrators with $\Phi_n = \Phi_n(\bar{X}_n)$ form a preferable base for adequate construction of numerical methods from a dynamical point of view. They may preserve the monotone character of norms along scalar products.

Definition 1.8.2. A numerical method Z is called **exact norm-monotone** iff the following implications while discretizing ODEs $dx/dt = f(t,x)$ with Caratheodory functions f can be established

$$
\begin{aligned}
&\forall x \in \mathbb{R}^d, t \in \mathbb{R}^1 :< f(t,x), x > \leq 0 \\
&\Longrightarrow \forall n \in \mathbb{N} : ||X_0|| \geq ||X_1|| \geq \ldots \geq ||X_n|| \geq ||X_{n+1}|| \geq \ldots
\end{aligned}
$$

and

$$
\begin{aligned}
&\forall x \in \mathbb{R}^d, t \in \mathbb{R}^1 :< f(t,x), x > \geq 0 \\
&\Longrightarrow \forall n \in \mathbb{N} : ||X_0|| \leq ||X_1|| \leq \ldots \leq ||X_n|| \leq ||X_{n+1}|| \leq \ldots
\end{aligned}
$$

for all adapted random initial values $X_0 \in \mathbb{R}^d$.

Theorem 1.8.4 (Exact Monotonicity of Midpoint Methods). *All midpoint-type methods X with increments $\Phi(\bar{X}_n) = f(t_n^*, \bar{X}_n)h_n$ and any sample time-points $t_n^* \in \mathbb{R}^1$ are exact norm-monotone for all ODEs with adapted random initial values X_0 and any choice of step sizes h_n.*

Proof. Apply the energy identity (1.188) to midpoint methods with increments $\Phi(\bar{X}_n) = f(t_n^*, \bar{X}_n)h_n$ and any sample time-points $t_n^* \in \mathbb{R}^1$. For them, this identity reads as

$$
||X_{n+1}||^2 = ||X_n||^2 + 2 < \bar{X}_n, f(t_n^*, \bar{X}_n) > h_n.
$$

Obviously, by taking the square root, this relation is equivalent to

$$
||X_{n+1}|| = \sqrt{||X_n||^2 + 2 < \bar{X}_n, f(t_n^*, \bar{X}_n) > h_n}.
$$

Now, the uniform monotonicity of Euclidean scalar product $< x, f(t,x) >$ with respect to x and positivity of h_n imply the exact norm-monotonicity of related midpoint methods X. For example, if $< x, f(t,x) > \leq 0$ then we have

$$||X_{n+1}|| \leq ||X_n||$$

for all $n \in \mathbb{N}$. Complete induction on $n \in \mathbb{N}$ yields the non-increasing evolution of Euclidean norms $||X_n||$ in n. Similarly, we can verify the monotonicity for $< x, f(t,x) > \geq 0$. Thus, the proof of Theorem 1.8.4 is completed. ◇

Remark 1.8.5. The situation with fully random increment functionals Φ_n (e.g. with $\Phi_n(X) = a(t_n, X_n)h_n + b(t_n, X_n)\Delta W_n$ for Euler methods) is somewhat more complicated (due to the non-monotone character of Wiener processes W) and requires further research.

1.8.4 Asymptotically Exact Methods

To relax the requirement of being exact at finite times, we may only require that numerical methods provide exact values as the integration time t_n tends to infinity or step sizes h_n tend to 0. In view of S. [118,120–122], midpoint and trapezoidal methods are good candidates. Consider (for brevity, autonomous) linear system of SDEs

$$dX(t) = AX(t)dt + \sum_{j=1}^{m} b^j dW^j(t) \tag{1.189}$$

with additive noise and constant drift matrix $A \in \mathbb{R}^{d \times d}$ Assume that there is a stationary solution X_∞ of (1.189). Then, for stationarity of autonomous systems (1.189) with additive white noise, it is a necessary and sufficient requirement that all eigenvalues $Re(\lambda_i(A)) < 0$. Under this condition, we also find the almost sure limit $\lim_{t\to+\infty} X(t) = X_\infty$.

Definition 1.8.3. The random sequence $(Y_n)_{n\in\mathbb{N}}$ is said to be **asymptotically p-th mean preserving** iff

$$\lim_{n\to+\infty} \mathbb{E}\|Y_n\|^p = \mathbb{E}\|X_\infty\|^p,$$

(asymptotically) mean preserving iff

$$\lim_{n\to+\infty} \mathbb{E}Y_n = \mathbb{E}X_\infty,$$

(asymptotically) in-law preserving iff

$$\mathcal{L}aw(Y_\infty) = \mathcal{L}aw(X_\infty),$$

(asymptotically) variance-preserving iff

$$\lim_{n\to+\infty} \mathbb{E}\|Y_n - \mathbb{E}[Y_n]\|^2 = \mathbb{E}\|X_\infty - \mathbb{E}[X_\infty]\|^2$$

with respect to systems (1.189).

This definition has been originally introduced by S. [120,121] (actually first time in WIAS Report No. 112, WIAS, Berlin, 1994).

Now, consider the family of drift-implicit Theta methods

$$Y_{n+1} = Y_n + (\theta A\, Y_{n+1} + (1-\theta) A\, Y_n)\, h_n + \sum_{j=1}^{m} b^j\, \Delta W_n^j \tag{1.190}$$

with independently Gaussian distributed increments $\Delta W_n^j = W^j(t_{n+1}) - W^j(t_n)$ and implicitness parameter $\theta \in [0,1] \subset \mathbb{R}^1$.

Theorem 1.8.5 (Asymptotic preservation by trapezoidal BTMs [120]–[122]). *Assume that:*

(i) $\forall \lambda(\textit{eigenvalue}\,(A))\ Re(\lambda(A)) < 0$.
(ii) (X_0, Y_0) *independent of* $\mathcal{F}_\infty^j = \sigma\{W^j(s) : 0 \le s < +\infty\}$.
(iii) $\mathbb{E}\|X_0\|^p + \mathbb{E}\|Y_0\|^p < +\infty$ *for* $p \ge 2$.
(iv) $A \in \mathbb{R}^{d\times d}, b^j \in \mathbb{R}^d$ *are deterministic.*

Then, the **trapezoidal method** *(i.e. (1.3) with* $\theta = 0.5$*) applied to system (1.189) with any equidistant step size* $h = h_n$ *is asymptotically mean, p-th mean, variance- and in-law preserving. Moreover, it is* **the only method from the entire family of drift-implicit Theta methods** *(1.190) with that behavior (i.e.,* $\Longrightarrow$ **asymptotic equivalence** *for systems with additive noise).*

Proof (Only main ideas with diagonalizable matrix A). First, the limit distribution of Y_n governed by (1.190) exists (for all implicitness parameters $\theta \ge 0.5$) (since one may apply standard fixed point principles of contractive mappings). Second, the limit as linear transformation of Gaussian random variables is Gaussian for all $\theta \in [0.5, +\infty)$. Third, $\mathbb{E}Y_n \longrightarrow 0$ as n tends to $+\infty$ (as in deterministics if $\theta \ge 0.5$). Fourth, due to uniqueness of Gaussian laws, it remains to to study the second moments for all constant step sizes $h > 0$. We arrive at

$$\lim_{n\to+\infty} \mathbb{E}\left([Y_n Y_n^T]\right)_{i,k} = -\sum_{j=1}^{m} \frac{b_i^j b_k^j}{\lambda_i + \lambda_k + (1-2\theta)h\lambda_i\lambda_k}$$

and

$$\mathbb{E}\left([X_\infty X_\infty^T]\right)_{i,k} = -\sum_{j=1}^{m} \frac{b_i^j b_k^j}{\lambda_i + \lambda_k}$$

Then

$$\lim_{n \to +\infty} \mathbb{E}\left([Y_n Y_n^T]\right)_{i,k} = \mathbb{E}\left([X_\infty X_\infty^T]\right)_{i,k} \quad \Longleftrightarrow \quad \theta = 0.5.$$

Thus, the stationarity with exact stationary Gaussian probability law including the preservation of all moments and variance under discretization is obvious. $\diamond$

Remark 1.8.6. For the proof with non-diagonalizable matrices A, one may apply fixed point principles to verify the same conclusion of Theorem 1.8.5. Note that, for linear autonomous systems of SDEs (1.189), the schemes of trapezoidal and midpoint methods coincide. Thus, midpoint methods possess the same preserving character of asymptotic exactness under discretization.

1.9 Order Bounds: Why Not Higher Order Methods?

As commonly known from deterministic numerical analysis, there are order bounds for convergence of numerical methods. Sometimes they are also called **Dahlquist barriers**.

1.9.1 Order Bounds of Clark-Cameron

Clark and Cameron [22] could prove the following very remarkable result on maximum order bounds of partition $\mathscr{F}_T^N$-measurable approximations.

Definition 1.9.1. The stochastic process $Y = (Y(t))_{0 \le t \le T}$ is called **partition $\mathscr{F}_T^N$-measurable** iff all values $Y(t_n)$ $(t_n \in [0,T], 0 \le n \le N)$ are $\mathscr{F}_{t_n}^N$-measurable with

$$\mathscr{F}_{t_n}^N = \sigma\left\{W^j(t_k) : k = 0,1,\ldots,n; j = 1,2,\ldots,m\right\}$$

for all $n = 0,1,\ldots,N$, along a given $\mathscr{F}_{t_n}^N$-measurable discretization $0 = t_0 \le t_1 \le \ldots \le t_N = T$ for the fixed deterministic time-interval $[0,T]$.

Remark 1.9.1. The conditional expectations $\mathbb{E}[X(t_{n+1})|\mathscr{F}_{t_n}^N]$ provide the partition $\mathscr{F}_T^N$-measurable stochastic approximations with the minimal mean square error due to their inherent projection property in Hilbert spaces $L^2(\Omega, \mathscr{F}, \mathbb{P})$. Thus it is natural to study their error and practical implementation at first.

Theorem 1.9.1 (Clark-Cameron Order-Bound Theorem). *Assume that $X=(X(t))_0 \le t \le T$ satisfies the one-dimensional autonomous SDE*

$$dX(t) = a(X(t))dt + dW(t) \tag{1.191}$$

where $a \in C^3(\mathbb{R})$ and all derivatives of a are uniformly bounded. Then

$$\mathbb{E}[(X(T) - \mathbb{E}[X(T)|\mathscr{F}_T^N])^2] = \frac{cT}{N^2} + o\left(\frac{1}{N^2}\right)$$

where

$$c = \frac{T^3}{12} \int_0^T \mathbb{E}\left[\exp\left(2\int_s^T a'(X(u))du\right)[a'(X(s))]^2\right] ds.$$

Thus, for systems with additive noise, we obtain the general mean square order bound 1.0 for numerical approximations using only the increments of underlying Wiener process. A similar result holds also for diffusions with variable diffusion coefficients $b(x)$ when

$$c(x) := a(x) - \frac{1}{2}b(x)b'(x) \not\equiv Kb(x)$$

for any real constant K, see Clark and Cameron [22]. They also provide a constructive example with multiplicative noise. Consider the two-dimensional SDE

$$\begin{aligned} dX^{(1)}(t) &= dW^1(t) \\ dX^{(2)}(t) &= X^{(1)}(t)\, dW^2(t) \end{aligned}$$

driven by two independent scalar Wiener processes W^1, W^2. This system obviously has the solutions $X^{(1)}(t) = W^1(t)$ and

$$X^{(2)}(t) = \int_0^t W^1(s)dW^2(s) = \int_0^t \int_0^s dW^1(u)\, dW^2(s)$$

(in fact it is an one-dimensional example with multi-dimensional "Wiener process differentials" (i.e. $m = 2$)). Then they compute the best convergence rate $\gamma_2 = 0.5$ (in mean square sense) for partition $\mathscr{F}_T^N$-measurable approximations using any set of N equidistant, $\mathscr{F}_{t_n}^N$-measurable time-instants $t_n = n\frac{T}{N}$, and the mean square minimally attainable approximation error

$$\left(\mathbb{E}\left[\left|X^{(2)}(T) - \mathbb{E}[X^{(2)}(T)|\mathscr{F}_T^N]\right|^2\right]\right)^{\frac{1}{2}} = \frac{T}{2}\left[\frac{1}{N}\right]^{\frac{1}{2}}.$$

It is worth noting that $X^{(2)}$ represents the simplest nontrivial multiple integral with length $l(\alpha) > 1$. Liske [89] has studied its joint distribution with $(W^1(t), W^2(t))$. In this case the error order bound for $\mathscr{F}_T^N$-measurable approximations of X_T^2 is already attained with 0.5, since X^2 cannot be expanded in a linear combination of W^1, W^2. This system also exhibits an interesting test equation for the qualitative behavior of numerical methods (e.g. compare the numerically estimated distribution with that of the exact solution derived by Liske [89]). Since in the L^2 sense one cannot provide better partition $\mathscr{F}_T^N$-measurable approximations than that of the projection done by conditional expectations, there are natural (convergence) order restrictions for

$\mathscr{F}_T^N$-measurable approximations. Thus we cannot exceed the order 1 in L^2-sense for $\mathscr{F}_T^N$-measurable approximations. On the other hand, if one wants higher order of convergence in general, one has to enlarge the condition σ-field substantially (actually done by higher order multiple integrals and Levy areas). Note also this is not always necessary for approximations of functionals $V(t, X(t))$ of diffusion processes X with V-commutativity, see S. [126]. In fact, for example for pure one-dimensional diffusions X (i.e. when drift a is zero), the x-commutativity condition (i.e. $V(x) = x$), which is then identical with the condition of *commutative noise* (in short: noise-commutativity) under the absence of drift terms

$$b^j(x)\frac{db^k(x)}{dx} = b^k(x)\frac{db^j(x)}{dx}$$

for all $j, k = 0, 1, 2, \ldots, m$. This requirement, together with $b^j \in C^\infty(\mathbb{R})$, effects that $b^j(x) = K_{j,k}b^k(x)$ with some deterministic real constants $K_{j,k}$. In this trivial case one could even obtain any order of p-th mean convergence ($p \leq 1$). (This is no surprise after one has carefully analyzed the observation of Clark and Cameron [22] which implies the approximation error 0 by the projection operator of conditional expectation under $a\prime(x) \equiv 0$ and the noise-commutativity assumption in the situation $d = 1$). Unfortunately, the situation in view of convergence order bounds is much more complicated in the fully multi-dimensional framework and needs more care in the near future.

1.9.2 Exact Mean Square Order Bounds of Cambanis and Hu

Cambanis and Hu [20] have established the following result concerning exact mean square convergence error bounds (i.e. for the asymptotic behavior of leading error coefficients of numerical schemes with respect to mean square convergence criteria). For the statement, we introduce the following definition of partition density.

Definition 1.9.2. A strictly positive, differentiable function $h \in C^0([0,T]^2, \mathbb{R}_+)$ with uniformly bounded derivatives is said to be a **regular partition density** of the time-interval $[0, T]$ iff

$$\int_{t_n^{N(t)}}^{t_{n+1}^{N(t)}} h(t,s)ds = \frac{1}{N(t)}$$

for $n = 0, 1, \ldots, N(t) - 1$, $t_0 = 0$, where $N = N(t)$ denotes the number of sub-intervals $[t_n^{N(t)}, t_{n+1}^{N(t)}]$ for a total time interval $[0, t]$ with terminal times $t \leq T$.

Regular partition densities possess the property that

$$\lim_{N(t)\to+\infty, t_n\to s} N(t)(t_{n+1}^{N(t)} - t_n^{N(t)}) = \frac{1}{h(t,s)}.$$

Therefore they describe the distribution of time-instants in discretizations of intervals $[0, T]$ in a fancy manner. Since the conditional approximation provides the mean square $\mathscr{F}^N$-measurable approximation (with $N = N(t)$) with minimal mean square error, one arrives at

Theorem 1.9.2. *Assume that X satisfies an one-dimensional SDE (1.2) with coefficients $a, b \in C^3(\mathbb{R}, \mathbb{R})$ possessing bounded derivatives up to third order, $\mathbb{E}|X(0)|^2 < +\infty$, and all time-discretizations are exclusively done along a given regular partition density h on $[0,T]^2$.*
Then, there exists a Gaussian process $\eta = (\eta(t))_{0\leq t\leq T}$ on $(\Omega, \mathscr{F}, (\mathscr{F}_t)_{0\leq t\leq T}, \mathbb{P})$ such that

$$\lim_{N(t)\to+\infty} N(t)\Big(X(t) - \mathbb{E}[X(t)|\mathscr{F}_t^{N(t)}]\Big) = \eta(t)$$

with mean 0 and covariance matrix $C(t) =$

$$\int_0^t \frac{[(\mathscr{L}^1 a - \mathscr{L}^0 b)(X(s))]^2}{6[h(s)]^2} \exp\left(\int_s^t (2a\prime(X_u) - [b\prime(X(u))]^2)du + 2\int_s^t b\prime(X(u))dW(u)\right)ds$$

which is the unique solution of $dC(t) =$

$$\left((2a\prime(X(t)) + [b\prime(X(t))]^2)C(t) + \frac{[(\mathscr{L}^1 a - \mathscr{L}^0 b)(X(t))]^2}{12}\right)dt + 2b\prime(X(t))C(t)dW(t)$$

with $\eta_0 = 0$ and has the property

$$\lim_{N(t)\to+\infty} N(t)\mathbb{E}\Big[X(t) - \mathbb{E}[X(t)|\mathscr{F}_t^{N(t)}]\Big]^2 = \mathbb{E}\eta(t) = \int_0^t \frac{H(t,s)}{h(t,s)}ds$$

where $H(t,s) =$

$$\frac{1}{6}\mathbb{E}\left[[(\mathscr{L}^1 a - \mathscr{L}^0 b)(X(s))]^2 \exp\left(\int_s^t (2a\prime(X(u)) - [b\prime(X(u))]^2)du + 2\int_s^t b\prime(X(u))dW(u)\right)\right].$$

The optimal double mean square approximation error satisfies a similar relation. For more details, see Cambanis and Hu [20]. Also their results can be generalized to multidimensional diffusions with some care. This result is fundamental with respect to asymptotically optimal mean square discretizations. This fact can be seen from the fact that the function $h^* \in C^0([0,T], \mathbb{R}_+)$ established by

$$h^*(t,s) = \frac{[H(t,s)]^{1/3}}{\int_0^t [H(t,s)]^{1/3}ds}$$

minimizes the functional $\int_s^t \frac{H(t,s)}{[h(t,s)]^2} ds$ where $H(t,s) \geq 0$ among all regular partition densities h with $h(t,s) > 0$ and $\int_0^t h(t,s)ds = 1$. Therefore, any asymptotically mean square optimal approximation has to use a discretization following that optimal partition law. However, the practical value is still in doubt, since it will be hard to evaluate those expressions in the fully multidimensional framework or has any reader another suggestion in the case $m, d > 1$?

1.9.3 Lower Order Bounds of Hofmann-Gronbach-Ritter

Hofmann, Müller-Gronbach and Ritter [50] have noticed in one dimension (i.e. $d = m = 1$) that, for continuously differentiable b and fixed terminal time $T = 1$, there are efficient estimates for the lower bounds of order of strong (mean square) convergence for all $\mathscr{F}_{t_n}^N$-measurable approximations Y^Δ for SDEs in $\mathbb{R}^1$ with additive noise. More precisely, for the error e_2 of mean square convergence and partitions with maximum step size Δ and $N + 1$ time-instants, we have

$$\lim_{\Delta \to 0} \sqrt{N} \inf_{Y^\Delta} e_2(Y^\Delta, a, b, X_0, T) = K_2 ||b||_{L^1}$$

with constants $K_2 = \frac{1}{\sqrt{6}}$ and $T = 1$. Therefore, the standard forward Euler method with an adaptive strategy of step size selection already produces asymptotically mean square optimal $\mathscr{F}_{t_n}^N$-measurable numerical approximations. Note that the number N of observations of the underlying driving noise W may be determined by any measurable termination criterion. Moreover, one can carry the qualitative assertion on lower order bounds over to d-dimensional SDEs with additive noise and L^p-integrable b as well (i.e. $p \geq 1$). It is worth noting that that step size selection suggested originally by Hofmann, Müller-Gronbach and Ritter [50] is only designed to control large diffusion fluctuations, and it does not seem to be very appropriate as one takes the limit as b goes to zero (i.e. incomplete adaptability is obtained in the presence of significant drift parts - an approach which leads to inconsistent results in view of deterministic limit equations, however a procedure which indeed might be appropriate for pure diffusion processes with very large diffusion coefficients $b(t) >> 1$). Furthermore, they did not study the dependence of the order and error bounds on variable terminal times T or as T tends to $+\infty$. Such as study would be important for adequate asymptotic analysis through numerical dynamical systems as discretizations.

In the paper [51], the authors Hofmann, Müller-Gronbach and Ritter analyze the double $L^2([0,1])$-error associated to the approximation of a scalar stochastic differential equation by numerical methods based on multiple Itô integrals as suggested in Kloeden and Platen [72]. Namely, for a scalar diffusion process XX, they consider approximations of the type $\overline{X} = f(I_{\alpha_1,s_1,t_1}, \ldots, I_{\alpha_k,s_k,t_k})$ where $f : \mathbb{R}^k \to L^2([0,1])$ is assumed to be measurable and $I_{\alpha,s,t}$ denotes the multiple

Itô integral corresponding to a zero-one-multi-index α and an interval $[s,t] \subset [0,1]$. They are interested in controlling $\mathbb{E}[\int_0^1 |X(s) - \overline{X}(s)|^2 ds]^{1/2}$ in terms of a function of the number N of observations for different integrals in the approximated process. It is shown that the double$L^2([0,1])$-error for linear interpolated Itô-Taylor approximations is at most of order $1/2$ with respect to $1/N$. For some special discretization grids and multi-indices sets, the authors provide an equivalent of the error. This result is in sharp contrast with the well known fact that high orders can be achieved by these methods with respect to the error at the discretization points [see, e.g. as claimed in Chap. 10 in Kloeden and Platen [72].

In Müller-Gronbach and Ritter [100], the authors give a survey on minimal errors for strong and weak approximation of stochastic differential equations. They investigate asymptotic optimality of numerical methods for integrating systems of stochastic differential equations (with Lipschitz continuous coefficients) in both the weak and strong sense, restricted to finite time-intervals $[0,1]$. The main emphasis is on algorithms with point evaluations of the driving Brownian motion at N time-instants. The number N of observations may be determined by any measurable termination criterion. Some (optimal) algorithms with variable step sizes (i.e. with varying cardinality) may have superior behaviour with respect to convergence and related costs. In some cases one even obtains an exponential speed-up by using (optimal) methods of varying cardinality compared to methods with fixed cardinality N. As an example, the same authors have studied the linear equation

$$dX(t) = bX(t)dW(t)$$

with a constant b. Then the cost of optimal methods is linear in b while the cost of optimal methods with fixed cardinality N is exponential in b. If $b = 3$ then the speed-up is already of factor 900. All in all, estimates on lower order bounds exist and are very important contributions which show that the "run for higher order" or any order higher than 0.5 can be a "run into the vain" (following famous words of D. Hilbert). That is also why we concentrate rather on the qualitative analysis of lower order methods and their qualitative improvements with constant and variable step sizes.

1.9.4 Convergence is an Asymptotic Property

It is clear that convergence is an asymptotic property. Such asymptotic properties can be never achieved in a finite time. This requirement represents only a quantitative assertion of asymptotic nature. Other qualitative properties such as longterm stability, monotonicity, boundedness, stationarity, energy and / or oscillatory behaviour seem to be of more interest than raising hypothetical convergence rates under non-realistic or non-practical assumptions in a real world scenario since we live in a finite-time constrained world.

1.10 Contractivity and Two-Point Behaviour

For simplicity, we shall exclusively illustrate the concepts of contractivity and non-expansivity by the drift-implicit Theta methods (1.1) without balancing terms c^j, i.e. all weight matrices $c^j \equiv 0$.

1.10.1 Two-Point Motion Process and General Contraction-Identity

One is able to establish a general contraction identity for numerical methods. This will show that midpoint-type methods are designed to adequately replicate the increasing or decreasing evolution of perturbations for random initial data. Recall that any difference method for the approximation of any differential equation is constructed from general scheme-structure (1.187) (i.e. $X_{n+1} = X_n + \Phi_n(X)$) with $\Phi_n(X)$ representing the **increment functional** of related numerical method. Recall $\bar{X}_n = (X_{n+1}+X_n)/2$. Let X and Y denote the stochastic processes belonging to the same numerical scheme (1.187) and started at $X_0 \in \mathbb{R}^d$ and $Y_0 \in \mathbb{R}^d$, respectively. We shall study the dynamic behaviour of the **two-point motion process** (X, Y) along the same numerical method governed by the schemes

$$X_{n+1} = X_n + \Phi_n(X)$$
$$Y_{n+1} = Y_n + \Phi_n(Y)$$

with one and the same increment functional Φ_n along one and the same partition $(t_n)_{n\in\mathbb{N}}$. Recall that $||.||$ denotes the Euclidean vector norm in $\mathbb{R}^d$ and $< .,. >$ the Euclidean scalar product.

Theorem 1.10.1 (General Contraction Identity). *For all numerical methods in $\mathbb{R}^d$ satisfying (1.187) with increment functional Φ_n, we have*

$$||X_{n+1} - Y_{n+1}||^2 = ||X_n - Y_n||^2 + 2 < \bar{X}_n - \bar{Y}_n, \Phi_n(X) - \Phi_n(Y) > \quad (1.192)$$

where $||.||$ is the Euclidean norm in $\mathbb{R}^d$ and $< .,. >$ the Euclidean scalar product.

Proof. First, for the Euclidean norm $||.||$ and scalar product $< .,. >$, recall the identity $||u+v||^2 = ||u||^2 + 2 < u, v > +||v||^2$ for all vectors $u, v \in \mathbb{R}^d$. Note that, for numerical methods (1.187) with increment functional Φ_n, we have

$$\Phi_n(X) = X_{n+1} - X_n.$$

Second, analyzing the Euclidean norm of the two-point motion process gives the identities

$$
\begin{aligned}
||X_{n+1} - Y_{n+1}||^2 &= ||X_n - Y_n + \Phi_n(X) - \Phi_n(Y)||^2 \\
&= ||X_n - Y_n||^2 + 2 < X_n - Y_n, \Phi_n(X) - \Phi_n(Y) > \\
&\quad + ||\Phi_n(X) - \Phi_n(Y)||^2 \\
&= ||X_n - Y_n||^2 + < X_n - Y_n, \Phi_n(X) - \Phi_n(Y) > \\
&\quad + < X_{n+1} - \Phi_n(X) - Y_{n+1} + \Phi_n(Y), \Phi_n(X) - \Phi_n(Y) > \\
&\quad + ||\Phi_n(X) - \Phi_n(Y)||^2 \\
&= ||X_n - Y_n||^2 + < X_{n+1} + X_n - Y_{n+1} - Y_n, \Phi_n(X) - \Phi_n(Y) > \\
&\quad - < \Phi_n(X) - \Phi_n(Y), \Phi_n(X) - \Phi_n(Y) > + ||\Phi_n(X) - \Phi_n(Y)||^2 \\
&= ||X_n - Y_n||^2 + 2 < \bar{X}_n - \bar{Y}_n, \Phi_n(X) - \Phi_n(Y) >
\end{aligned}
$$

which confirm the validity of contraction identity (1.192). ◇

Remark 1.10.1. For stochastic numerical methods, the contraction identity (1.192) holds almost surely too (with their increment functional Φ_n which is random). This identity (1.192) also explains why midpoint-type numerical integrators with $\Phi_n = \Phi_n(\bar{X}_n)$ form a preferable base for adequate construction of numerical methods from a dynamical point of view. They may preserve the monotone character of contractions (perturbations) of two-point motion process along scalar products.

Definition 1.10.1. A numerical method Z is called **exact contraction-monotone** iff the following implications while discretizing ODEs $dx/dt = f(t,x)$ with Caratheodory functions f can be established

$$
\begin{aligned}
&\forall x, y \in \mathbb{R}^d, t \in \mathbb{R}^1 :< f(t,x) - f(t,y), x - y > \leq 0 \\
&\Longrightarrow \forall n \in \mathbb{N}: ||X_0 - Y_0|| \geq ||X_1 - Y_1|| \geq \ldots \geq ||X_n - Y_n|| \geq ||X_{n+1} - Y_{n+1}|| \geq \ldots
\end{aligned}
$$

and

$$
\begin{aligned}
&\forall x, y \in \mathbb{R}^d, t \in \mathbb{R}^1 :< f(t,x) - f(t,y), x - y > \geq 0 \\
&\Longrightarrow \forall n \in \mathbb{N} : ||X_0 - Y_0|| \leq ||X_1 - Y_1|| \leq \ldots \leq ||X_n - Y_n|| \leq ||X_{n+1} - Y_{n+1}|| \leq \ldots
\end{aligned}
$$

for all adapted random initial values $X_0, Y_0 \in \mathbb{R}^d$.

Theorem 1.10.2 (Exact Contraction-Property of Midpoint Methods). *All midpoint-type methods X with increments $\Phi(\bar{X}_n) = f(t_n^*, \bar{X}_n)h_n$ and any sample time-points $t_n^* \in \mathbb{R}^1$ are exact contraction-monotone for all ODEs with adapted random initial values X_0 and any choice of step sizes h_n.*

Proof. Apply the contraction identity (1.192) to midpoint methods with increment functional $\Phi(\bar{X}_n) = f(t_n^*, \bar{X}_n)h_n$ and any sample time-points $t_n^* \in \mathbb{R}^1$. For them, this identity reads as

$$
||X_{n+1} - Y_{n+1}||^2 = ||X_n - Y_n||^2 + 2 < \bar{X}_n - \bar{Y}_n, f(t_n^*, \bar{X}_n) - f(t_n^*, \bar{Y}_n > h_n.
$$

Obviously, by taking the square root, this relation is equivalent to

$$||X_{n+1} - Y_n|| = \sqrt{||X_n - Y_n||^2 + 2 < \bar{X}_n - \bar{Y}_n, f(t_n^*, \bar{X}_n) - f(t_n^*, \bar{Y}_n) > h_n}.$$

Now, the uniform monotonicity of Euclidean scalar product

$$< x - y, f(t,x) - f(t,y) >$$

with respect to $x, y \in \mathbb{R}^d$ and positivity of h_n imply the exact contraction-monotonicity of related midpoint methods X. For example, if $< x - y, f(t,x) - f(t,y) > \leq 0$ for all $x, y \in \mathbb{R}^d$ then we have

$$||X_{n+1} - Y_{n+1}|| \leq ||X_n - Y_n||$$

for all $n \in \mathbb{N}$. Complete induction on $n \in \mathbb{N}$ yields the non-increasing evolution of Euclidean norms $||X_n - Y_n||$ in n. Similarly, we can verify the monotonicity for $< x - y, f(t,x) - f(t,y) > \geq 0$ for all $x, y \in \mathbb{R}^d$. Thus, the proof of Theorem 1.10.2 is completed. ◇

Remark 1.10.2. The situation with fully random increment functionals Φ_n (e.g. with $\Phi_n(X) = a(t_n, X_n)h_n + b(t_n, X_n)\Delta W_n$ for Euler methods) is somewhat more complicated (due to the non-monotone character of Wiener processes W) and requires further research. However, an extension to p-th mean contractions (appropriate for the concept of B-stability) gives some more insight for stochastic Theta methods (see next subsections).

1.10.2 P-th Mean Contractivity and Non-expansivity of Backward Euler Methods

Let $X_{s,x}(t)$ denote the value of the stochastic process X at time $t \geq s$, provided that it has started at the value $X_{s,x}(s) = x$ at prior time s. x and y are supposed to be adapted initial values. Let Π denote an ordered time-scale (discrete ($\Pi = \mathbb{N}$) or continuous ($\Pi = [0, +\infty)$)) and $p \neq 0$ be a nonrandom constant.

Definition 1.10.2. A stochastic process $X = (X(t))_{t \in \Pi}$ with basis $(\Omega, \mathcal{F}, (\mathcal{F}_t)_{t \in \Pi}, \mathbb{P})$ is said to be uniformly **p-th mean (forward) contractive** on $\mathbb{R}^d$ iff $\exists K_C^X \in \mathbb{R}\ \forall t \geq s \in \Pi\ \forall x, y \in \mathbb{R}^d$

$$\mathbb{E}\Big[||X_{s,x}(t) - X_{s,y}(t)||^p \Big| \mathcal{F}_s\Big] \leq \exp\Big(pK_C^X(t-s)\Big)||x - y||^p \qquad (1.193)$$

with **p-th mean contractivity constant** K_C^X. In the case $K_C^X < 0$, we speak of **strict p-th mean contractivity**. Moreover, X is said to be a process with **p-th**

mean non-expansive perturbations iff $\forall t \geq s \in \Pi\ \forall x, y \in \mathbb{R}^d$

$$\mathbb{E}\Big[||X_{s,x}(t) - X_{s,y}(t)||^p \Big| \mathscr{F}_s\Big] \leq ||x - y||^p. \tag{1.194}$$

If $p = 2$ then we speak of **mean square contractivity** and **mean square non-expansivity**.

For strictly contractive processes, adapted perturbations in the initial data have no significant impact on its longterm dynamic behaviour. Adapted perturbations of non-expansive processes are under control along the entire time-scale Π. These concepts are important for the control of error propagation through numerical methods. They are meaningful to test numerical methods while applying to SDEs with monotone coefficient systems.

Let $p > 0$ be a nonrandom constant.

Definition 1.10.3. A coefficient system (a, b^j) of SDEs (1.2) and its SDE are said to be **strictly uniformly p-th mean monotone** on $\mathbb{R}^d$ iff $\exists K_{UC} \in \mathbb{R}\ \forall t \in \mathbb{R} \forall x, y \in \mathbb{R}^d$

$$< a(t,x) - a(t,y), x - y >_d + \frac{1}{2} \sum_{j=1}^{m} ||b^j(t,x) - b^j(t,y)||^2$$
$$+ \frac{p-2}{2} \sum_{j=1}^{m} \frac{< b^j(t,x) - b^j(t,y), x - y >_d^2}{||x - y||^2} \leq K_{UC} ||x - y||^2. \tag{1.195}$$

If $p = 2$ then we speak of **mean square monotonicity**.

Lemma 1.10.1. *Assume that X satisfies SDE (1.2) with p-th mean monotone coefficient system (a, b^j).*
Then X is p-th mean contractive for all $p \geq 2$ and its p-th mean contractivity constant K_C^X can be estimated by

$$K_C^X \leq K_{UC}.$$

This lemma can be proved by Dynkin's formula (averaged Itô formula). Let us discuss the possible "worst case effects" on perturbations of numerical methods under condition (1.195) with $p = 2$. It turns out that the drift-implicit backward Euler methods are mean square contractive.

Theorem 1.10.3. *Assume that*

(i) $\theta_n = 1$.
(ii) $0 < \inf_{n \in \mathbb{N}} h_n \leq \sup_{n \in \mathbb{N}} h_n < +\infty$*, all h_n nonrandom (i.e. only admissible step sizes).*
(iii) $\exists K_a \leq 0 \forall x, y \in R^d\ \forall t \geq 0 \ : < a(t,x) - a(t,y), x - y > \leq K_a ||x - y||^2$.
(iv) $\exists K_b \forall x, y \in R^d\ \forall t \geq 0 \ : \ \sum_{j=1}^{m} ||b^j(t,x) - b^j(t,y)||^2 \leq K_b ||x - y||^2$.

Then, the drift-implicit Euler methods (1.1) with scalar implicitness $\theta_n = 1$ and vanishing $c^j = 0$ have mean square contractive perturbations when applied to SDEs (1.2) with mean square monotone coefficients (a, b^j) with contractivity constant

$$K_C^X = \sup_{n \in I\!N} \frac{2K_a + K_b}{1 - 2h_n K_a}. \tag{1.196}$$

If additionally $2K_a + K_b \leq 0$ then they are mean square non-expansive and

$$K_C^X = \frac{2K_a + K_b}{1 - 2K_a \sup_{n \in I\!N} h_n}. \tag{1.197}$$

Proof. Rearrange the scheme (1.1) for the drift-implicit Theta methods with nonrandom scalar implicitness $(\Theta_n) = \theta_n I$ to separate implicit from explicit part such that

$$X_{n+1} - \theta_n h_n a(t_{n+1}, X_{n+1}) = X_n + (1 - \theta_n) h_n a(t_n, X_n) + \sum_{j=1}^{m} b^j(t_n, X_n) \Delta W_n^j. \tag{1.198}$$

Recall that X and Y denote the values of the same scheme (1.1) started at values X_0 and Y_0, respectively. Now, take the square of Euclidean norms on both sides. By taking the expectation on both sides we arrive at

$$\begin{aligned} &I\!E||X_{n+1} - Y_{n+1}||^2 - 2\theta_n h_n I\!E < X_{n+1} - Y_{n+1}, a(t_{n+1}, X_{n+1}) - a(t_{n+1}, Y_{n+1}) > \\ &\quad + \theta_n^2 h_n^2 I\!E||a(t_{n+1}, X_{n+1}) - a(t_{n+1}, Y_{n+1})||^2 \\ &= I\!E||X_n - Y_n||^2 + 2(1 - \theta_n) h_n I\!E < X_n - Y_n, a(t_n, X_n) - a(t_n, Y_n) > \\ &\quad + (1-\theta_n)^2 h_n^2 I\!E||a(t_n, X_n) - a(t_n, Y_n)||^2 + h_n \sum_{j=1}^{m} I\!E||b^j(t_n, X_n) - b^j(t_n, Y_n)||^2. \end{aligned}$$

Under the assumption (iii) we have

$$-2\theta_n h_n < a(t, x) - a(t, y), x - y > \geq -2\theta_n h_n K_a ||x - y||^2 \geq 0$$

for all $x, y \in I\!R^d$ and $t \geq 0$. Consequently, under (iii) and (iv), we may estimate

$$\begin{aligned} &(1 - 2\theta_n h_n K_a) I\!E||X_{n+1} - Y_{n+1}||^2 \\ &\leq [1 + (2(1-\theta_n) K_a + K_b) h_n]_+ I\!E||X_n - Y_n||^2 \\ &\quad + (1-\theta_n)^2 h_n^2 I\!E||a(t_n, X_n) - a(t_n, Y_n)||^2. \end{aligned}$$

for all $n \in \mathbb{N}$. This leads to the estimate

$$\begin{aligned}
&\mathbb{E}||X_{n+1} - Y_{n+1}||^2 \\
&\le \frac{[1 + (2(1-\theta_n)K_a + K_b)h_n]+}{1 - 2\theta_n h_n K_a}\mathbb{E}||X_n - Y_n||^2 \\
&= \left(1+\frac{(2K_a+K_b)h_n}{1 - 2\theta_n h_n K_a}\right)\mathbb{E}||X_n-Y_n||^2+\frac{(1-\theta_n)^2h_n^2}{1-2\theta_n h_n K_a}\mathbb{E}||a(t_n, X_n) - a(t_n, Y_n)||^2 \\
&\le \exp\left(\frac{(2K_a+K_b)h_n}{1 - 2\theta_n h_n K_a}\right)\mathbb{E}||X_n-Y_n||^2+\frac{(1-\theta_n)^2h_n^2}{1-2\theta_n h_n K_a}\mathbb{E}||a(t_n, X_n)-a(t_n, Y_n)||^2
\end{aligned}$$

since $1 + z \le \exp(z)$ for $z \ge -1$. Now, set all parameters $\theta_n = 1$ in the above inequality. In this case one encounters

$$\mathbb{E}||X_{n+1} - Y_{n+1}||^2 \le \exp\left(\frac{2K_a + K_b}{1 - 2h_n K_a}h_n\right)\mathbb{E}||X_n - Y_n||^2.$$

Therefore, the drift-implicit backward Euler methods have mean square contractive perturbations with contractivity constant

$$\begin{aligned}
K_C^X &= \sup_{n\in\mathbb{N}} \frac{2K_a + K_b}{1 - 2h_n K_a} \\
&= \frac{2K_a + K_b}{1 - 2\sup_{n\in\mathbb{N}} h_n K_a} \quad \text{if } 2K_a + K_b \le 0.
\end{aligned}$$

If additionally $2K_a + K_b \le 0$, then the perturbations are mean square non-expansive. ◇

1.10.3 P-th Mean Non-contractivity and Expansivity of Euler Methods

Let $X_{s,x}(t)$ denote the value of the stochastic process X at time $t \ge s$, provided that it has started at the value $X_{s,x}(s) = x$ at prior time s. x and y are supposed to be adapted initial values. Let Π denote an ordered time-scale (discrete ($\Pi = \mathbb{N}$) or continuous ($\Pi = [0, +\infty)$)) and $p > 0$ be a nonrandom constant.

Definition 1.10.4. A stochastic process $X=(X(t))_{t\in\Pi}$ with basis $(\Omega, \mathcal{F}, (\mathcal{F}_t)_{t\in\Pi}, \mathbb{P})$ is said to be **p-th mean (forward) non-contractive** (in the strict sense) on $\mathbb{R}^d$ iff $\forall t \ge s \in \Pi$ $\forall x, y \in \mathbb{R}^d$ (adapted)

$$\mathbb{E}\left[||X_{s,x}(t) - X_{s,y}(t)||^p \middle| \mathcal{F}_s\right] \ge ||x - y||^p. \tag{1.199}$$

X is said to be a process with **p-th mean expansive perturbations** iff $\forall t > s \in \Pi \ \forall x, y \in \mathbb{R}^d (x \neq y)$ (adapted)

$$\mathbb{E}\Big[||X_{s,x}(t) - X_{s,y}(t)||^p \Big| \mathscr{F}_s\Big] > ||x - y||^p. \qquad (1.200)$$

If $p = 2$ then we speak of **mean square non-contractivity** and **mean square expansivity**, respectively.

For non-contractive processes, perturbations in the initial data may have significant impact on its longterm dynamic behaviour. Adapted perturbations of expansive processes lead to chaotic, sensitive dynamic behaviour along the entire time-scale Π. These concepts are important for the control of error propagation through numerical methods. They are meaningful to test numerical methods while applying to SDEs with non-contractive coefficient systems.

Let $p > 0$ be a nonrandom constant.

Definition 1.10.5. A coefficient system (a, b^j) of SDEs (1.2) and its SDE are said to be **strictly uniformly p-th mean non-decreasing** on $\mathbb{R}^d$ iff $\exists K_{UC} \geq 0 \in \mathbb{R} \ \forall t \in \mathbb{R} \ \forall x, y \in \mathbb{R}^d$

$$< a(t,x) - a(t,y), x - y >_d + \frac{1}{2} \sum_{j=1}^{m} ||b^j(t,x) - b^j(t,y)||^2$$
$$+ \frac{p-2}{2} \sum_{j=1}^{m} \frac{< b^j(t,x) - b^j(t,y), x - y >_d^2}{||x - y||^2}$$
$$\geq K_{UC} \ ||x - y||^2. \qquad (1.201)$$

If $K_{UC} > 0$ in (1.201) then the coefficient system (a, b^j) is said to be **p-th mean expansive** and its SDE has p-th mean expansive perturbations. Moreover, if $p = 2$ then we speak of **mean square non-decreasing** and **mean square expansive** perturbations and systems, respectively.

Lemma 1.10.2. *Assume that X satisfies SDE (1.2) with p-th mean non-decreasing coefficient system (a, b^j).*
Then X has p-th mean non-decreasing perturbations for all $p \geq 2$. If additionally $K_{UC} > 0$ in (1.201) then X possesses p-th mean expansive perturbations.

This lemma can be proved by Dynkin's formula (averaged Itô formula). Let us discuss the possible "worst case effects" on perturbations of numerical methods under condition (1.201) with $p = 2$. It turns out that the drift-implicit forward Euler methods are mean square non-contractive under this condition and may have even mean square expansive perturbations.

Theorem 1.10.4. *Assume that:*

(i) $\theta_n = 0$.
(ii) $0 < \inf_{n \in I\!N} h_n \leq \sup_{n \in I\!N} h_n < +\infty$, *all* h_n *nonrandom (i.e. only admissible step sizes).*
(iii) $\exists K_a \forall x, y \in R^d \forall t \geq 0 :< a(t,x) - a(t,y), x - y >\geq K_a ||x - y||^2$.
(iv) $\exists K_b \forall x, y \in R^d \forall t \geq 0 : \sum_{j=1}^{m} ||b^j(t,x) - b^j(t,y)||^2 \geq K_b ||x - y||^2$.

Then, the drift-implicit (forward) Euler methods (1.1) with scalar implicitness $\theta_n = 0$ *and vanishing* $c^j = 0$ *have mean square non-contractive perturbations when applied to SDEs (1.2) with coefficients* (a, b^j) *satisfying* $2K_a + K_b \geq 0$. *If additionally* $2K_a + K_b > 0$ *then they are mean square expansive.*

Proof. Consider the scheme (1.1) for the drift-implicit Theta methods with nonrandom scalar implicitness $(\Theta_n) = \theta_n I$ and separate implicit from explicit part such that

$$X_{n+1} - \theta_n h_n a(t_{n+1}, X_{n+1}) = X_n + (1 - \theta_n) h_n a(t_n, X_n) + \sum_{j=1}^{m} b^j(t_n, X_n) \Delta W_n^j. \quad (1.202)$$

Recall that X and Y denote the values of the same scheme (1.1) started at values X_0 and Y_0, respectively. Now, take the square of Euclidean norms on both sides. By taking the expectation on both sides we arrive at

$$\begin{aligned}
&\mathbb{E}||X_{n+1} - Y_{n+1}||^2 - 2\theta_n h_n \mathbb{E} < X_{n+1} - Y_{n+1}, a(t_{n+1}, X_{n+1}) - a(t_{n+1}, Y_{n+1}) > \\
&\quad + \theta_n^2 h_n^2 \mathbb{E}||a(t_{n+1}, X_{n+1}) - a(t_{n+1}, Y_{n+1})||^2 \\
&\quad = \mathbb{E}||X_n - Y_n||^2 + 2(1 - \theta_n) h_n \mathbb{E} < X_n - Y_n, a(t_n, X_n) - a(t_n, Y_n) > \\
&\quad + (1-\theta_n)^2 h_n^2 \mathbb{E}||a(t_n, X_n) - a(t_n, Y_n)||^2 + h_n \sum_{j=1}^{m} \mathbb{E}||b^j(t_n, X_n) - b^j(t_n, Y_n)||^2.
\end{aligned}$$

Under the assumption (iii) and $\theta_n \leq 1$ we have

$$\begin{aligned}
&2(1 - \theta_n) h_n < a(t,x) - a(t,y), x - y > + h_n \sum_{j=1}^{m} ||b^j(t,x) - b^j(t,y)||^2 \\
&\quad \geq [2(1 - \theta_n) K_a + K_b] h_n ||x - y||^2
\end{aligned}$$

for all $x, y \in \mathbb{R}^d$ and $t \geq 0$. Consequently, under (ii)–(iv), $\theta_n \leq 1$ and $2K_a + K_b \geq 0$, we may estimate

$$(1-2\theta_n h_n K_a)\mathbb{E}||X_{n+1}-Y_{n+1}||^2+\theta_n^2 h_n^2\mathbb{E}||a(t_{n+1},X_{n+1})-a(t_{n+1},Y_{n+1})||^2$$
$$\geq [1+(2(1-\theta_n)K_a+K_b)h_n]\mathbb{E}||X_n-Y_n||^2+(1-\theta_n)^2 h_n^2\mathbb{E}||a(t_n,X_n)-a(t_n,Y_n)||^2.$$

for all $n \in \mathbb{N}$. Now, set $\theta_n = 0$. This leads to the estimate

$$\mathbb{E}||X_{n+1}-Y_{n+1}||^2 \geq [1+(2K_a+K_b)h_n]\mathbb{E}||X_n-Y_n||^2 \geq \mathbb{E}||X_n-Y_n||^2.$$

Therefore, the forward Euler methods have mean square non-contractive perturbations under the condition $2K_a + K_b \geq 0$. After returning to the latter inequality above, one clearly recognizes that, if additionally $2K_a + K_b > 0$, then the perturbations are mean square expansive. $\diamond$

1.10.4 Mean Square BN- and B-stability of Backward Euler Methods

It is natural to ask for transferring the deterministic concept of B-stability to the stochastic case. This can be done in the p-th mean moment sense fairly straight–forward, and it has been studied by S. [120] at first.

Definition 1.10.6. A numerical sequence $Z = (Z_n)_{n\in\mathbb{N}}$ (method, scheme, approximation, etc.) is called **p-th mean B-stable** iff it is p-th mean contractive for all autonomous SDEs (1.2) with p-th mean monotone coefficient systems (a, b^j) and for all admissible step sizes. It is said to be **p-th mean BN-stable** iff it is p-th mean contractive for all non-autonomous SDEs (1.2) with p-th mean monotone coefficient systems (a, b^j) for all admissible step sizes. If $p = 2$ then we also speak of **mean square B-** and **BN-stability**.

Indeed, the drift-implicit backward Euler methods are appropriate to control the growth of its perturbations as long as the underlying SDE does. This fact is documented by the mean square B-stability of these methods in the following Theorem.

Theorem 1.10.5 (Mean Square BN-, B-Stability of Backward Euler Methods). *Assume that*

$$\forall n \in \mathbb{N} : \Theta_n = \theta_n I, \ \forall j = 1,2,\ldots,m : c^j \equiv 0.$$

Then, the drift-implicit backward Euler method (1.1) applied to Itô SDEs (1.2) with scalar implicitness parameters $\theta_n = 1$ and nonrandom step sizes h_n is mean square BN-stable and B-stable.

Proof. Combine the main assertions of Lemma 1.10.1 and Theorem 1.10.3 $\diamond$

1.11 On a First Definition of Dynamic Consistency in Stochastics

The previous course of sections presenting several concepts is unified in the following concept of dynamic consistency.

1.11.1 The Definition of Dynamic Consistency

Definition 1.11.1. A class of numerical methods Y discretizing SDEs (1.2) is said to be **dynamically consistent** iff the following properties hold for their representatives:

(1) Y is p-th mean consistent.
(2) Y is p-th mean stable for all nonrandom step sizes h_n whenever the underlying SDE has a p-th mean stable fixed point.
(3) Y is p-th mean contractive for all nonrandom step sizes h_n whenever the underlying SDE is p-th mean contractive.
(4) The limit $\lim_{n\to+\infty} Y_n = \lim_{t\to+\infty} X(t)$ whenever the latter limit exist (in the sense of moments or in the sense of probability law).
(5) Y has positive outcomes for all nonrandom step sizes h_n whenever the underlying SDE has positive solutions (a.s.).
(6) Y is p-th mean convergent to the unique solution of underlying SDE with Lipschitz-continuous coefficients.
(7) Y is an energy-exact numerical method for all nonrandom step sizes h_n and all adapted initial data.
(8) Y is exact norm-monotone for all nonrandom step sizes h_n and all adapted initial data.
(9) Y is exact contraction-monotone for all nonrandom step sizes h_n and all adapted initial data.
(10) Y does not admit spurious solutions of any kind.

Remark 1.11.1. This list of requirements for dynamic consistency can be easily extended, hence it can be considered as incomplete at this stage. These requirements act like those requirements in statistical hypothesis testing which leads to an acceptance or non-acceptance of test objects (here numerical methods). These requirements should be understood as a sort of minimum fair requirements for "qualitative goodness of numerical approximations".

The concept of dynamic consistency leads to the problem of construction and verification of nonstandard numerical methods (as initiated in Mickens [92] in deterministic analysis).

1.11.2 Class of Balanced Theta Methods is Dynamically Consistent

As we could recognize from previous sections, the class of balanced (improved) Theta methods provides us dynamically consistent numerical approximations of SDEs. In fact, it is rich enough to replicate qualitatively the dynamic behaviour of underlying SDEs. We gain control on stability (cf. Sect. 1.3), positivity (cf. Sect. 1.5), boundedness (cf. Sect. 1.6), oscillations (cf. Sect. 1.7), contractivity (cf. Sect. 1.10), consistency (cf. Sect. 1.2) and convergence (cf. Sect. 1.4) through their representatives. The balanced terms c^j with $j \geq 1$ are needed to control the pathwise behaviour of Theta methods (1.1) (e.g. for almost sure positivity, boundedness and absence of oscillations). The parameters Θ or balanced terms c^0 in them are used to control the moment behaviour of Theta methods (1.1) (e.g. for moment contractivity, moment stability, moment boundedness, etc.). In the previous sections we have reported numerous facts supporting the **preference of midpoint-type numerical methods** (or trapezoidal methods) since they avoid spurious solutions, are simplectic, exact norm-monotone, exact contraction-monotone and a stochastically improved version of them are energy- and asymptotically exact. They can practically be implemented by predictor-corrector procedures, linear- or partial-implicit versions to avoid the time-consuming resolution of implicit algebraic equations at each integration step. Dynamically consistent methods of higher order of convergence than midpoint-type methods are not known so far. This is a challenging subject for further research.

1.11.3 Remarks and Practical Guideline on Optimal Choice of θ, c^j

The optimal choice of implicitness parameters θ and weights c^j is a fairly complex problem. Its choice depends on the qualitative properties which one is aiming at to be guaranteed by numerical approximations. Here a shortlist of main recommendations and conclusion on its choice is given, based on the prior observations in previous sections:

1. $\theta = 0.5$ is optimal for linear systems (see S. [121, 122]) with respect to stability (cf. Sect. 1.3), contractivity (cf. Sect. 1.10), monotonicity, energy-exactness (cf. Sect. 1.8), etc.
2. Equivalent choices with appropriate c^0 are possible (see S. [120]).
3. Positive semi-definite matrices c^j are preferrable since they imply no additional step size restrictions (guaranteeing the inverses of "stabilizers" M (matrices) in local one step representation).

4. $\theta > 0$ and positive negative-definite part $c^0 = \frac{1}{2}[-\frac{\partial a(t,x)}{\partial x}]_+$ of Jacobian matrix $\frac{\partial a(t,x)}{\partial x}$ are recommendable choices to achieve "optimal numerical stabilization" based on standard linearization techniques.
5. $c^j = 0 (j = 1, 2, \ldots, m)$ suffices to control p-th moments (stability, boundedness of moments).
6. $c^j \neq 0 (j = 1, 2, \ldots, m)$ is needed to control pathwise behaviour such as a.s. stability, a.s. boundedness or a.s. positivity (cf. Sect. 1.5).
7. $c^j = b^j(t,x)/x$ (componentwise) is recommended to be chosen for systems with $b^j(t,x) = B^j(t,x)x$ as commonly met in population models in ecology, biology, finance, marketing.

Of course, this represents an incomplete list of observations and further studies are needed. As commonly known, optimal choices must depend on the specific structure of drift and diffusion coefficients, and, above all, on the goal one is aiming at by the approximation. This clearly depends on the knowledge on qualitative behaviour of underlying class of SDEs.

Acknowledgements We sincerely thank the anonymous reviewers for their useful comments and Mounir Zili for his continuous support.

References

1. Allen, E.: Modeling with Itô Stochastic Differential Equations. Springer, New York (2007)
2. Appleby, J.A.D., Rodkina, A., Schurz, H.: Non-positivity and oscillations of solutions of nonlinear stochastic difference equations. J. Difference Equ. Appl. **16**(7), 807–830 (2010)
3. Arnold, L.: Random Dynamical Systems. Springer, Berlin (1998)
4. Arous, G.B.: Flots et séries de taylor stochastiques. Probab. Theory Rel. **81**(1), 29–77 (1989)
5. Artemiev, S.S., Averina, T.A.: Numerical Analysis of Systems of Ordinary and Stochastic Differential Equations. VSP, Utrecht (1997)
6. Bally, V., Protter, P., Talay, D.: The law of the Euler scheme for stochastic differential equations, Z. Angew. Math. Mech. **76**(3), 207–210 (1996)
7. Bally, V., Talay, D.: The Euler scheme for stochastic differential equations: Error analysis with Malliavin calculus. Math. Comput. Simul. **38**(1–3), 35–41 (1995)
8. Bally, V., Talay, D.: The law of the Euler scheme for stochastic differential equations I. Convergence rate of the distribution function. Probab. Theory Relat. Fields **104**, 43–60 (1996)
9. Bally, V., Talay, D.: The law of the Euler scheme for stochastic differential equations. II: Convergence rate of the density. Monte Carlo Methods Appl. **2**(2), 93–128 (1996)
10. Bashkirtseva, I., Ryashko, L., Schurz, H.: Analysis of noise-induced transitions for Hopf system with additive and multiplicative random disturbances. Chaos Solitons Fractals **39**(1), 72–82 (2009)
11. Bernard, P., Fleury, G.: Convergence of numerical schemes for stochastic differential equations. Monte Carlo Methods Appl. **7**(1–2), 35–44 (2001)
12. Bihari, T.: A generalization of a lemma of Bellman and its application to uniqueness problems of differential equations. Acta Math. Acad. Sci. Hung. **7**, 81–94 (1956)
13. Biscay, R., Jimenez, J.C., Riera, J.J., Valdes, P.A.: Local linearization method for the numerical solution of stochastic differential equations. Ann. Inst. Statist. Math. **48**(4), 631–644 (1996)
14. Bouleau, N., Lépingle, D.: Numerical Methods for Stochastic Processes. Wiley, New York (1993)

15. Bryden, A., Higham, D.J.: On the boundedness of asymptotic stability regions for the stochastic theta method. BIT **43**(1), 1–6 (2003)
16. Burrage, K., Platen, E.: Runge-Kutta methods for stochastic differential equations. Ann. Numer. Math. **1**, 63–78 (1994)
17. Burrage, K., Burrage, P.M., Mitsui, T.: Numerical solutions of stochastic differential equations – implementation and stability issues. J. Comput. Appl. Math. **125**(1–2), 171–182 (2000)
18. Burrage, K., Tian, T.: A note on the stability properties of the Euler methods for solving stochastic differential equations. New Zealand J. Math. **29**(2), 115–127 (2000)
19. Burrage, K., Tian, T.: The composite Euler method for stiff stochastic differential equations. J. Comput. Appl. Math. **131**(1-2), 407–426 (2001)
20. Cambanis, S., Hu, Y.: Exact convergence rate of the Euler-Maruyama scheme with application to sampling design. Stochastics Stochastics Rep. **59**, 211–240 (1996)
21. Chang, C.C.: Numerical solution of stochastic differential equations with constant diffusion coefficients. Math. Comp. **49**(180), 523–542 (1987)
22. Clark, J.M.C., Cameron, R.J.: The maximum rate of convergence of discrete approximations for stochastic differential equations. In: Stochastic Differential Systems (Proc. IFIP-WG 7/1 Working Conf., Vilnius, 1978), pp. 162–171, Lecture Notes in Control and Information Sci. **25**, Springer, Berlin-New York (1980)
23. Dahlquist, G.: A special stability problem for linear multistep methods. BIT **3**, 27–43 (1963)
24. Deelstra, G., Delbaen, F.: Convergence of discretized stochastic (interest rate) processes with stochastic drift term. Appl. Stochastic Models Data Anal. **14**(1), 77–84 (1998)
25. Del Moral, P.: A uniform convergence theorem for the numerical solving of the nonlinear filtering problem. J. Appl. Probab. **35**(4), 873–884 (1998)
26. Ermakov, S.M., Mikhajlov, G.A.: Statistical Simulation (in Russian). Nauka, Moscow (1982)
27. Faure, O.: Simulation de mouvement brownien et des diffusions. Thèse ENPC, Paris (1992)
28. Gaines, J.G.: The algebra of iterated stochastic integrals. Stochastics Stoch. Reports **49**, 169–179 (1994)
29. Gaines, J.G.: A basis for iterated stochastic integrals. Math. Comput. Simulation **38**(1–3), 7–11 (1995)
30. Gaines, J.G., Lyons, T.J.: Random generation of stochastic area integrals. SIAM J. Appl. Math. **54**(4), 1132–1146 (1994)
31. Gaines, J.G., Lyons, T.J.: Variable step size control in the numerical solution of stochastic differential equations. SIAM J. Appl. Math. **57**(5), 1455–1484 (1997)
32. Gard, T.C.: Introduction to Stochastic Differential Equations. Marcel Dekker, Basel (1988)
33. Gikhman, I.I., Skorochod, A.V.: Stochastische Differentialgleichungen. Akademie-Verlag, Berlin (1971)
34. Girsanov, I.V.: An example of non-uniqueness of the solution of the stochastic equation of K. Itô. Theor. Probab. Appl. **7**, 325–331 (1962)
35. Golec, J.: Averaging Euler-type difference schemes. Stochastic Anal. Appl. **15**(5), 751–758 (1997)
36. Golec, J., Ladde, G.S.: Euler–type approximation for systems of stochastic differential equations. J. Appl. Math. Simul. **28**, 357–385 (1989)
37. Golec, J., Sathananthan, S.: Sample path approximation for stochastic integro-differential equations. Stoch. Anal. Appl. **17**(4), 579–588 (1999)
38. Golec, J., Sathananthan, S.: Strong approximations of stochastic integro-differential equations, Dyn. Contin. Discrete Impuls. Syst. Ser. B Appl. Algorithms **8**(1), 139–151 (2001)
39. Guo, S.J.: On the mollifier approximation for solutions of stochastic differential equations. J. Math. Kyoto Univ. **22**, 243–254 (1982)
40. Guo, S.J.: Approximation theorems based on random partitions for stochastic differential equations and applications. Chinese Ann. Math. **5**, 169–183 (1984)
41. Gyöngy, I.: On approximation of Itô stochastic equations. Math. SSR Sbornik. **70**, 165–173 (1991)
42. Gyöngy, I.: A note on Euler's approximations. Potential Anal. **8**(3), 205–216 (1998)

43. Gyöngy, I., Nualart, D.: Implicit scheme for quasi-linear parabolic partial differential equations perturbed by space-time white noise. Stoch. Processes Appl. **58**(1), 57–72 (1995)
44. Gyöngy, I., Nualart, D.: Implicit scheme for stochastic partial differential equations driven by space-time white noise. Potential Anal. **7**, 725–757 (1997)
45. Hernandez, D.B., Spigler, R.: A–stability of implicit Runge–Kutta methods for systems with additive noise. BIT **32**, 620–633 (1992)
46. Hernandez, D.B., Spigler, R.: Convergence and stability of implicit Runge–Kutta methods for systems with multiplicative noise. BIT **33**, 654–669 (1993)
47. Higham, D.J.: A-stability and stochastic mean-square stability. BIT **40**(2), 404–409 (2000)
48. Higham, D.J.: Mean-square and asymptotic stability of the stochastic Theta method. SIAM J. Numer. Anal. **38**(3), 753–769 (2000)
49. Higham, D.J., Mao, X., Stuart, A.M.: Strong convergence of Euler-type methods for nonlinear stochastic differential equations. SIAM J. Numer. Anal. **40**(3), 1041–1063 (2002)
50. Hofmann, N., Müller-Gronbach, T., Ritter, K.: Optimal approximation of stochastic differential equations by adaptive step-size control. Math. Comp. **69**(231), 1017–1034 (2000)
51. Hofmann, N., Müller-Gronbach, T., Ritter, K.: On the global error of Itô-Taylor schemes for strong approximation of scalar stochastic differential equations. J. Complexity **20**(5), 732–752 (2004)
52. Hong, J., Scherer, R., Wang, L.: Midpoint rule for a linear stochastic oscillator with additive noise. Neural, Parallel Sci. Comput. **14**(1), 1–12 (2006)
53. Horváth-Bokor, R.: Convergence and stability properties for numerical approximations of stochastic ordinary differential equations. Ph.D. thesis, University of Zagreb, Zagreb (2000)
54. Hu, Y.Z.: Series de Taylor stochastique et formule de Campbell-Hausdorff d'apres Ben Arous. In: Séminaire de Probabilites XXVI; Lecture Notes in Math. **1526**, pp. 579–586, Springer, New York (1992)
55. Hu, Y.: Semi-implicit Euler-Maruyama scheme for stiff stochastic equations, In: Koerezlioglu (ed.) Stochastic Analysis and Related Topics V: The Silvri Workshop, Progr. Probab. **38**, pp. 183–202, Birkhauser, Boston (1996)
56. Itô, K.: Stochastic integral, Proc. Imp. Acad. Tokyo **20**, 519–524 (1944)
57. Itô, K.: On a formula concerning stochastic differential equations. Nagoya Math. J. **3**, 55–65 (1951)
58. Jacod, J., Protter, P.: Asymptotic error distributions for the Euler method for stochastic differential equations. Ann. Probab. **26**(1), 267–307 (1998)
59. Jacod, J., Jakubowski, A., Mémin, J.: About asymptotic errors in discretization of processes. PMA **661**, Preprint, University of Paris 6 & 7, Paris (2001)
60. Kahl, C., Schurz, H.: Balanced implicit Milstein methods for SDEs. Monte Carlo Meth. Appl. **12**(2), 143–170 (2006)
61. Kanagawa, S.: On the rate of convergence for Maruyama's approximation solutions of stochastic differential equations. Yokohama Math. J. **36**(1), 79–86 (1988)
62. Kanagawa, S.: The rate of convergence for approximate solutions of stochastic differential equations. Tokyo J. Math. **12**, 33–48 (1989)
63. Kanagawa, S.: Error estimations for the Euler-Maruyama approximate solutions of stochastic differential equations. Monte Carlo Meth. Appl. **1**(3), 165–171 (1995)
64. Kanagawa, S.: Convergence rates for the Euler-Maruyama type approximate solutions of stochastic differential equations. In: Probability Theory and Mathematical Statistics (Tokyo, 1995), pp. 183–192, World Sci. Publishing, River Edge, NJ (1996)
65. Kanagawa, S.: Confidence intervals of discretized Euler-Maruyama approximate solutions of SDE's. In: Proceedings of the Second World Congress of Nonlinear Analysts, Part 7 (Athens, 1996), Nonlinear Anal. **30**(7), 4101–4104 (1997)
66. Kanagawa, S., Ogawa, S.: Numerical solution of stochastic differential equations and their applications. Sugaku Expositions **18**(1), 75–99 (2005)
67. Kannan, D., De Ting Wu: A numerical study of the additive functionals of solutions of stochastic differential equations. Dynam. Syst. Appl. **2**(3), 291–310 (1993)

68. Kannan, D., Zhang, Q.: Nonlinear filtering of an interactive multiple model with small observation noise: Numerical methods. Stoch. Anal. Appl. **16**(4), 631–659 (1998)
69. Kantorovič, L.V.: Functional analysis and applied mathematics (in Russian). Uspehi Matem. Nauk (N.S.) **3**(6 (28)), 89–185 (1948) and Vestnik Leningrad. Univ. **3**(6), 3–18 (1948)
70. Karatzas, I., Shreve, S.: Brownian motion and stochastic calculus. Springer, New York (1988)
71. Khas'minskiǐ, R.Z.: Stochastic Stability of Differential equations. Sijthoff & Noordhoff, Alphen aan den Rijn (1980)
72. Kloeden, P.E., Platen, E.: Numerical Solution of Stochastic Differential Equations. Springer, Berlin (1992)
73. Kloeden, P.E., Platen, E., Schurz, H.: Numerical Solution of Stochastic Differential Equations through Computer Experiments. Universitext, Springer, New York (1994)
74. Kohatsu-Higa, A.: High order Itô-Taylor approximations to heat kernels. J. Math. Kyoto Univ. **37**, 129–150 (1997)
75. Kohatsu-Higa, A.: Weak approximations. A Malliavin calculus approach. Math. Comput. **70**(233), 135–172 (2001)
76. Kohatsu-Higa, A., Ogawa, S.: Weak rate of convergence for an Euler scheme of nonlinear SDE's. Monte Carlo Meth. Appl. **3**(4), 327–345 (1997)
77. Kohatsu-Higa, A., Protter, P.P.: The Euler scheme for SDE's driven by semimartingales, In: Kunita, H. and Kuo, H.H. (eds.) Stochastic Analysis on Infinite-dimensional Spaces (Baton Rouge, LA, 1994), pp. 141–151, Pitman Res. Notes Math. Ser. **310**, Longman Sci. Tech., Harlow (1994)
78. Komori, Y., Mitsui, T.: Stable ROW-type weak scheme for stochastic differential equations. Monte Carlo Meth. Appl. **1**(4), 279–300 (1995)
79. Komori, Y., Saito, Y., Mitsui, T.: Some issues in discrete approximate solution for stochastic differential equations. Sūrikaisekikenkyūsho Kōkyūroku **850**, 1–13 (1993)
80. Korzeniowski, A.: Stability of stochastic differential equations under discretization. Stoch. Anal. Appl. **26**(6), 1267–1273 (2008)
81. Krylov, N.: Introduction to the Theory of Diffusion Processes. AMS, Providence (1995)
82. Küchler, U., Platen, E.: Strong discrete time approximation of stochastic differential equations with time delay. Math. Comput. Simulation **54**(1–3), 189–205 (2000)
83. Kurtz, T., Protter, P.: Wong–Zakai corrections, random evolutions and numerical schemes for SDE's, in Meyer-Wolf, E.M.E. and Schwartz, A. (eds.) Stochastic Analysis: Liber Amicorum for Moshe Zakai, pp. 331–346, Academic Press, Boston (1991)
84. Kushner, H.J.: On the weak convergence of interpolated Markov chains to a diffusion. Ann. Probab. **2**, 40–50 (1974)
85. Kushner, H.J., Dupuis, P.G.: Numerical Methods for Stochastic Control Problems in Continuous Time. Springer, New York (1992)
86. Kuznetzov, D.F.: Stochastic Differential Equations: Theory and Practice of Numerical Solutions (in Russian). Polytechnical State University, St. Petersburg (2010)
87. Ladde, G.S., Lawrence, B.A.: Stability and convergence of stochastic approximation procedures under Markovian structural perturbations. Dynam. Syst. Appl. **10**(2), 145–175 (2001)
88. Ladde, G.S., Sathananthan, S.: Error estimates and stability of Itô-type systems of nonlinear stochastic integro-differential equations. Appl. Anal. **43**(3–4), 163–189 (1992)
89. Liske, H.: On the distribution of some functional of the Wiener process (in Russian). Theory of Random Processes **10**, 50–54 (1982)
90. Maruyama, G.: Continuous Markov processes and stochastic equations. Rend. Circ. Mat. Palermo **4**, 48–90 (1955)
91. Medhin, N.G., Sambandham, M.: A numerical study of random integro-differential equations. Dynam. Syst. Appl. **1**(2), 171–185 (1992)
92. Mickens, R.E.: Nonstandard Finite Difference Models of Differential Equations. World Scientific, Singapore (1994)
93. Milstein, G.N.: Weak approximation of solutions of systems of stochastic differential equations. Theor. Probab. Applic. **30**, 750–766 (1985)

94. Milstein, G.N.: A theorem on the order of convergence of mean square approximations of solutions of systems of stochastic differential equations. Theor. Probab. Applic. **32**, 738–741 (1988)
95. Milstein, G.N.: Numerical Integration of Stochastic Differential Equations. Kluwer, Dordrecht (1995)
96. Milstein, G.N., Platen, E., Schurz, H.: Balanced implicit methods for stiff stochastic systems. SIAM J. Numer. Anal. **35**(3), 1010–1019 (1998)
97. Mitsui, T.: Computational issues in numerical solution of stochastic differential equations. In: Kawarada, H. et al. (eds.) Advances in Numerical Mathematics. Proceedings of the 4th Japan-China Joint Seminar on Numerical Mathematics, Chiba Univ., Chiba, Japan, August 24–28, 1998, Gakkotosho, Tokyo, GAKUTO Int. Ser., Math. Sci. Appl. **12**, 75–84 (1999)
98. Mitsui, T., Saito, Y.: Stability analysis of numerical schemes for stochastic differential equations. SIAM J. Numer. Anal. **33**(6), 2254–2267 (1996)
99. Moro, E., Schurz, H.: Boundary preserving semi-analytical numerical algorithms for stochastic differential equations. SIAM J. Sci. Comput. **29**(4), 1525–1549 (2007)
100. Müller-Gronbach, T., Ritter, K.: Minimal errors for strong and weak approximation of stochastic differential equations. In: Monte Carlo and quasi-Monte Carlo Methods 2006, pp. 53–82, Springer, Berlin (2008)
101. Newton, N.J.: Asymptotically efficient Runge–Kutta methods for a class of Itô and Stratonovich equations. SIAM J. Appl. Math. **51**(2), 542–567 (1991)
102. Øksendal, B.: Stochastic Differential Equations. Springer, New York (1985)
103. Ogawa, S.: Monte Carlo simulation of nonlinear diffusion processes. Jpn. J. Ind. Appl. Math. **9**(1), 22–33 (1992)
104. Ogawa, S.: Some problems in the simulation of nonlinear diffusion processes. Math. Comput. Simul. **38**, 217–223 (1995)
105. Ogawa, S.: On a deterministic approach to the numerical solution of the SDE. Math. Comput. Simul. **55**(1–3), 209–214 (2001)
106. Pardoux, E., Talay, D.: Discretization and simulation of stochastic differential equations. Acta Appl. Math. **3**(1), 23–47 (1985)
107. Petersen, W.P.: Some experiments on numerical simulations of stochastic differential equations and a new algorithm. J. Comput. Phys. **113**(1), 75–81 (1994) (see also Petersen, W.P.: Stability and accuracy of simulations for stochastic differential equations. IPS Report **90-02**, ETH, Zürich (1990))
108. Protter, P.: Stochastic Integration and Differential Equations. Springer, New York (1990)
109. Protter, P., Talay, D.: The Euler scheme for Levy driven stochastic differential equations. Ann. Probab. **25**(1), 393–423 (1997)
110. Rodkina, A., Schurz, H.: A theorem on global asymptotic stability of solutions to nonlinear stochastic difference equations with Volterra type noises. Stab. Control Theory Appl. (SACTA) **6**(1), 23–34 (2004)
111. Rodkina, A., Schurz, H.: Global asymptotic stability of solutions to cubic stochastic difference equations. Adv. Diff. Equat. **1**(3), 249–260 (2004)
112. Rodkina, A., Schurz, H.: Almost sure asymptotic stability of drift-implicit theta-methods for bilinear ordinary stochastic differential equations in $\mathbb{R}^1$. J. Comput. Appl. Math. **180**(1), 13–31 (2005)
113. Rodkina, A., Schurz, H.: On global asymptotic stability of solutions to some in-arithmetic-mean-sense drift-monotone stochastic difference equations. Int. J. Numer. Anal. Model. **2**(3), 355–366 (2005)
114. Rodkina, A., Schurz, H., Shaikhet, L.: Almost sure stability of some stochastic dynamical systems with memory. Discrete Contin. Dyn. Syst. Ser. A. **21**(2), 571–593 (2008)
115. Römisch, W., Wakolbinger, A.: On the convergence rates of approximate solutions of stochastic equations. Lect. Notes Contr. Inform. Sci. **96**, 204–212 (1987)
116. Ryashko, L.B., Schurz, H.: Mean square stability analysis of some linear stochastic systems. Dynam. Syst. Appl. **6**(2), 165–190 (1997)

117. Saito, Y., Mitsui, T.: T-stability of numerical scheme for stochastic differential equations. In: "Contributions in Numerical Mathematics", pp. 333–344, World Sci. Ser. Appl. Anal. **2**, World Sci. Publishing, River Edge, NJ (1993)
118. Schurz, H.: Asymptotical mean square stability of an equilibrium point of some linear numerical solutions with multiplicative noise, Stochastic Anal. Appl. **14**(3), 313–354 (1996)
119. Schurz, H.: Numerical regularization for SDEs: construction of nonnegative solutions. Dynam. Syst. Appl. **5**(1), 323–352 (1996)
120. Schurz, H.: Stability, Stationarity, and Boundedness of Some Implicit Numerical Methods for Stochastic Differential Equations and Applications. Logos-Verlag, Berlin (1997)
121. Schurz, H.: Preservation of asymptotical laws through Euler methods for Ornstein-Uhlenbeck process. Stoch. Anal. Appl. **17**(3), 463–486 (1999)
122. Schurz, H.: The invariance of asymptotic laws of linear stochastic systems under discretization. Z. Angew. Math. Mech. **79**(6), 375–382 (1999)
123. Schurz, H.: Moment attractivity, stability and contractivity exponents of stochastic dynamical systems. Discrete Cont. Dynam. Syst. **7**(3), 487–515 (2001)
124. Schurz, H.: On moment-dissipative stochastic dynamical systems. Dynam. Syst. Appl. **10**(1), 11–44 (2001)
125. Schurz, H.: Numerical Analysis of SDEs without tears, In: Kannan, D., Lakshmikantham, V.: (eds.) Handbook of Stochastic Analysis and Applications, pp. 237–359, Marcel Dekker, Basel (2002) (see also Schurz, H.: Applications of numerical methods and its analysis for systems of stochastic differential equations. Bull. Karela Math. Soc. **4**(1), 1–85 (2007))
126. Schurz, H.: Remarks on Taylor series expansions and conditional expectations for Stratonovich SDEs with complete V-commutativity. Stoch. Anal. Appl. **21**(4), 865–894 (2003)
127. Schurz, H.: Stability of numerical methods for ordinary SDEs along Lyapunov-type and other functions with variable step sizes. Electron. Trans. Numer. Anal. **20**, 27–49 (2005)
128. Schurz, H.: Convergence and stability of balanced implicit methods for SDEs with variable step sizes. Int. J. Numer. Anal. Model. **2**(2), 197–220 (2005)
129. Schurz, H.: An axiomatic approach to numerical approximations of stochastic processes. Int. J. Numer. Anal. Model. **3**(4), 459–480 (2006)
130. Schurz, H.: Modeling, analysis and discretization of stochastic logistic equations. Int. J. Numer. Anal. Model. **4**(2), 180–199 (2007)
131. Schurz, H.: New stochastic integrals, oscillation theorems and energy identities. Preprint m-07-005, Department of Mathematics, SIU, Carbondale, 2007 (Revised version published in Commun. Appl. Anal. **13**(2), 181–194 (2009))
132. Schurz, H.: Nonlinear stochastic wave equations in $\mathbb{R}^1$ with power-law nonlinearity and additive space-time noise. Contemp. Math. **440**, 223–242 (2007)
133. Schurz, H.: How rates of L^p-convergence carry over to numerical approximations of some convex, non-smooth functionals of SDEs. Int. J. Numer. Anal. Model. **5**(1), 55–72 (2008)
134. Shiryaev, A.N.: Probability (2nd Ed.). Springer, New York (1996)
135. Strmmen–Melb, A.H., Higham, D.J.: Numerical simulation of a linear stochastic oscillator with additive noise, Appl. Numer. Math. **51**(1), 89–99 (2004)
136. Stuart, A.M., Peplow, A.T.: The dynamics of the theta method. SIAM J. Sci. Statist. Comput. **12**(6), 1351–1372 (1991)
137. Sussmann, H.J.: Product expansions of exponential Lie series and the discretization of stochastic differential equations. In: Fleming, W., Lions, J.: (eds.) Stochastic Differential Systems, Stochastic Control Theory, and Applications (Minneapolis, Minn., 1986), IMA Vol. Math. Appl. **10**, 563–582. Springer, New York 563–582 (1988)
138. Talay, D.: Analyse Numérique des Equations Différentielles Stochastiques. Thèse 3ème Cycle, Universite de Provence, Centre Saint Charles (1982)
139. Talay, D.: Efficient numerical schemes for the approximation of expectations of functionals of the solution of an SDE and applications. Springer Lect. Notes Contr. Inf. Sci. **61**, 294–313 (1984)

140. Talay, D.: Simulation of stochastic differential systems. In: "Probabilistic Methods in Applied Physics" (P. Kré and W. Wedig, Eds.), Lecture Notes in Physics **451**, pp. 54–96, Springer, Berlin (1995)
141. Talay, D., Tubaro, L.: Expansion of the global error for numerical schemes solving stochastic differential equations. Stoch. Anal. Appl. **8**, 483–509 (1990)
142. Tudor, M.: Approximation schemes for two-parameter stochastic equations. Probab. Math. Statist. **13**(2), 177–189 (1992)
143. Tudor, C., Tudor, M.: Approximation schemes for Itô-Volterra stochastic equations. Bol. Soc. Mat. Mexicana **3**(1), 73–85 (1995)
144. Wagner, W., Platen, E.: Approximation of Itô integral equations. Preprint ZIMM, Akad. Wiss. DDR, Berlin (1978)

Chapter 2
Kernel Density Estimation and Local Time

Ciprian A. Tudor

Abstract In this paper we develop an asymptotic theory for some regression models involving standard Brownian motion and standard Brownian sheet.

2.1 Introduction

The motivation of our work comes from the econometric theory. Consider a regression model of the form

$$y_i = f(x_i) + u_i, \qquad i \geq 0 \tag{2.1}$$

where $(u_i)_{i\geq 0}$ is the "error" and $(x_i)_{i\geq 0}$ is the regressor. The purpose is to estimate the function f based on the observation of the random variables y_i, $i \geq 0$. The conventional kernel estimate of $f(x)$ is

$$\hat{f}(x) = \frac{\sum_{i=0}^{n} K_h(x_i - x) y_i}{\sum_{i=0}^{n} K_h(x_i - x)}$$

where K is a nonnegative real kernel function satisfying $\int_{\mathbb{R}} K^2(y)dy = 1$ and $\int_{\mathbb{R}} yK(y)dy = 0$ and $K_h(s) = \frac{1}{h}K(\frac{s}{n})$. The bandwidth parameter $h \equiv h_n$ satisfies $h_n \to 0$ as $n \to \infty$. We will choose in our work $h_n = n^{\alpha}$ with $0 < \alpha < \frac{1}{2}$. The asymptotic behavior of the estimator $\hat{f}$ is usually related to the behavior of the sequence

$$V_n = \sum_{i=1}^{n} K_h(x_i - x)u_i.$$

C.A. Tudor (✉)
Laboratoire Paul Painlevé, Université de Lille 1, F-59655 Villeneuve d'Ascq, France
e-mail: tudor@math.univ-lille1.fr.

M. Zili and D.V. Filatova (eds.), *Stochastic Differential Equations and Processes*, Springer Proceedings in Mathematics 7, DOI 10.1007/978-3-642-22368-6_2,

The limit in distribution as $n \to \infty$ of the sequence V_n has been widely studied in the literature in various situations. We refer, among others, to [5] and [6] for the case where x_t is a recurrent Markov chain, to [12] for the case where x_t is a partial sum of a general linear process, and [13] for a more general situation. See also [9] or [10]. An important assumption in the main part of the above references is the fact that u_i is a martingale difference sequence. In our work we will consider the following situation: first the error u_i is chosen to be $u_i = W_{i+1} - W_i$ for every $i \geq 0$, where $(W_t)_{t\geq 0}$ denotes a standard Wiener process and $x_i = W_i$ for $i \geq 0$. Note that in this case, although for every i the random variables u_i and x_i are independent, there is not global independence between the regressor $(x_i)_{i\geq 0}$ and $(u_i)_{i\geq 0}$. However, this case has been already treated in previous works (see e.g. [12, 13]). See also [2] for models related with fractional Brownian motion. In this case, the sequence V_n reduces to (we will also restrict to the case $x = 0$ because the estimation part is not addressed in this paper)

$$S_n = \sum_{i=0}^{n-1} K(n^\alpha W_i)\,(W_{i+1} - W_i)\,. \tag{2.2}$$

The second case we consider concerns a two-parameter model:

$$y_{i,j} = f(x_{i,j}) + e_{i,j}, \qquad i,j \geq 0 \tag{2.3}$$

where $e_{i,j} = W^{(2)}_{i+1,j+1} - W^{(2)}_{i+1,j} - W^{(2)}_{i,j+1} + W^{(2)}_{i,j}$ are the rectangular increments of a Wiener sheet $W^{(2)}$ (see Sect. 2 for the definition of the Wiener sheet). This case seems to be new in the literature. But in this situation, because of the complexity of the stochastic calculus for two-parameter processes, we will restrict ourselves to case when the regressor $x_{i,j}$ is independent by the error $u_{i,j}$. That is, we assume that $x_{i,j} = W^{(1)}_{i,j}$ where $W^{(1)}$ is a Wiener sheet independent by $W^{(2)}$. The model (2.3) leads to the study of the sequence

$$T_n = \sum_{i,j=0}^{n-1} K\left(n^\alpha W^{(1)}_{i,j}\right)\left(W^{(2)}_{i+1,j+1} - W^{(2)}_{i+1,j} - W^{(2)}_{i,j+1} + W^{(2)}_{i,j}\right).$$

We will assume that the kernel K is the standard Gaussian kernel

$$K(x) = \frac{1}{\sqrt{2\pi}} e^{-\frac{x^2}{2}}. \tag{2.4}$$

The limits in distribution of S_n and T_n will be $c_1\beta_{L^W(1,0)}$ and $c_2\beta_{L^{W^{(1)}}(\underline{1},0)}$ respectively, where L^W and $L^{W^{(1)}}$ denote the local time of W and $W^{(1)}$ respectively, β is a Brownian motion independent by W and $W^{(1)}$ and c_1, c_2 are explicit positive constants.

2.2 The One Parameter Case

Let $(W_t)_{t\geq 0}$ be a standard Brownian motion on a standard probability space $(\Omega, \mathscr{F}, P)$ and let us consider the sequence S_n given by (2.2) with $0 < \alpha < \frac{1}{2}$ and the kernel function K given by (2.4). Denote by $\mathscr{F}_t$ the filtration generated by W. Our first step is to estimate the L^2 mean of S_n.

Lemma 2.1. *As $n \to \infty$ it holds that*

$$n^{\alpha-\frac{1}{2}}\mathbf{E}S_n^2 \to C = \frac{\sqrt{2}}{2\pi}.$$

Proof. Recall that, if Z is a standard normal random variable, and if $1 + 2c > 0$

$$\mathbf{E}\left(e^{-cZ^2}\right) = \frac{1}{\sqrt{1+2c}}. \tag{2.5}$$

Since the increments of the Brownian motion are independent and $W_{i+1} - W_i$ is independent by $\mathscr{F}_i$ for every i, it holds that (here Z denotes a standard normal random variable)

$$\begin{aligned}
\mathbf{E}S_n^2 &= \mathbf{E}\sum_{i=0}^{n-1} K^2(n^\alpha W_i)\,(W_{i+1} - W_i)^2 = \mathbf{E}\sum_{i=0}^{n-1} K^2(n^\alpha W_i) \\
&= \frac{1}{2\pi}\sum_{i=0}^{n-1} \mathbf{E}e^{-n^{2\alpha} i Z^2} = \frac{1}{2\pi}\sum_{i=0}^{n-1}\left(1 + 2n^{2\alpha} i\right)^{-\frac{1}{2}}
\end{aligned}$$

and this behaves as $\frac{\sqrt{2}}{2\pi} n^{-\alpha+\frac{1}{2}}$ when n tends to infinity. □

In the following our aim is to prove that the sequence $n^{-\frac{1}{4}+\frac{\alpha}{2}} S_n$ converges in distribution to a non-trivial limit. Note that the sequence S_n can be written as

$$\begin{aligned}
S_n &= \sum_{i=0}^{n-1} K(n^\alpha W_i)(W_{i+1} - W_i) = \sum_{i=0}^{n-1}\int_i^{i+1} K(n^\alpha W_i)dW_s \\
&= \sum_{i=0}^{n-1}\int_i^{i+1} K(n^\alpha W_{[s]})dW_s = \int_0^n K(n^\alpha W_{[s]})dW_s
\end{aligned}$$

where $[s]$ denotes the integer part of the real number s. Define, for every $t \geq 0$,

$$S_t^n = \int_0^t K(n^\alpha W_{[s]})dW_s. \tag{2.6}$$

Then for every $n \geq 1$ the process $(S_t^n)_{t\geq 0}$ is a $\mathscr{F}_t$ martingale (recall that $\mathscr{F}_t$ denotes the sigma algebra generated by the Wiener process W). The bracket of the martingale $(S_t)_{t\geq 0}$ will be given by, for every $t \geq 0$

$$\langle S^n \rangle_t = \int_0^t K^2(n^\alpha W_{[s]})ds.$$

This bracket plays a key role in order to understand the behavior of S_n. Let us first understand the limit of the sequence $\langle S^n \rangle_n$. Its asymptotic behavior is related to the local time of the Brownian motion. We recall its definition. For any $t \geq 0$ and $x \in \mathbb{R}$ we define $L^W(t,x)$ as the density of the occupation measure (see [1, 21])

$$\mu_t(A) = \int_0^t 1_A(W_s)ds, \qquad A \in \mathscr{B}(\mathbb{R}).$$

The local time $L^W(t,x)$ satisfies the occupation time formula

$$\int_0^t f(W_s)ds = \int_{\mathbb{R}} L^W(t,x) f(x)dx \tag{2.7}$$

for any measurable function f. The local time is Hölder continuous with respect to t and with respect to x. Moreover, it admits a bicontinuous version with respect to (t,x).

We will denote by p_ε the Gaussian kernel with variance $\varepsilon > 0$ given by $p_\varepsilon(x) = \frac{1}{\sqrt{2\pi\varepsilon}} e^{-\frac{x^2}{2\varepsilon}}$. Note that

$$n^{\alpha+\frac{1}{2}} K^2(n^{\alpha+\frac{1}{2}} W_t) = \frac{1}{2\sqrt{\pi}} p_{\frac{1}{2}n^{-\alpha-\frac{1}{2}}}(W_t)$$

and by the scaling property of the Brownian motion

$$\begin{aligned} n^{-\frac{1}{2}+\alpha}\langle S^n \rangle_n &= n^{-\frac{1}{2}+\alpha} \sum_{i=0}^{n-1} K^2(n^\alpha W_i) \\ &=_{(d)} n^{-\frac{1}{2}+\alpha} \sum_{i=0}^{n-1} K^2\left(n^{\alpha+\frac{1}{2}} W_{\frac{i}{n}}\right) = \frac{1}{n}\sum_{i=0}^{n-1} \frac{1}{2\sqrt{\pi}} p_{\frac{1}{2}n^{-2\alpha-1}}\left(W_{\frac{i}{n}}\right). \end{aligned}$$

where $=_{(d)}$ means the equality in distribution. A key point of our paper is the following result which gives the convergence of the "bracket."

Lemma 2.2. *The sequence* $\frac{1}{n}\sum_{i=0}^{n-1} p_{\frac{1}{2}n^{-\alpha-\frac{1}{2}}}(W_{\frac{i}{n}})$ *converges in* $L^2(\Omega)$, *as* $n \to \infty$ *to* $L^W(1,0)$.

Proof. Let us recall that $\int_0^1 p_\varepsilon(W_s)ds$ converges as $\varepsilon \to 0$ to $L^W(1,0)$ in $L^2(\Omega)$ and almost surely (see e.g. [8]). Using this fact, it suffices to show that the quantity

$$I_n := \mathbf{E}\left(\int_0^1 \left(p_{\alpha_n}(W_s) - p_{\alpha_n}\left(W_{\frac{[ns]}{n}}\right)\right)ds\right)^2 \tag{2.8}$$

converges to zero as $n \to \infty$, where we denoted by $\alpha_n = \frac{1}{2}n^{-\alpha-\frac{1}{2}}$. We have

$$\begin{aligned}
I_n &= \mathbf{E}\int_0^1\int_0^1 dsdt\left(p_{\alpha_n}(W_s) - p_{\alpha_n}\left(W_{\frac{[ns]}{n}}\right)\right)\left(p_{\alpha_n}(W_t) - p_{\alpha_n}\left(W_{\frac{[nt]}{n}}\right)\right)\\
&= 2\mathbf{E}\int_0^1 dt\int_0^t ds\left(p_{\alpha_n}(W_s) - p_{\alpha_n}\left(W_{\frac{[ns]}{n}}\right)\right)\left(p_{\alpha_n}(W_t) - p_{\alpha_n}\left(W_{\frac{[nt]}{n}}\right)\right).
\end{aligned}$$

Notice that for every $s,t \in [0,1]$, $s \leq t$,

$$\begin{aligned}
\mathbf{E}p_\varepsilon(W_s)p_\varepsilon(W_t) &= \mathbf{E}\left(\mathbf{E}\left[p_\varepsilon(W_s)p_\varepsilon(W_t)|\mathscr{F}_s\right]\right)\\
&= \mathbf{E}\left(p_\varepsilon(W_s)\mathbf{E}\left[p_\varepsilon(W_t)|\mathscr{F}_s\right]\right)\\
&= \mathbf{E}\left(p_\varepsilon(W_s)\mathbf{E}\left[p_\varepsilon(W_t - W_s + W_s)|\mathscr{F}_s\right]\right).
\end{aligned}$$

By the independence of $W_t - W_s$ and $\mathscr{F}_s^W$ we get

$$\mathbf{E}\left[p_\varepsilon(W_t - W_s + W_s)|\mathscr{F}_s\right] = \left(\mathbf{E}p_\varepsilon(W_t - W_s + x)\right)_{x=W_s} = p_{\varepsilon+t-s}(W_s).$$

We will obtain

$$\begin{aligned}
\mathbf{E}p_\varepsilon(W_s)p_\varepsilon(W_t) &= \mathbf{E}p_\varepsilon(W_s)p_{\varepsilon+t-s}(W_s)\\
&= \frac{1}{\sqrt{2\pi}}\left(\frac{s}{\varepsilon s + \varepsilon(t-s+\varepsilon) + s(t-s+\varepsilon)}\right)^{\frac{1}{2}}.
\end{aligned} \tag{2.9}$$

This sequence converges to $\frac{1}{\sqrt{2\pi s(t-s)}}$ as $\varepsilon \to 0$. If we replace s or t by $\frac{[ns]}{s}$ or $\frac{[nt]}{n}$ respectively, we get the same limit.

As a consequence of the Lemma 2.2 we obtain

Proposition 1. *The sequence $n^{-\frac{1}{2}+\alpha}\langle S^n\rangle_n$ converges in distribution, as $n \to \infty$, to*

$$\left(\int_{\mathbb{R}} K^2(y)dy\right)L^W(1,0) = \frac{1}{2\sqrt{\pi}}L^W(1,0)$$

where L^W denotes the local time of the Brownian motion W.

Proof. The conclusion follows because

$$n^{-\frac{1}{2}+\alpha}\langle S^n\rangle_n = n^{-\frac{1}{2}+\alpha}\sum_{i=0}^{n-1} K^2(n^\alpha W_i) =_{(d)} n^{-\frac{1}{2}+\alpha}\sum_{i=0}^{n-1} K^2(n^{\alpha+\frac{1}{2}} W_{\frac{i}{n}})$$

and this converges to $L^W(1,0)$ in $L^2(\Omega)$ from Lemma 2.2.

Remark 2.1. Intuitively, the result in Proposition 1 follows because

$$n^{-\frac{1}{2}+\alpha}\sum_{i=0}^{n-1} K^2(n^\alpha W_i) =_{(d)} n^{-\frac{1}{2}+\alpha}\sum_{i=0}^{n-1} K^2(n^{\alpha+\frac{1}{2}} W_{\frac{i}{n}})$$

$$\sim n^{\frac{1}{2}+\alpha}\int_0^1 K^2(n^{\alpha+\frac{1}{2}} W_s)ds = \int_{\mathbb{R}} K^2(n^{\alpha+\frac{1}{2}} x)L^W(1,x)dx$$

$$= \int_{\mathbb{R}} dy K^2(y) L^W\left(1, \frac{y}{n^{\alpha+\frac{1}{2}}}\right)$$

where we used the occupation time formula (2.7). The bicontinuity of the local time implies that this last expression converges to the limit in Proposition 1.

We state the main result of this part.

Theorem 2.1. *Let S_n be given by (2.2). Then as $n \to \infty$, the sequence $n^{\frac{\alpha}{2}-\frac{1}{4}} S_n$ converges in distribution to*

$$\left(\left(\int_{\mathbb{R}} K^2(y)dy\right) L^W(1,0)\right)^{\frac{1}{2}} Z$$

where Z is a standard normal random variable independent by $L^W(1,0)$.

Proof. A similar argument has already been used in [4]. Obviously,

$$S_n =_{(d)} n^{\frac{1}{4}+\frac{\alpha}{2}}\int_0^1 K\left(n^{\alpha+\frac{1}{2}} W_{\frac{[ns]}{n}}\right) dW_s := T_n.$$

Let

$$T_t^n = n^{\frac{1}{4}+\frac{\alpha}{2}}\int_0^t K\left(n^{\alpha+\frac{1}{2}} W_{\frac{[ns]}{n}}\right) dW_s.$$

Then T_t^n is a martingale with respect to the filtration of W. We can show that $\langle T^n, W\rangle$ converges to zero in probability as $n \to \infty$. Indeed,

$$\langle T^n, W\rangle_t = n^{\frac{1}{4}+\frac{\alpha}{2}}\int_0^t K\left(n^{\alpha+\frac{1}{2}} W_{\frac{[ns]}{n}}\right) ds$$

and this clearly goes to zero using formula (2.5). It is not difficult to see that the convergence is uniform on compact sets. On the other hand $\langle T^n\rangle_1$ converges to $\left(\int_{\mathbb{R}} K^2(y)dy\right) L^W(1,0)$ in $L^2(\Omega)$ from Lemma 2. The result follows immediately from the asymptotic Knight theorem (see [11], Theorem 2.3 page 524, see also [4]).

□

2.3 The Multiparameter Settings

This part concerns the two-parameter model (2.3) defined in the introduction. Let $W^{(1)}$ and $W^{(2)}$ denote two independent Wiener sheets on a probability space $(\Omega, \mathscr{F}, P)$. Recall that a Brownian sheet $(W_{u,v})_{u,v\geq 0}$ is defined as a centered two-parameter Gaussian process with covariance function

$$\mathbf{E}\left(W_{s,t} W_{u,v}\right) = (s \wedge u)(t \wedge v)$$

for every $s, t, u, v \geq 0$. The model (2.3) leads to the study of the sequence

$$T_n = \sum_{i,j=0}^{n-1} K\left(n^{\alpha} W^{(1)}_{i,j}\right)\left(W^{(2)}_{i+1,j+1} - W^{(2)}_{i+1,j} - W^{(2)}_{i,j+1} + W^{(2)}_{i,j}\right). \tag{2.10}$$

As in the previous section, we will first give the renormalization of the L^2 norm of T_n as $n \to \infty$.

Proposition 2. *We have*

$$\mathbf{E}\left(n^{\alpha-1} T_n\right)^2 \to_{n\to\infty} \frac{\sqrt{2}}{\pi}.$$

Proof. By the independence of $W^{(1)}$ and $W^{(2)}$ and by the independence of the increments of the Brownian sheet $W^{(2)}$ we have, using (2.5)

$$\mathbf{E}T_n^2 = \sum_{i,j=0}^{n-1} \mathbf{E}\left(K^2(n^{\alpha} W^{(1)}_{i,j})\right) = \frac{1}{2\pi}\sum_{i,j=0}^{n-1} \mathbf{E}\left(e^{-n^{2\alpha}(W^{(1)}_{i,j})^2}\right) = \frac{1}{2\pi}\sum_{i,j=0}^{n-1} \frac{1}{\sqrt{1+2n^{2\alpha} ij}}$$

and the conclusion follows because $\sum_{i=0}^{n-1} \frac{1}{\sqrt{i}}$ behaves, when $n \to \infty$ as $2\sqrt{n}$.

□

We will first study the "bracket" $\langle T\rangle_n = \sum_{i,j=0}^{n-1} K^2(n^{\alpha} W^{(1)}_{i,j})$ which is in some sense the analogous of the bracket of S_n defined in the one-dimensional model. For simplicity, we will still use the notation $\langle T\rangle_n$ even if it is not anymore a true martingale bracket (the stochastic calculus for two parameter martingales is more complex, see e.g. [7]). By the scaling property of the Brownian sheet, the sequence $n^{\alpha-1}\langle T\rangle_n$ has the same distribution as

$$n^{\alpha-1}\sum_{i,j=0}^{n-1} K^2\left(n^{\alpha+1}W^{(1)}_{\frac{i}{n},\frac{j}{n}}\right).$$

Note that for every $u,v\geq 0$ we can write

$$\sqrt{\pi}n^{\alpha+1}K^2\left(n^{\alpha+1}W^{(1)}_{u,v}\right)=\frac{1}{2}p_{\frac{1}{2n^{2(1+\alpha)}}}(W^{(1)}_{u,v}).$$

As a consequence $n^{\alpha-1}\langle T\rangle_n$ has the same law as

$$\frac{1}{2\sqrt{\pi}}\frac{1}{n^2}\sum_{i,j=0}^{n-1} p_{\frac{1}{2n^{2(1+\alpha)}}}\left(W^{(1)}_{\frac{i}{n},\frac{j}{n}}\right).$$

In the limit of the above sequence, the local time of the Brownian sheet $W^{(1)}$ will be involved. This local time can be defined as in the one-dimensional case. More precisely, for any $s,t\geq 0$ and $x\in\mathbb{R}$ the local time $L^{W^{(1)}}((s,t),x)$ is defined as the density of the occupation measure (see [1,21])

$$\mu_{s,t}(A)=\int_0^t\int_0^s 1_A(W_{u,v})dudv, \qquad A\in\mathcal{B}(\mathbb{R}).$$

and it satisfies the occupation time formula: for any measurable function f

$$\int_0^t\int_0^s f(W^{(1)}_{u,v})dudv=\int_{\mathbb{R}} L^{W^{(1)}}((s,t),x)f(x)dx. \tag{2.11}$$

The following lemma is the two-dimensional counterpart of Lemma 2.2.

Lemma 2.3. *The sequence* $\frac{1}{n^2}\sum_{i,j=0}^{n-1} p_{\frac{1}{2n^{2\alpha+2}}}\left(W^{(1)}_{\frac{i}{n},\frac{j}{n}}\right)$ *converges in* $L^2(\Omega)$ *an* $n\to\infty$ *to* $L^{W^{(1)}}(\underline{1},0)$ *where* $L^{W^{(1)}}(\underline{1},0)$ *denotes the local time of the Brownian sheet* $W^{(1)}$*, where* $\underline{1}=(1,1)$.

Proof. This proof follows the lines of the proof of Lemma 2.2. Since $\int_0^1\int_0^1 p_\varepsilon(W_{u,v})duv$ converges to $L^{W^{(1)}}(\underline{1},0)$ as $\varepsilon\to 0$ (in $L^2(\Omega)$ and almost surely, it suffices to check that

$$J_n:=\mathbf{E}\left(\int_0^1\int_0^1\left(p_{\alpha_n}(W_{u,v})-p_{\alpha_n}\left(W_{\frac{[un]}{n},\frac{[vn]}{n}}\right)\right)dudv\right)^2$$

converges to zero as $n \to \infty$ with $\alpha_n = \frac{1}{2}n^{-2\alpha-2}$. This follows from the formula, for every $a \geq u$ and $b \geq v$

$$\mathbf{E}\left(p_\varepsilon(W_{a,b} - W_{u,v})p_\varepsilon(W_{u,v})\right) = \mathbf{E}\left(p_\varepsilon(W_{u,v})p_{\varepsilon+ab-uv}(W_{u,v})\right)$$

and relation (2.9). □

Let us now state our main result of this section.

Theorem 2.2. *As $n \to \infty$, the sequence $n^{\frac{\alpha}{2}-\frac{1}{2}}T_n$ converges in distribution to $\left(c_0 L^{W^{(1)}}(\underline{1},0)\right)^{\frac{1}{2}} Z$ where $L^W(1,0)$ is the local time of the Brownian sheet $W^{(1)}$, $c_0 = \frac{1}{2\sqrt{\pi}}$ and Z is a standard normal random variable independent by $W^{(1)}$.*

Proof. We will compute the characteristic function of the T_n. Let $\lambda \in \mathbb{R}$. Since the conditional law of T_n given $W^{(1)}$ is Gaussian with variance $\sum_{i,j=0}^{n-1} K^2\left(n^\alpha W^{(1)}_{i,j}\right)$ we can write

$$\begin{aligned}\mathbf{E}\left(e^{i\lambda n^{\frac{\alpha}{2}-\frac{1}{2}}T_n}\right) &= \mathbf{E}\left(\mathbf{E}\left(e^{i\lambda n^{\frac{\alpha}{2}-\frac{1}{2}}T_n}|W^{(1)}\right)\right)\\ &= \mathbf{E}\left(e^{-\frac{\lambda^2}{2}n^{\alpha-1}\sum_{i,j=0}^{n-1}K^2\left(n^\alpha W^{(1)}_{i,j}\right)}\right) = \mathbf{E}\left(e^{-\frac{\lambda^2}{2}n^{\alpha-1}\langle T\rangle_n}\right).\end{aligned}$$

By the scaling property of the Brownian sheet, the sequence

$$n^{\alpha-1}\langle T\rangle_n =_{(d)} n^{\alpha-1}\sum_{i,j=0}^{n-1} K^2\left(n^{\alpha+1}W^{(1)}_{\frac{i}{n},\frac{j}{n}}\right) = \frac{1}{2\sqrt{\pi}}\frac{1}{n^2}\sum_{i,j=0}^{n-1} p_{\frac{1}{2n^{2(1+\alpha)}}}\left(W^{(1)}_{\frac{i}{n},\frac{j}{n}}\right).$$

The result follows from Lemma 2.3. □

Remark 2.2. A similar remark as Remark 1 is available in the two-parameter settings. Indeed, the basic idea of the result is that

$$\begin{aligned}n^{\alpha-1}T_n &=_{(d)} n^{\alpha-1}\sum_{i,j=0}^{n-1} K^2\left(n^{\alpha+1}W^{(1)}_{\frac{i}{n},\frac{j}{n}}\right) \sim n^{\alpha+1}\int_0^1\int_0^1 K^2\left(n^{\alpha+1}W^{(1)}_{u,v}\right)du dv\\ &= n^{\alpha+1}\int_{\mathbb{R}} K^2\left(n^{\alpha+1}x\right)L^{W^{(1)}}(\underline{1},x)dx\\ &= \int_{\mathbb{R}} K^2(y)L^{W^{(1)}}\left(1,\frac{y}{n^{\alpha+1}}\right)dy \to_{n\to\infty} \int_{\mathbb{R}} K^2(y)dy L^{W^{(1)}}(\underline{1},0)\end{aligned}$$

by using (2.11) and the bicontinuity of the local time.

As a final remark, let us mention that above result (and) the model (2.3) can be relatively easily extended to the case of N-parameter Brownian motion, with $N \geq 2$.

References

1. Berman, S. Local nondeterminism and local times of Gaussian processes. Indiana Univ. Math. J. **23**, 69–94 (1973)
2. Bourguin, S., Tudor, C.A.: Asymptotic theory for fractional regression models via Malliavin calculus. J. Theoretical Probab., to appear (2010)
3. Geman, D., Horowitz, J.: Occupation densities. Ann. Probab. **8**, 1-67 (1980)
4. Hu, Y., Nualart, D.: Stochastic integral representation of the L^2 modulus of local time and a centram limit theorem. Electron. Comm. Probab. **14**, 529–539 (2009)
5. Karlsen, H.A., Tjostheim, D.: Nonparametric estimation in null recurrent time series. Ann. Stat. **29**, 372–416 (2001)
6. Karlsen, H.A., Mykklebust, T., Tjostheim, D.: Non parametric estimation in a nonlinear cointegrated model. Ann. Stat. **35**, 252–299 (2007)
7. Nualart, D.: On the Quadratic Variation of Two-Parameter Continuous Martingales. Ann. Probab. **12**(2), 445–457 (1984)
8. Nualart, D., Vives, J.: Chaos expansion and local time. Publications Matemàtiques. **36**, 827–836 (1992b)
9. Park, J.Y., Phillips, P.C.B. Nonlinear regression with integrated time series. Economet. **74**, 117–161 (2001)
10. Phillips, P.C.B.: Regression theory for near-integrated time series. Economet. **56**, 1021–1044 (1988)
11. Revuz, D., Yor, M.: Continuous martingales and Brownian motion. 3rd edn. Springer, Berlin (1999)
12. Wang, Q., Phillips, P.: Asymptotic theory for the local time density estimation and nonparametric cointegrated regression. Economet. Theor. **25**, 710–738 (2009)
13. Wang, Q., Phillips, P.: Structural nonparametric cointegrating regression. Economet. **77**(6), 1901–1948 (2009)

Chapter 3
General Shot Noise Processes and Functional Convergence to Stable Processes

Wissem Jedidi, Jalal Almhana, Vartan Choulakian, and Robert McGorman

Abstract In traffic modeling theory, many authors present models based on particular shot noise representations. We propose here a model based on a general Poisson shot noise representation. Under minimal assumptions, we obtain an approximation of the cumulative input process by a stable Lévy motion via a functional limit theorem.

3.1 Introduction

In the present paper, we consider a unique server dealing with an infinite sized source. The source sends data to the server over independent transmissions and according to a Poisson process. Our aim is to study the traffic generated by the transmissions over an interval of time $[0, t]$ and denoted by A_t. The cumulative input process $A = (A_t)_{t \geq 0}$ has a structure of a Poisson shot noise (see [3] and [13]) that is a natural generalization of the compound Poisson process when the summands are stochastic processes starting at the points of the underlying Poisson process. This has become popular in modeling traffic, computer failure times, geophysical phenomena and finance. Here, we do not assume any particular mechanism of

W. Jedidi (✉)
Department of Mathematics, Faculty of Sciences of Tunis, Campus Universitaire, 1060 Tunis, Tunisia
e-mail: wissem.jedidi@fst.rnu.tn

J. Almhana · V. Choulakian
GRETI group, University of Moncton, Moncton, NB E1A3E9, Canada
e-mail: almhanaj@umoncton.ca; choulav@umoncton.ca

R. McGorman
NORTEL Networks, 4001 E. Chapel Hill-Nelson Hwy, research Triangle Park, North Carolina, USA 27709
e-mail: mcgorman@nortelnetworks.com

M. Zili and D.V. Filatova (eds.), *Stochastic Differential Equations and Processes*, Springer Proceedings in Mathematics 7, DOI 10.1007/978-3-642-22368-6_3,

evolution in time of this process (i.e. on the summands). In Sect. 3.5, we will compare with the specialized literature where some mechanisms are assumed : Kaj [11], Konstantopoulos and Lin [14] Maulik et al. [15], Maulik and Resnick [16], Mikosch et al. [18], Resnick and Van den Berg [20]. Of course, when specifying the mechanism, some random quantities take a particular importance. We will show later that only two data are relevant for us, namely:

1. The size of each transmission (denoted X_∞ in the sequel).
2. The length (i.e. the duration) of each transmission (denoted τ in the sequel).

in order to show that the cumulative process A is approximated in a strong sense by a stable process. The strong sense means that the process A, when adequately drifted, rescaled in time and normalized, functionally converges in law to a stable (non-Gaussian) process (see Sect. 3.2 below) for the precise definition of functional convergence in law for stochastic processes). This convergence is obtained under two crucial assumptions: (1) the size of each transmission has a distribution tail regularly varying and (2) an assumption very close (and sometimes weaker) to what is called, in the literature cited below, a *slow connection rate* or *slow input rate* condition on the length of the transmission. Notice that if condition (3) is substituted by (4) a *fast connection rate* (resp. *intermediate connection rate*), the limit is no longer stable but selfsimilar Gaussian (resp. non-Gaussian and non-stable stochastic process with stationary increments). See Çağlar [4], Kaj [11], Kaj and Taqqu [12], Maulik and Resnick [16], K. Maulik and Resnick [17], Mikosch et al. [18] for non-Stable limits. Some extensions in the context of cluster Poisson point processes are studied in Fasen and Samorodnitsky [7] and Fasen [6] (with superpositions of input processes) and analogous results are shown. A generalization of the popular ON/OFF model can be found in [19] and also analogous results are shown.

We stress, here, that we are only interested by a simple formalism leading to stable limits. The main idea of the paper is simple and is stated in theorem 5: we observe that if the total traffic of sessions which have been active up to time t is taken into account, then the cumulative input process would be compound Poisson. In the scaling we consider, we prove that the difference goes to 0 and the compound Poisson is approximated by a stable process totally skewed to the right with index of stability in $\alpha \in (0, 2)$.

To the best of our knowledge, apart the model of Resnick and Van Den Berg [20], all the results available in the literature only treat the finite-dimensional part of the attraction (expressed in 3.21 below) by a stable process totally skewed to the right with index of stability $\alpha \in (1, 2)$.

The model proposed by Klüppelberg et al. [13] deals with the Poisson shot noise process B in (3.2) below. Their main result says that the convergence of the finite-dimensional distributions of a (rescaled) Poisson shot noise process to a stable process S with index in $\alpha \in [1, 2)$ is equivalent to what they call a condition of *regular variation in the mean*. The limiting stable process S is not necessarily a stable Lévy motion (we relax the independence of the increments in S) and one has to describe it according to the mechanism of B. We consider that our approach intercepts the one of [13], but only on the finite-dimensional point of view. Then we reinforce the result of [13] when S is a stable Lévy motion.

As already announced, the main result is stated in theorem 5. Surprisingly, the assumptions required in theorem 5 are weaker than those required in the literature (see Sect. 3.5 for a comparison), but they provide a stronger conclusion: the functional convergence to a stable process totally skewed to the right with index of stability $\alpha \in (0, 2)$.

3.2 The Topology

All the convergence of processes shown in this paper hold in the functional (or weak) mode i.e. in the space $\mathbb{D} = \mathbb{D}(\mathbb{R}+, \mathbb{R})$ of càdlàg functions equipped with the M_1-Skorohod topology. We will not get into details about Skorohod topologies, we just say that, according to 3.20 p.350 [8], a sequence Z^n of stochastic processes is said to functionally (or weakly) converge in law to a process Z (we denote $Z^n \Longrightarrow Z$) if and only if we have the two statements: (1) the limit process Z is well identified via finite-dimensional convergence, i.e. for all $d \geq 1$ and $t_1, t_2, \cdots, t_d \geq 0$, we have convergence of the d-dimensional random variable $(Z^n_{t_1}, Z^n_{t_2}, \cdots Z^n_{t_d}) \longrightarrow (Z_{t_1}, Z_{t_2}, \cdots Z_{t_d})$ and (2) the sequence of processes Z^n is tight which is a technical condition (strongly related to the modulus of continuity of the topology) ensuring the existence of the limit. This paper mostly uses the powerful tools on the M_1 topology presented in the book of Jacod and Shiryaev [8]. Another important reference on the topic is the book of Whitt [22]. Notice that many natural and important functionals on $\mathbb{D}$ (like the supremum functional) are continuous in M_1 and for the other stronger Skorohod topology namely J_1 (see Billingsley [1]). Our choice to use M_1 topology relies on this fact that it is desirable to use as weak topology as possible, since it requires weak conditions for tightness. We are going to make some significant discontinuous element of $(\mathbb{D}, M_1)$ (namely, the processes ε^1 and ε^2 introduced in (3.6) and (3.7) below) converge to 0. See also remark 6 below.

Observe that finite-dimensional convergence of a sequence Z^n to 0 is equivalent to the unidimensional convergence $(Z^n_t \to 0, \ \forall t > 0)$. In the following, this simple constatation will be useful.

3.3 The Model

Through all the following, we assume that a unique server deals with an infinite number of sources. Following Resnick's formalism [20], the transmissions arrive to the server according to an homogeneous Poisson process on $\mathbb{R}$ labeled by its points $(T_k)_{k\in\mathbb{Z}}$ so that $\cdots T_{-1} < T_0 < 0 < T_1 \cdots$ and hence $\{-T_0, T_1, (T_{k+1} - T_k, \ k \neq 0)\}$ are i.i.d. exponential random variables with parameter λ.

T_k *is considered as the time of initiation of the* k-th *transmission.*

Let the counting measure $N(\mathrm{d}u) = \sum_{k\in\mathbb{Z}} \varepsilon_{T_k}(\mathrm{d}u)$ and define the Poisson process N by

$$N_t = N[0,t] \text{ if } t \geq 0 \quad \text{and} \quad N_t = N[t,0) \text{ if } t < 0.$$

The quantity N_t represents the number of transmissions started between time $s = 0$ and time $s = t$.

We are interested in the *cumulative input* of work to the server (also called *total accumulated work*) over an interval of time $[0,t]$ and denoted by A_t. It corresponds to the size of the files transmitted by the source. There are many ways to model it (from the most trivial way to the most sophisticated). Specification of the source behavior could be taken into account adding more and more parameters. In order to avoid this intricacy, most authors (Kaj [11], Konstantopoulos and Lin [14] Maulik et al. [15], Maulik and Resnick [16], Maulik and Resnick [17], Mikosch et al. [18], Resnick and Van den Berg [20]) have a macroscopic approach strongly connected with times of initiation of the transmissions, their duration and their rate. As we will see, this paper confirms the pertinence of this approach, and we show that it is sufficient for having the required control on the cumulative input process.

Our aim is to describe, in the more general setting, the cumulative input process and to give an approximation of its law. Notice that the cumulative input process describes the work generated over the interval $[0,t]$. Time $s = 0$ is when our "observation starts" and $s = t$ is when our "observation finishes." Observe that times of initiation of transmissions are either negative or positive (before or after our observation starts, respectively). The k-th transmission starts at time T_k and continues over the interval of time $[T_k, +\infty)$. Suppose we observed the transmissions for all times and until time s and we want to calculate the work generated by the k-th transmission. This work occurs over the random interval $(-\infty, s] \cap [T_k, +\infty)$. The length of this interval is the r.v.

$$(s - T_k)_+ = Max\{0, s - T_k\}.$$

We deduce that the work generated by the k-th transmission is given by a quantity which depends on the length $(s - T_k)_+$. We will denote this work

$$X^k(s - T_k)$$

where $\big(X^k(r)\big)_{r\in\mathbb{R}}$ is a stochastic process, the random variable $X^k(r)$ is an increasing function of r, vanishing if $r \leq 0$ and describing the quantity of work potentially generated by the k-th transmission over an interval of time of length $r > 0$.

If we had observed only over the interval $(s', s]$ (instead of $(-\infty, s]$), the work potentially generated by the k-th transmission should be written as the difference

$$X^k(s - T_k) - X^k(s' - T_k).$$

Notice that $t \geq T_l \geq T_k \geq 0 \geq T_j \geq T_i$ is equivalent to $N_t \geq l \geq k \geq 0 \geq j \geq i$. Because of the above considerations, we propose the following general model

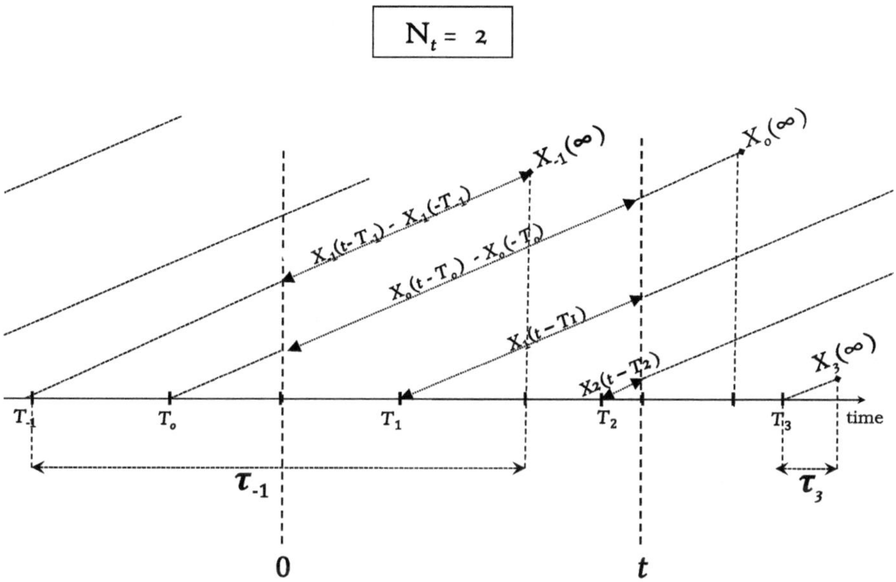

Fig. 3.1 The quantity A_t is formed by the four slanted segments between 0 and t while B_t is formed by the two slanted segments on the right between 0 and t

describing the cumulative input over $[0, t]$ by a "moving average" type (Fig. 3.1):

$$A_t = \sum_{k=-\infty}^{N_t} [X^k(t - T_k) - X^k(-T_k)] = \sum_{k=-\infty}^{\infty} [X^k(t - T_k) - X^k(-T_k)] 1_{(T_k \leq t)}, t \geq 0. \tag{3.1}$$

The processes $(X^k)_{k \in \mathbb{Z}}$ describe the evolution in time of the job completion of the transmissions. They are assumed to form an i.i.d. sequence independent from the arrival process $(T_n)_{n \in \mathbb{Z}}$. For each $k \in \mathbb{Z}$, the process X^k is càdlàg (right continuous with left limits), vanishing on the negative real axis and increasing to a finite r.v. $X^k(\infty)$.

We will see in the sequel that the distributional behavior of the process $(A_t)_{t \geq 0}$ at large times scales is the same as the one of its "finite memory" part B which is a Poisson shot noise (see Bremaud [3] and Klüppelberg [13]):

$$B_t = \sum_{k=1}^{N_t} X^k(t - T_k), \quad t \geq 0. \tag{3.2}$$

Observe that the process A has stationary increments, while B has not, nevertheless B is of special interest because it only takes into account the transmissions started after time $s = 0$. The special structure of the process B merits some other comments. The problem is that at any fixed time t we can not "see" if the k-th transmission has finished or not and moreover we are unable to calculate during the time $t - T_k$ the accumulated work $X^k(t - T_k)$. The only available information is the quantity

$X^k(\infty)$ *which is the total work required by the* k*-th transmission.*

It is then natural to introduce the process

$$C_t = \sum_{k=1}^{N_t} X^k(\infty), \quad t \geq 0, \tag{3.3}$$

which characterizes the total work required by all the transmissions started within the interval $[0, t]$. This process enjoys a very special property: it is a Lévy process, i.e. it has independent and stationary increments and, more precisely, is a compound Poisson process (see the book of Sato [21] for an account of Lévy processes or appendix 3.7). This process turns out to be the principal component of the processes A and B, a component which will give the right approximation by a stable process, as stated in theorem 5 below.

Observe that the processes B and C defined in (3.2) and (8.5) are very well defined because they are finite sums while the process A introduces in (3.1) may have problems of definition, since the r.v. A_t, $t > 0$, could be infinite. We will see in lemma 2 that they are actually finite under our assumptions. There are many ways to represent these processes. Recall the random Poisson measure $N(du)$ associated with the sequence $(T_k)_{k\in\mathbb{Z}}$ and define the integer random measure M on $(\mathbb{R} \times \mathbb{D}, \mathscr{B} \otimes \mathscr{D})$ by

$$M(\mathrm{d}u, \mathrm{d}y) := \sum_{k\in\mathbb{Z}} \epsilon_{(T_k, X^k)}(\mathrm{d}u, \mathrm{d}y), \tag{3.4}$$

which is actually a marked Poisson measure. Then, the process A has a Poisson integral representation of the *moving average* type

$$A_t = \int_{(-\infty,t]\times\mathbb{D}} \left[\, y((t-u)_+) - y((-u)_+) \,\right] M(\mathrm{d}u, \mathrm{d}y).$$

Notice also that the process A is increasing and, as shown later on, is locally integrable if $1 < \alpha < 2$. Then, the process A may have the structure of a special semimartingale (see definition 4.21 p. 43 [8]). We will only invoke some more familiar aspects in the structure of the process A, namely, it is represented as follows:

$$A_t = \sum_{k=-\infty}^{N_t} \left[\, X^k(t - T_k) - X^k(-T_k) \,\right] = B_t + \epsilon_t^1 \tag{3.5}$$

where

$$B_t = \sum_{k=1}^{N_t} X^k(t - T_k), \qquad \epsilon_t^1 = \sum_{k=-\infty}^{0} [X^k(t - T_k) - X^k(-T_k)]. \tag{3.6}$$

We can also decompose B in $B = C - \epsilon^2$ with

$$C_t = \sum_{k=1}^{N_t} X^k(\infty), \qquad \epsilon_t^2 = \sum_{k=1}^{N_t} [X^k(\infty) - X^k(t - T_k)]. \tag{3.7}$$

All the processes presented above are closely connected to the load of the transmissions, this is the reason why we will call them *load processes*. The lemma 2 below unveils some of their properties.

3.4 The Results

We will show that the process $Z = A,\ B$ or C, after being correctly drifted rescaled and normalized, satisfy a limit theorem and share a common limiting process, a strict stable Lévy motion totally skewed to the right (see appendix 3.7 for more on stable Lévy motions).

3.4.1 Assumptions and Notations

Through all the following we adopt the following notations for all real numbers a, b

$$a \wedge b = \min\{a, b\}, \quad a \vee b = \max\{a, b\} \quad \text{and} \quad (a-b)_+ = (a-b) \vee 0 = a - (a \wedge b).$$

Recall the processes X^k are i.i.d. Let

$$(X_t)_{t \geq 0} := (X^1(t))_{t \geq 0}, \quad X_\infty := \lim_{t \to \infty} X_t, \tag{3.8}$$

and assume the stopping time

$$\tau := \inf\{ t \ : \ X_t = X_\infty\} \quad \textit{is finite}. \tag{3.9}$$

Notice that the event $(X_\infty > X_u)$, $u \geq 0$, is given by $(\tau > u)$. Then, the behavior of the r.v. τ describes the way the cumulative process of each transmission reaches its maximum, or in other words the way each transmission is completed. The r.v. X_∞ and τ are actually versions of the size and of the length of any transmissions.

3.4.1.1 Assumption on the Size of the Transmissions

The r.v. X_∞ is finite and has a distribution with regularly varying tail of order $-\alpha$, with $\alpha \in (0,2)$ i.e. there exists a deterministic increasing function r such that for each $x > 0$,

$$\lim_{u\to\infty} u\, P(X_\infty > r(u)\,x) = \frac{1}{x^\alpha} \tag{3.10}$$

which is equivalent to $r(u) = u^{\frac{1}{\alpha}} l(u)$ for some slowly varying function l i.e. satisfying $lim_{u\to\infty} l(u\,x)/l(u) = 1$, for all $x > 0$. It is commonly said that X_∞ has heavy tails. The reader is referred to the book of Bingham et al. [2] or to appendix 3.7 for more details on regular variation theory.

3.4.1.2 Assumption on the Arrivals

Notice that the intensity parameter λ of the arrival process N is not necessarily constant,

it may depend on a scale c and $\lambda = \lambda_c$ may depend on a parameter $c > 0$ as studied in Mikosch et al. [17] and Kaj [11]. Through all the following, we consider the case where λ_c increases to a value $\lambda_\infty \in (0,\infty]$ as c goes to infinity. Of course, the constant case $\lambda_c = \lambda_\infty$ is included in the latter. Sometimes, we will simply write λ instead of λ_c.

3.4.1.3 Technical Assumptions on the Length of the Transmissions: Connection with the Intensity and with the Size

At large scales, we will need one of these four assumptions expressed with a function $r : (0,\infty) \to (0,\infty)$ such that:

$$\lim_{c\to\infty} \lambda_c\, \mathbb{E}\left[(\tau \wedge c) \left(\frac{X_\infty}{r(c\lambda_c)} \wedge 1 \right) \right] = 0 \tag{3.11}$$

or

$$\lim_{c\to\infty} \lambda_c\, \mathbb{E}\left[\tau \left(\frac{X_\infty}{r(c\lambda_c)} \wedge 1 \right) \right] = 0 \tag{3.12}$$

or

$$\mathbb{E}[X_\infty] < \infty \quad \text{and} \quad \lim_{c\to\infty} \frac{\lambda_c}{r(c\,\lambda_c)} \int_0^c \mathbb{E}[X_\infty - X_u]\,du = 0 \tag{3.13}$$

or

$$\mathbb{E}[X_\infty] < \infty \quad \text{and} \quad \lim_{c\to\infty} \frac{\lambda_c}{r(c\,\lambda_c)} \int_0^{ct} \mathbb{E}[X_\infty - X_u]\,du = 0,\ \forall t > 0. \tag{3.14}$$

Remark 1. (a) The technical lemma 3.32 below is the key of the proof of theorem 5 and works under the assumptions (3.10), (3.11),(3.12),(3.13). We stress that apart assertions (2) and (3)(iii), the function r used in lemma 3.32 needs not to be the one introduced in (3.10). Trivially, assumption (3.12) implies (3.11) and assumption (3.14) implies (3.13). We will see in the proof of lemma 3.32 below, that assumption (3.10), with $\alpha \in (1,2)$, together with (3.13) implies (3.14). Assume that $\limsup_{c\to\infty} r(cT)/r(c) < \infty,\ \forall T > 0$, which is the case if the function r is the one of (3.10), and assume we are in the case of finite intensity ($\lambda_c = cst$ or $0 < \lambda_\infty < \infty$), then it is also obvious that each of the conditions (3.11), (3.12), (3.13) and (3.14) is equivalently expressed with $\lambda_c = 1$. Finally, notice that if (3.10) is satisfied and τ has finite expectation, then (3.12) or (3.14) are satisfied by many simple conditions on the distribution of τ. For more details, the reader is invited to see the comments 1 and 3 of Sect. 3.6. Another point is that (3.10) with $\alpha \in (1,2)$ implies $\mathbb{E}[X_\infty] < \infty$.

(b) Assumption (3.12), when coupled with (3.10), is the one that makes the functional convergence work for A, B, C in theorem 5 below. There are two simple conditions implying (3.12). The first one is τ regularly varying distribution tail with index $-\beta > 0$ such that $\frac{1}{\alpha} + \frac{1}{\beta} > 1$. The second condition is called in the literature a *slow connection rate* or *slow input rate* condition, precisely

$$\lim_{c\to\infty} c\,\lambda_c\,\mathbb{P}(\tau > c) = 0,$$

See the discussion right after (3.22).

The following lemma treats the problem of existence and infinite divisibility of the different load process involved in the cumulative input process A.

Lemma 2. Infinite divisibility of the load processes. *For all $t \geq 0$ we have:*

(i) The process C is a a compound Poisson process with Lévy exponent given for all $\theta \in \mathbb{C}$, $\mathrm{Re}(\theta) \geq 0$, by

$$\Psi_t(\theta) := \log \mathbb{E}\big[\, e^{-\theta\, C_t}\,\big] = \lambda\, t \int_{\mathbb{R}_+} \big(e^{-\theta\, x} - 1\big)\, \mathbb{P}(X_\infty \in \mathrm{d}x). \tag{3.15}$$

(ii) The r.v. B_t and ϵ_t^2 are finite and infinitely divisible.

(iii) Assume $\mathbb{E}[\tau\,(X_\infty \wedge 1)] < \infty$, then the r.v. A_t and ϵ_t^1 are finite and infinitely divisible.

(iv) Assume $\mathbb{E}[X_\infty] < \infty$, then the r.v. A_t and ϵ_t^1 are finite, infinitely divisible and all the load quantities have finite expectations given by

$$\mathbb{E}[A_t] = \mathbb{E}[C_t] = \lambda\, t\, \mathbb{E}[X_\infty],$$

$$\mathbb{E}[B_t] = \lambda \int_0^t \mathbb{E}[X_u]\,\mathrm{d}u$$

$$\mathbb{E}[\epsilon_t^1] = \mathbb{E}[\epsilon_t^2] = \lambda \int_0^t \mathbb{E}[X_\infty - X_u]\,\mathrm{d}u.$$

When proving our main result (theorem 5 below), we will see that the central point is that the processes $Z = A,\ B$ or C we are dealing with,

share a principal common Lévy component which is the Lévy process C

and functional limit theorems for Lévy processes are quite easy to check. Proposition 3 gives a taste on the kind of controls we obtain, and theorem 5 will extend it on the processes A and B. All these processes will share the same limit, a stable motion totally skewed to the right.

Proposition 3. Independently from the behavior of the intensity of the arrivals, the process C is functionally attracted in law by a stable process. *Assume (3.10). Let C^c be the process defined for each $t \geq 0$ by*

$$C_t^c = \frac{C_{ct} - t\, d(c\lambda_c)}{r(c\,\lambda_c)}, \tag{3.16}$$

where the function r is given by (3.10) and the function d is defined by

$$d(u) = u\, r(u)\, \mathbb{E}\left[h\left(\frac{X_\infty}{r(u)} \right) \right] \quad \text{and} \quad h(x) = \begin{cases} 0 & \text{if } 0 < \alpha < 1 \\ \sin x & \text{if } \alpha = 1 \\ x & \text{if } 1 < \alpha < 2. \end{cases} \tag{3.17}$$

Let S be a strict α-stable Lévy motion totally skewed to the right. Letting $c \uparrow \infty$, we get the functional convergence

$$(C_t^c)_{t\geq 0} \Longrightarrow (S_t)_{t\geq 0}. \tag{3.18}$$

Now, we propose a lemma which will be the central key for proving that the cumulative process is attracted by a stable process. Lemma 4 below says that the residual processes $\epsilon^{1,c}$ and $\epsilon^{2,c}$ go to 0 functionally in law when c goes to infinity. The latter can be interpreted as follows: at large time scales (1) the contribution of the past (before time 0) in the cumulative process is negligible and (2) within the interval of observation the difference between the total requirement of work and the accumulated work is also negligible.

Lemma 4. Technical controls on the residual processes.

Let $i = 1, 2,\ c > 0$ and the processes $\epsilon^{i,c}$ defined by $\epsilon_t^{i,c} = \dfrac{\epsilon_{ct}^i}{r(c\,\lambda_c)}$.

1. *If assumption (3.14) holds, then $\epsilon^{i,c}$ converge to 0 in the finite-dimensional sense.*
2. *If assumption (3.10) with $\alpha \in (1, 2)$ holds together with (3.13), then $\epsilon^{i,c}$ converge to 0 in the finite-dimensional sense.*
3. *Furthermore, assume* $\limsup_{c\to\infty} \frac{r(cT)}{r(c)} < \infty,\ \forall T > 0$. *Then $\epsilon^{i,c}$ satisfies*

$$\forall T, \eta > 0, \quad \lim_{c \to \infty} \mathbb{P}\left(\sup_{t \le T} \epsilon_t^{i,c} > \eta \right) = 0 \tag{3.19}$$

(which certainly implies the $\epsilon^{i,c}$ *functionally converges in law to the null process) in any of the following situations:*

(i) $i = 1$, *and assumption (3.13) holds;*
(ii) $i = 1$ *and assumption (3.12) holds;*
(iii) $i = 2$ *and assumption (3.10) holds together with (3.11) (recall (3.11) is implied by (3.12)).*

After the previous preliminaries, we tackle our main result dealing with the attraction of the load processes by a stable process.

Theorem 5. *The processes* A, B, C *are attracted functionally in law by a common stable process. Assume (3.10). Let* $Z = A$, B *or* C *and define for each* $t \ge 0$

$$Z_t^c = \frac{Z_{ct} - t\, d(c\lambda_c)}{r(c\lambda_c)}, \tag{3.20}$$

where $r(c\lambda_c)$ *is given by (3.10) and* $d(c\lambda_c)$ *is defined in (3.17). Let* S *be a strict* α*-stable Lévy motion totally skewed to the right. Letting* $c \uparrow \infty$, *we get the functional convergence*

$$(Z_t^c)_{t \ge 0} \Longrightarrow (S_t)_{t \ge 0}, \tag{3.21}$$

in any of the following situations:

(a) $Z = C$;
(b) $Z = B$ *and (3.11) holds;*
(c) $Z = A$ *and (3.12) holds;*
(d) $Z = A, B,$ $\alpha \in (1, 2)$ *and (3.13) holds. Then, the convergence (3.21) is true in the finite-dimensional sense, and the drift "*$t\, d(c\lambda_c) = c\, \lambda_c\, t\, \mathbb{E}[X_\infty]$*" in (3.20) can be replaced by the expectation* $\mathbb{E}[B_{ct}]$.

Remark 6. Our main result stated in the las theorem merits some comments:

(i) We would like to call the attention of the reader that we had allowed the intensity of the arrival process λ_c to increase to infinity with the time scale. The condition $\alpha \in (1, 2)$ in the statement d) of theorem 5 is connected to the finiteness of the moments $\mathbb{E}[Z_t]$, $t > 0$.
(ii) The reader should also notice that in the case that the intensity $\lambda_c \to \lambda_\infty \in (0, \infty)$ and τ has a finite expectation then the functional convergence in theorem 5 simply holds under condition (3.10) (See the details in Sect. 3.6–3 below).
(iii) As already noticed in Sect. 3.2, the mode of convergence (functional in $(\mathbb{D}, M_1)$) is extremely useful because it implies that some special functionals of the sequence of processes in $\mathbb{D}$ converge to the functionals of the limit process. For instance, consider the inverse in time of Z

$$\tau(Z)_a = \inf\{t \geq 0 :, Z_t > a\}, \quad a \geq 0.$$

It corresponds to the *first time the cumulative input crosses the critical barrier* a. Correctly normalized, and with convergence (3.21), it is easy to show that the distribution of the process $\tau(Z)$ can be approximated by $\tau(S)$ which is well studied in probability literature. Functionals such as supremum and reflection maps are treated similarly. See the monograph of Whitt [23] for these considerations.

3.5 Overview of Some Related Work

We will describe the different infinite source Poisson models we encountered in the literature according to their chronological order of appearance. Of course, our list is not exhaustive. We will see that all the models we present are particular forms of (3.1) or (3.2) when the processes X^k is specified. In what follows, we denote by $\tau_k = \inf\{t : X^k(t) = X^k(\infty)\}$ the length (or duration) of the k-th transmission and R_k its rate (R_k could be interpreted as the size of the transmission per time unit, i.e. $R_k = X^k(\infty)/\tau_k$). In all the following models, the sequence $(\tau_k, R_k)_{k\in\mathbb{Z}}$ is assumed to be i.i.d. and independent from the arrival process N. As done previously, we denote $(\tau, R) = (\tau_1, R_1)$ and $X = X^1$. The assumptions used in the works cited below imply (3.10) and (3.11) according to the discussions in Sect. 3.6 below. At the end of each model presented, we explain why the last claim is true by referring to the corresponding argument in Sect. 3.6.

1. Konstantopoulos et Lin [14]:The intensity λ is constant. The model is in the form (3.1) with
$$X^k(r) = \xi(r \wedge \tau_k)$$
where $\xi(.)$ is a deterministic regularly varying function with index $a > 0$ and τ has regularly varying tails with index $-\beta$ where $\beta \in (1,2)$ and $\beta > a$, so that $\mathbb{E}[\tau] < \infty$ and $X_\infty = \xi(\tau)$ has a regularly varying tail of index $\alpha = \beta/a \in (1,2)$. The authors assumed many other constraints on the indexes a and β and on the increments of the function ξ. They also conjectured that the convergence might hold with the M_1 topology for their model. See Sect. 3.6–3 below.
2. Resnick and Van Den Berg [20]: The intensity λ is constant. The model of Konstantopoulos et al. [14] is revisited under the form (3.2) by keeping some identical assumptions cited in [14] and relaxing the extraneous ones. The convergence (3.21) is shown to hold in the weak Skorohod sense, i.e. in $(\mathbb{D}, M_1)$. See Sect. 3.6–3 below.
3. Mikosch et al. [18]: The intensity λ is constant. Assuming that τ has a regularly varying tail with index $-\alpha$, $\alpha \in (1,2)$, the authors proposed a model in the form (3.2) with
$$X^k(r) = (r_+ \wedge \tau_k).$$
Then $X_\infty = \tau$ and $\mathbb{E}[\tau] < \infty$. See Sect. 3.6–3 below.

4. Maulik et al. [15]: The intensity λ is constant. The model is of the form (3.2) with

$$X^k(r) = (r_+ \wedge \tau_k)\, R_k .$$

The authors imposed a kind of asymptotic independence with a bivariate regularly varying nature on the tails of the pair (τ, R): τ and R have both regularly varying distribution tail with indexes $-\alpha^\tau$, $-\alpha^R$, $1 < \alpha^\tau, \alpha^R < 2$. There exists two regularly varying function $r^\tau(c)$, $r^R(c)$ with indexes $1/\alpha^\tau, 1/\alpha^R$ and a probability measure G on $\mathbb{R}_+$ with α^τ-th finite moment such that the following vague convergence holds: if $\alpha^\tau < \alpha^R$, then

$$c\,\mathbb{P}\left(\left(\frac{\tau}{r^\tau(c)}, R\right) \in .\right) \longrightarrow \frac{\mathrm{d}x}{x^{\alpha^\tau+1}} \times G(.) \quad \text{on } (0,\infty] \times [0,\infty],$$

and

$$\lim_{\epsilon\to 0} \limsup_{c\to\infty} \mathbb{E}\left[\left(\frac{\tau}{r^\tau(c)}, R\right)^\delta 1_{(\tau < r^\tau(c)\,\epsilon)}\right] = 0 \quad \text{for some } \delta > 0.$$

If $\alpha^\tau > \alpha^R$ the same convergence hold when exchanging τ by R Mutatis Mutandis. It is shown in this paper that these assumptions imply: $X_\infty = \tau R$ has a regularly varying distribution tail with index $-\alpha$, $\alpha = \alpha^\tau \wedge \alpha^R \in (1,2)$ and because τ has a regularly varying distribution tail with index $-\alpha$, $\alpha > 1$ we have $\mathbb{E}[\tau] < \infty$. See Sect. 3.6–3 below.
5. Maulik and Resnick [16]: The intensity λ is constant. The model has the form (3.2) with $X^k(t) = Y^k(t) \wedge X^k(\infty)$ and $(Y_k, X_k(\infty))_k$ independent form N. The authors assumed a kind of asymptotic independence type between the processes Y^k and the r.v. $X^k(\infty)$ with regularly varying nature on the tails: denoting $Y_{c,t} = Y^1_{ct}$, the authors assumed the existence of a regularly varying function $\sigma(c)$ with index $H > 1$ and a process χ in $\mathbb{D}$ with stationary increments such that $\mathbb{E}[(\chi_1)^{-\alpha}] < \infty$ for some $\alpha \in (1,2)$ and for each $\epsilon > 0$, the following convergence of finite measures is true

$$\frac{1}{\mathbb{P}(X_\infty > \sigma(c))}\,\mathbb{P}\left(\frac{X_\infty}{\sigma(c)} > \epsilon,\ \frac{Y_{c,.}}{\sigma(c)} \in .\right) \longrightarrow \frac{1}{\epsilon^\alpha}\,\mathbb{P}(\chi \in .)$$

and for each $\gamma > 0$

$$\lim_{\epsilon\to 0} \limsup_{c\to\infty} \frac{1}{\mathbb{P}(X_\infty > \sigma(c))}\,\mathbb{P}\left(\frac{X_\infty}{\sigma(c)} \le \epsilon,\ \frac{\tau}{c} > \gamma\right) = 0.$$

It is shown in this paper that these assumptions imply: χ is self-similar, X_∞ has regularly varying distribution tail with index $-\alpha$ and τ has regularly varying distribution tail with index $-\beta$, $\beta = H\alpha > 1$ which implies $\mathbb{E}[\tau] < \infty$. See Sect. 3.6–3 below.

6. Maulik and Resnick [17]: The model partly extends [16] by $\lambda_c \to \infty$ by letting $\lambda = \lambda_c \to \infty$ but Y^1 is H-self-similar. The authors studied two rates of growth: a slow connection growth condition ($\lim_{c\to\infty} c\,\lambda_c\,\mathbb{P}(\tau > c) = 0$ giving a stable limit process) and a fast connection growth condition ($\lim_{c\to\infty} c\,\lambda_c\,\mathbb{P}(\tau{>}c){=}\infty$) giving a self-similar Gaussian limit process). We compare only with the slow connection setting. The authors assumed (3.10) with $\alpha \in (1,2)$, $\mathbb{E}[(Y_1^1)^{-\alpha}] < \infty$, $\mathbb{E}[(Y_1^1)^{2+\delta-\alpha}] < \infty$ for some $\delta > 0$ and $\beta = H\alpha > 1$ satisfies $\frac{1}{\alpha} + \frac{1}{\beta} > \frac{3-\alpha}{\alpha} \vee 1$. It is shown in this paper that these assumptions imply: τ has a regularly varying distribution tail with index β. See Sect. 3.6–4 below.
7. Kaj [11]: The intensity $\lambda = \lambda_c \to \infty$. The model is expressed with the processes B in (3.2) with
$$X^k(r) = r_+ \wedge \tau_k$$
and the cumulative process A is expressed as the sum $A = B + \epsilon^1$ where ϵ^1 is build in order to have a stationary version of the process A. Namely, $\epsilon_t^1 = \sum_{k=1}^{M_\nu} t \wedge \widetilde{\tau}_k$, where M_ν is a Poisson r.v with intensity parameter $\nu = E[\tau]$ independent from $(\tau_k)_{k\geq 1}$, the sequence $(\widetilde{\tau}_k)_{k\geq 1}$ is i.i.d., independent from M_ν and $(\tau_k)_{k\geq 1}$ and $\widetilde{\tau}_1$ has the equilibrium distribution $\mathbb{P}(\widetilde{\tau}_1 \leq u) = \frac{1}{\nu}\int_0^u \mathbb{P}(\tau{>}z)\,dz$. Trivial calculations show that the process ϵ^1 introduced by the author has the same law as the process ϵ^1 defined in (3.6) by
$$\epsilon_t^1 = \sum_{k=-\infty}^{0} (t - T_k) \wedge \tau_k - (-T_k) \wedge \tau_k.$$
The author assumed that the right tail of $X_\infty = \tau$ is regularly varying with index $-\alpha$, $\alpha \in (1,2)$, and connection rate holds: there exists a function $a(c)$ such that $a(c)\,\lambda_c\,P(\tau > a(c)) \to K \in [0,\infty]$. The slow (resp. intermediate, resp. fast) connection rate is for $K = 0$ (resp. > 0, resp $= \infty$), In the intermediate case, when, replacing the functions $r(c\,\lambda_c)$ and $d(c\,\lambda_c)$ in (3.20) by $a(c)$, the author show that (3.21) holds but the limit is a non-Gaussian and non-stable stochastic process with stationary increments Y. The limit process is a H-fractional Brownian motion (resp. stable Lévy motion) in the fast (resp.slow) connection rates and can be recovered from the new intermediate regime limit process by applying two different (extreme) rescaling options: taking $Y^c = c^{-H}\,Y(ct)$ (resp. $Y^c = c^{-1/\alpha}\,Y(ct)$), then Y^c converges functionally to the H-fractional Brownian motion when $c \to 0$ (resp. finite-dimensionally to a totally skewed to the right stable motion when $c \to \infty$).

3.6 Comments on Our Assumptions

The regular variation condition (3.10) is familiar in probability theory and will be the main argument for approximating the Lévy measure of the processes $Z = A, B, C$. It is explicitly assumed in this section. For more details on regular variation theory we refer to [2] or to appendix 3.7. Actually, the conditions leading to assumptions (3.11), (3.12) and (3.13) are simple and numerous. Some of them are presented here:

1. Assumptions (3.11) or (3.12) simply hold if we take any function $t \mapsto f(t)$ which is strictly increasing, less than 1 for all t, has limit 1 and

$$X_t = f(t)X_\infty, \quad \text{with } X_\infty \text{ satisfying (3.10).}$$

Furthermore, since $\alpha \in (1,2)$, we have $\mathbb{E}[X_\infty] < \infty$ and assumption (3.13) works if

$$\lim_{c\to\infty} \frac{\lambda_c}{r(c\,\lambda_c)} \int_0^c [1 - f(u)]\,du = 0.$$

2. In the finite intensity case ($\lambda_c = cst = \lambda$ or $\lambda_\infty \in (0,\infty)$), assumptions (3.11) and (3.12) are also simply implied by $\mathbb{E}[\tau] < \infty$. Recall we have chosen an increasing version of the regularly varying function $r(u)$, then applying the monotone convergence theorem (decreasing version) we have

$$\mathbb{E}\left[\tau\left(1 \wedge \frac{X_\infty}{r(c\,\lambda)}\right)\right] \longrightarrow 0.$$

In the infinite intensity case ($\lambda_\infty = \infty$), and in order to have (3.12) it suffices that τ is a bounded r.v or even if there exists $\beta \in (1, c_\alpha)$,

$$c_\alpha = +\infty 1_{(\alpha\le 1)} + \frac{\alpha}{\alpha-1} 1_{(\alpha>1)},$$

such that $\mathbb{E}[\tau^\beta] < \infty$ and $\lambda_c/c^{\beta-1} \to 0$. Indeed, the Hölder inequality and (3.42) in appendix 3.7 yield

$$\lambda_c\mathbb{E}\left[\tau\left(\frac{X_\infty}{r(c\lambda_c)} \wedge 1\right)\right] \le \mathbb{E}\left[\tau^\beta\right]^{1/\beta} \lambda_c\, \mathbb{E}\left[\left(\frac{X_\infty}{r(c\lambda_c)}\right)^{\beta/(\beta-1)} \wedge 1\right]^{(\beta-1)/\beta}$$

$$\overset{+\infty}{\sim} \mathbb{E}[\tau^\beta]^{1/\beta}\lambda_c(c\lambda_c)^{(1-\beta)/\beta} = \mathbb{E}[\tau^\beta]^{1/\beta}(\lambda_c/c^{\beta-1})^{1/\beta} \to 0.$$

3. Assume the length τ has also a regularly varying distribution tail with index $-\beta$ such that $\frac{1}{\alpha} + \frac{1}{\beta} > 1$. Assume also the *slow connection rate* condition

$$\lim_{c\to\infty} c\,\lambda_c\,\mathbb{P}(\tau > c) = 0, \tag{3.22}$$

then, assumption (3.11) is satisfied. Indeed, using the Hölder inequality with $p > \beta, q > \alpha$, we have

$$\lambda_c\mathbb{E}\left[(\tau\wedge c)\left(\frac{X_\infty}{r(c\lambda_c)}\wedge 1\right)\right] \le \left(c\lambda_c\mathbb{E}\left[\left(\frac{\tau}{c}\right)^p \wedge 1\right]\right)^{\frac{1}{p}} \left(c\lambda_c\mathbb{E}\left[\left(\frac{X_\infty}{r(c\lambda_c)}\right)^q \wedge 1\right]\right)^{\frac{1}{q}}.$$

Now, using (3.42,3.43, appendix 3.7), we have:

$$\left(c\lambda_c\mathbb{E}\left[\left(\frac{\tau}{c}\right)^p \wedge 1\right]\right)^{\frac{1}{p}} \left(c\lambda_c\mathbb{E}\left[\left(\frac{X_\infty}{r(c\lambda_c)}\right)^q \wedge 1\right)\right]\right)^{\frac{1}{q}}$$

$$\overset{+\infty}{\sim} \left(\frac{p}{p-\beta} c\lambda_c \mathbb{P}(\tau > c)\right)^{\frac{1}{p}} \left(\frac{q}{q-\alpha}\right)^{\frac{1}{q}}.$$

4. Assume we have a fixed intensity λ (the case $\lambda = \lambda_c \to \lambda_\infty \in (0,\infty)$ being treated similarly). Recall that $1 < \alpha < 2$ implies $\mathbb{E}[X_\infty] < \infty$. Then necessarily

$$\lim_{t\to\infty} \frac{1}{t} \int_0^t \mathbb{E}[X_\infty - X_u]\, du = 0$$

which this is weaker than assumption (3.13). Condition (3.13) is easily checked if the length τ satisfies $\mathbb{E}[\tau^\gamma] < \infty$ for some $\gamma > 1$ (of course the right tail of the distribution of τ can also be regularly varying with index $-\beta$, $\beta > \gamma$). In this case, necessarily $y \mapsto y^\gamma\, \mathbb{P}(\tau > y)$ is a bounded function (by some bound K). Indeed, choose

$$\gamma < q < \gamma\alpha/(\alpha - 1) \text{ (which implies } 1 - 1/\alpha < \gamma/q < 1)$$

and take $p = q/(q-1)$. By the Hölder inequality, we have

$$0 \le \int_0^t \mathbb{E}[X_\infty - X_u] du = \mathbb{E}[\int_0^{\tau\wedge t} (X_\infty - X_u) du] \le \mathbb{E}[(\tau\wedge t)X_\infty]$$

$$\le \mathbb{E}[X_\infty^p]^{\frac{1}{p}}\, \mathbb{E}[(\tau\wedge t)^q]^{\frac{1}{q}}.$$

Now, write

$$\mathbb{E}[(\tau\wedge t)^q] = t^q \int_0^1 \mathbb{P}(\tau > t\, y^{1/q})\, dy = t^{q-\gamma} \int_0^1 \frac{(t\, y^{1/q})^\gamma\ \mathbb{P}(\tau > t\, y^{1/q})}{y^{\gamma/q}}\, dy$$

$$\le K\, t^{q-\gamma} \int_0^1 \frac{1}{y^{\gamma/q}}\, du = K'\, t^{q-\gamma},$$

and finally because $r_t = t^{1/\alpha}\, l(t)$, we get for some constant $K'' > 0$

$$0 \le \frac{1}{r_t} \int_0^t \mathbb{E}[X_\infty - X_u]\, du \le K'' \frac{t^{1-1/\alpha-\gamma/q}}{l(t)} \longrightarrow 0.$$

Assumption (3.13) is also obviously implied by $\mathbb{E}[\tau\, X_\infty] < \infty$.

5. *Example of a non trivial candidate for the process X describing the evolution in time of the job completion of the transmissions*

Take any positive r.v. X_∞ satisfying (3.10) and an independent increasing H-self-similar process Y such that $\mathbb{E}[Y_1^{-\rho}] < \infty$ when ρ is in the neighborhood of α. Then form the process

$$X_t = Y_t \wedge X_\infty.$$

We know according to (3.44, appendix 3.7) that $\mathbb{P}(\tau > u) \overset{+\infty}{\sim} \mathbb{E}[Y_1^{-\alpha}]\, \mathbb{P}(X_\infty > u^H)$. Then, going through the preceding situations for example, the adequate choice of H leads to our assumptions (3.11), (3.12) or (3.13). In the finite intensity case ($\lambda_c = cst$ or $\lambda_\infty \in (0,\infty)$), a candidate for Y could be a strict H–stable subordinator S^H with $0 < H < \alpha \wedge 1$. See appendix 3.7 below for a precise definition of stable subordinators, and for their distributional properties, the reader is referred to the to the book of Zolotarev [24] or to Jedidi [9, 10]. For self-similar processes, we suggest the book of Embrecht and Maejima [5].

3.7 Proofs of the Results

We start by proving lemma 2 justifying the existence of the load processes.

Proof of lemma 2. Recall the integer Poisson measure M defined in (3.4). For any positive measurable function f defined in $\mathbb{R}_+ \times \mathbb{R} \times \mathbb{D}$, the process $(Z_t)_{t\geq 0}$ defined by

$$Z_t = \int_{(u,y)\in\mathbb{R}\times\mathbb{D}} f(t,u,y)\, M(\mathrm{d}u,\mathrm{d}y) = \sum_{k\in\mathbb{Z}} f(t, T_k, X^k)$$

admits a Lévy-Khintchine representation: for all $\theta \in \mathbb{R}$, $\mathrm{Re}(\theta) \geq 0$ we have

$$\log \mathbb{E}[e^{-\theta Z_t}] = \lambda \int_{\mathbb{R}} \mathbb{E}[e^{-\theta f(t,u,X)} - 1]\, \mathrm{d}u. \tag{3.23}$$

Due to the Poisson representation of the load processes, we deduce that if $Z = A, B, C, \epsilon^1, \epsilon^2$ then for all $t \geq 0$ and $\theta \in \mathbb{R}$, $\mathrm{Re}(\theta) \geq 0$ we have a Lévy-Khintchine representation

$$\log \mathbb{E}[e^{-\theta Z_t}] = \lambda \int_{(0,\infty)} (e^{-\theta x} - 1)\, \nu_t^Z(\mathrm{d}x), \tag{3.24}$$

where

$$\nu_t^Z(\mathrm{d}x) = \begin{cases} \int_{-t}^{\infty} \mathbb{P}(X_{t+u} - X_u \in \mathrm{d}x)\, \mathrm{d}u & \text{if } Z = A \\ \displaystyle\int_0^t \mathbb{P}(X_u \in \mathrm{d}x)\, \mathrm{d}u & \text{if } Z = B \\ t\mathbb{P}(X_\infty \in \mathrm{d}x) & \text{if } Z = C \\ \displaystyle\int_0^\infty \mathbb{P}(X_{t+u} - X_u \in \mathrm{d}x)\, \mathrm{d}u & \text{if } Z = \epsilon^1 \\ \displaystyle\int_0^t \mathbb{P}(X_\infty - X_u \in \mathrm{d}x)\, \mathrm{d}u & \text{if } Z = \epsilon^2. \end{cases}$$

We recall the following useful and easily checkable inequality

$$|e^{-\theta x}-1| \le (2\wedge|\theta|\,x) \le (2\vee|\theta|)(x\wedge 1), \quad \forall x\ge 0,\ \theta\in\mathbb{C},\ Re(\theta)\ge 0. \quad (3.25)$$

Since $x^2\wedge 1 \le x\wedge 1$, then proving that $\nu_t^Z(\mathrm{d}x)$ integrates $x\wedge 1$ will insure that it is a Lévy measure and the r.v. Z_t is finite and infinitely divisible. For more account on infinite divisibility and Lévy processes, the reader is referred to appendix 3.7 or to the book of Sato [21].

(i) and (ii): If $Z = B, C, \epsilon^2$, the result is obvious since in these cases

$$\int_{(0,\infty)} (x\wedge 1)\nu_t^Z(\mathrm{d}x) \le t\,\mathbb{E}[X_\infty\wedge 1].$$

We make a special emphasis on the r.v. C_t. The process $(C_t)_{t\ge 0}$ has stationary and independent increments. This is easily seen from its Poisson integral representation: $\forall\, 0\le s\le t$, we have

$$\begin{aligned} C_t - C_s &= \int_{(s,t]\times\mathbb{D}} y(\infty)\; M(\mathrm{d}u,\mathrm{d}y)\\ &= \int_{(s,t]\times\mathbb{R}_+} y\sum_{k\ge 0}\epsilon_{(T_k,X^k(\infty))}(\mathrm{d}u,\mathrm{d}y) \end{aligned}$$

and because the sequence $(T_k, X^k(\infty))_{k\ge 0}$ forms a marked Poisson process.

(iii): Since $\nu_t^A(\mathrm{d}x) = \nu_t^B(\mathrm{d}x) + \nu_t^{\epsilon^1}(\mathrm{d}x)$, it suffices to check when $\nu_t^{\epsilon^1}(\mathrm{d}x)$ is a Lévy measure, but this is easy since

$$\begin{aligned} \int_{(0,\infty)} x\wedge 1\;\nu_t^{\epsilon^1}(\mathrm{d}x) &= \int_0^\infty \mathbb{E}[(X_{t+u}-X_u)\wedge 1]\,\mathrm{d}u\\ &= \int_0^\infty \mathbb{E}[1_{(\tau>u)}\;(X_{t+u}-X_u)\wedge 1]\mathrm{d}u \le \mathbb{E}[\tau\;(X_\infty\wedge 1)]<\infty. \end{aligned}$$

(iv): The assumption $E[X_\infty] < \infty$ implies $\int_{(0,\infty)} x\,\nu_t^Z(\mathrm{d}x) < \infty$ for $Z = B,\ C, \epsilon^2$ and allows to differentiate the Lévy-Khintchine formula (3.24). Then,

$$\mathbb{E}[B_t]=\int_0^t \mathbb{E}[X_u]\,\mathrm{d}u,\quad \mathbb{E}[C_t]=t\mathbb{E}[X_\infty] \quad\text{and}\quad \mathbb{E}[\epsilon_t^2]=\int_0^t \mathbb{E}[X_\infty - X_u]\,\mathrm{d}u.$$

Now, we tackle the processes A and ϵ^1. Noticing that $x^2\wedge 1 \le x\wedge 1 \le x$, we will show the stronger result: $\int_{(0,\infty)} x\,\nu_t^A(\mathrm{d}x) < \infty$. This result will provide in the same time two results: (1) finiteness and then infinite divisibility of A_t, ϵ_t^1 and (2) the equality $E[A_t] = \lambda\, t\, E[X_\infty]$. Write

$$\int_{(0,\infty)} x\, \nu_t^A(\mathrm{d}x) = \int_{-t}^{\infty} \mathbb{E}[X_{t+u} - X_u]\,\mathrm{d}u = \int_0^t \mathbb{E}[X_u]\,\mathrm{d}u + \int_0^{\infty} \mathbb{E}[X_{t+u} - X_u]\,\mathrm{d}u.$$

Denote the Stieljes function $\mathbb{E}[X_u]$ by $f(u)$. We have $f(\infty) := \lim_{u\to\infty} f(u) = \mathbb{E}[X_\infty] < \infty$. It is then obvious that

$$\lim_{t\to\infty} \frac{1}{t}\int_0^t (f(\infty) - f(u))\,\mathrm{d}u = 0.$$

Applying the Tonelli-Fubini result, we get

$$\begin{aligned}\int_0^{\infty} (f(t+u) - f(u))\mathrm{d}u &= \int_0^{\infty}\int_{[u,t+u]} \mathrm{d}f(r)\mathrm{d}u = \int_{(0,\infty)} (r\wedge t)\mathrm{d}f(r)\\ &= t[f(\infty) - f(t)] + \int_{(0,t]} r\mathrm{d}f(r).\end{aligned}$$

Then, integrating by parts the last integral, we get

$$\int_0^{\infty} (f(t+u) - f(u))\,\mathrm{d}u = \int_0^t (f(\infty) - f(u))\,\mathrm{d}u \tag{3.26}$$

and the last quantity is finite. We conclude by saying that the infinitely divisible r.v. A_t has finite expectation if its Lévy measure $\nu_t(x)$ integrates x in the neighborhood on ∞ (see theorem 25.3 [21]). Then, $\forall t > 0$,

$$\mathbb{E}[A_t]{=}\lambda\int_{(0,\infty)} x\nu_t^A(\mathrm{d}x) = \lambda\{\int_0^t f(u)\mathrm{d}u{+}\int_0^{\infty}(f(t{+}u){-}f(u))\mathrm{d}u\} = \lambda t\mathbb{E}[X_\infty].$$

Notice that this is an expected result since A_t has stationary increments, and then, its expectation is linear in t. We have also proved that

$$\mathbb{E}[\epsilon_t^2] = \lambda\int_0^t \mathbb{E}[X_\infty - X_u]\,\mathrm{d}u = \lambda\int_0^{\infty} \mathbb{E}[X_{t+u} - X_u]\,\mathrm{d}u = \mathbb{E}[\epsilon_t^1]. \tag{3.27}$$

□

Proof of proposition 3. Proving the convergence functional in law (3.18) is an easy problem since the process C and then C^c are Lévy processes. Due to corollary 3.6 p.415 [8], it is enough to show that

$$\textit{the r.v. } C_1^c = \frac{C_c - d(c\,\lambda_c)}{r(c\,\lambda_c)} \textit{ converges in distribution to } S_1$$

a strict totally skewed to the right α-stable variable. Thanks to (3.15), for all $\theta > 0$, the Laplace exponent of C_c is defined by

$$\Psi^c(\theta) = \log \mathbb{E}\big[\, e^{-\theta\, C_1^c}\,\big] = \int_{\mathbb{R}_+} \big(e^{-\theta\, x} - 1 + \theta\, h(x)\big)\; \nu^c(\mathrm{d}x).$$

where

$$h(x) = \begin{cases} 0 & \text{if } 0 < \alpha < 1 \\ \sin x & \text{if } \alpha = 1 \\ x & \text{if } 1 < \alpha < 2. \end{cases}$$

and

$$\nu^c(\mathrm{d}x\,) = c\,\lambda_c\;\mathbb{P}\left(\frac{X_\infty}{r(c\,\lambda_c)} \in \mathrm{d}x\right).$$

Now notice two things, the measure $\nu^c(\mathrm{d}x\,)$ integrates x, then integration by parts gives

$$\Psi^c(\theta) = \theta \int_{\mathbb{R}_+} \big(h'(x) - e^{-\theta\, x}\big)\; \nu^c((x,\infty))\,\mathrm{d}x.$$

Due to assumption (3.10) and then using property (3.40, appendix 3.7), we know that

$$\nu^c((x,\infty)) = \lambda_c\, c\;\mathbb{P}(\,X_\infty > r(c\,\lambda_c)\,x) = \frac{1}{x^\alpha}\,\frac{L(r(c\,\lambda_c))}{l(c\,\lambda_c)^\alpha}\,\frac{L(r(c\,\lambda_c)\,x)}{L(r(c\,\lambda_c))} \longrightarrow \frac{1}{x^\alpha}$$

and

$$L(r(c\,\lambda_c))/\,l(c\lambda_c)^\alpha \to 1.$$

Using the contribution of the function $(h'(x) - e^{-\theta x})/x^\alpha$, it remains to show that $L(r(c\lambda_c)x)/L(r(c\lambda_c))$ is appropriately dominated. This is due to Potter's bounds (3.41, appendix 3.7). Depending on the values of $\alpha \in (0,2)$, and using the flexible dominations (3.41), we conclude that

$$\lim_{c\to\infty} \Psi^c(\theta) = \theta \int_{\mathbb{R}_+} \big(h'(x) - e^{-\theta\, x}\big)\,\frac{1}{x^\alpha}\,\mathrm{d}x,$$

and notice that the latter is the Laplace exponent of a strict α-stable random variable totally skewed to the right (see appendix 3.7). □

Proof of lemma 4. (1) Assume (3.14). For $t > 0$, use the expression given by lemma 2:

$$\frac{\mathbb{E}[\epsilon_t^{1,c}]}{r(c\,\lambda_c)} = \frac{\mathbb{E}[\epsilon_t^{2,c}]}{r(c\,\lambda_c)} = \frac{\lambda_c}{r(c\,\lambda_c)} \int_0^{ct} \mathbb{E}[X_\infty - X_u]\,du \to 0.$$

The claimed result is a simple application of Markov inequality and the end of Sect. 3.2.

(2) We claim that (3.10), with $\alpha \in (1,2)$ and (3.13) together imply (3.14), i.e.

$$\lim_{c\to\infty} \frac{\lambda_c}{r(c\,\lambda_c)} \int_0^{ct} \mathbb{E}[X_\infty - X_u]\,du = 0, \quad \forall t > 0. \tag{3.28}$$

Indeed, (3.28) is immediate for $t \leq 1$. For $t > 1$, we have $\frac{\lambda_c}{\lambda_{ct}} \leq 1$ ($\lambda_.$ is increasing). Now, choose $\rho < 1 - \frac{1}{\alpha}$ and use Potter's bounds (3.41, appendix 3.7) in order to write

$$\begin{aligned}
\theta_{c,t} &= \frac{\lambda_c}{r(c\,\lambda_c)} \int_0^{ct} \mathbb{E}[X_\infty - X_u]\mathrm{d}u = \theta_{c,1}\,\frac{\lambda_c}{\lambda_{ct}}\,\frac{r(ct\lambda_{ct})}{r(c\lambda_c)} \\
&\leq \theta_{c,1}\ K(\rho)t^{\rho+\frac{1}{\alpha}}\,\left(\frac{\lambda_c}{\lambda_{ct}}\right)^{1-\rho-\frac{1}{\alpha}} \\
&\leq \theta_{c,1}\ K(\rho)t^{\rho+\frac{1}{\alpha}} \to 0, \quad \text{as } c \to \infty,
\end{aligned}$$

then, conclude by 1).

(3) Observe that $\epsilon^{1,c}$ is an increasing process while $\epsilon^{2,c}$ is not. In order to control the quantities $\mathbb{P}(\sup_{t\leq T} \epsilon_t^{i,c} > \eta)$ for $i = 1, 2$ and $T > 0$, observe also that

$$\sup_{t\leq T} \epsilon_t^{i,c} = \sup_{t\leq T} \frac{\epsilon^i_{c\,t}}{r(c\,\lambda_c)} = \sup_{s\leq 1} \frac{\epsilon^i_{c\,T\,s}}{r(c\,\lambda_c)} = \frac{r(c\,\lambda_c\,T)}{r(c\,\lambda_c)} \sup_{s\leq 1} \frac{\epsilon^i_{c\,T\,s}}{r(c\,\lambda_c\,T)}. \tag{3.29}$$

Since $\limsup_{c\to\infty} \dfrac{r(c\lambda_c T)}{r(c\lambda_c)} < \infty$, $\forall T > 0$, *it is enough to show that*

$\mathbb{P}(\sup_{s\leq 1} \epsilon_s^{i,c} > \eta) \to 0.$

(3)(i): The control of $\epsilon^{1,c}$ is easier than the one of $\epsilon^{2,c}$ because it is increasing and then

$$\sup_{s\leq 1} \epsilon_s^{1,c} = \epsilon_1^{1,c} = \epsilon_c^1 / r(c\,\lambda_c).$$

It is enough to reproduce the argument of (1).

(3)(ii) and (2)(iii): Recall τ_k is the length of the k-th transmission: $\tau_k = \inf\{t : X^k(t) = X^k(\infty)\}$. Due to the Lévy-Khintchine representation (3.24), the following stochastic inequality holds:

$$\begin{aligned}
\frac{\epsilon_c^1}{r(c\,\lambda_c)} &\stackrel{d}{=} \frac{1}{r(c\,\lambda_c)} \sum_{k=1}^{\infty} (X^k(c + T_k) - X^k(T_k))\, 1_{(T_k \leq \tau_k)} \\
&\leq \delta_c^1 := \frac{1}{r(c\,\lambda_c)} \sum_{k=1}^{\infty} X^k(\infty)\, 1_{(T_k \leq \tau_k)}.
\end{aligned} \tag{3.30}$$

For the process $\epsilon^{2,c}$, we use the more tractable Poisson integral representation. Let

$$y^{\leftarrow}(\infty) = \inf\{u : y(u) = y(\infty)\}$$

the first time an increasing function y on $\mathbb{D}$ reaches its maximum. We have

$$\begin{aligned}
\epsilon_s^{2,c} &= \frac{1}{r(c\,\lambda_c)} \int_{\left((cs-y^{\leftarrow}(\infty))_+,\, cs\right]\times\mathbb{D}} \big(y(\infty) - y(cs-u)\big)\, M(\mathrm{d}u, \mathrm{d}y) \\
&\leq \frac{1}{r(c\,\lambda_c)} \int_{\left((cs-y^{\leftarrow}(\infty))_+,\, cs\right]\times\mathbb{D}} y(\infty)\, M(\mathrm{d}u, \mathrm{d}y) \\
&= \delta_{cs}^2 := C_s^c - D_s^c
\end{aligned} \tag{3.31}$$

where the process C^c is same as the one defined in (3.16),

$$C_s^c = \frac{1}{r(c\,\lambda_c)} \Big\{ \int_{[0,\,cs]\times\mathbb{D}} y(\infty)\, M(\mathrm{d}u, \mathrm{d}y) \;-\; s\, d(c\,\lambda_c) \Big\}$$

and

$$D_s^c = \frac{1}{r(c\,\lambda_c)} \Big\{ \int_{\left[0,\,(cs-y^{\leftarrow}(\infty))_+\right]\times\mathbb{D}} y(\infty)\, M(\mathrm{d}u, \mathrm{d}y) \;-\; s\, d(c\,\lambda_c) \Big\}.$$

To resume, we have: for all $\eta > 0$,

$$\mathbb{P}(\sup_{s\leq 1} \epsilon_s^{i,c} > \eta) \leq \begin{cases} \mathbb{P}(\delta_c^1 > \eta) & \text{if } i = 1 \\ \mathbb{P}(\sup_{s\leq 1} \delta_{cs}^2 > \eta) & \text{if } i = 2. \end{cases} \tag{3.32}$$

We will give the proof of $(2)(ii)$ and $(2)(iii)$ through the following 5 steps:

Step (1) As $c \to \infty$, prove the convergence of δ_c^1, this will complete the assertion $(3)(ii)$ for $\epsilon^{1,c}$ and the finite-dimensional convergence of the processes $(\delta_{cs}^2)_{s\in[0,1]}$ to the null process.

Step (2) Prove tightness of the family $\{(\delta_{cs}^2)_{s\in[0,1]},\ c > 0\}$.

Step (3) Immediately deduce from steps (1) and (2) the convergence functionally in law of $(\delta_{cs}^2)_{s\in[0,1]}$ to the null process.

Step (4) Apply proposition 2.4 p. 339 [8] that shows the continuity of the functions

$$\begin{aligned}
\psi_a : \mathbb{D} &\longrightarrow \mathbb{R}_+ \\
z &\longmapsto \sup_{s\leq a} |z_s|
\end{aligned}$$

at each point z such that z is continuous in a.

Step (5) Knowing step (3), deduce from step (4) that $\mathbb{P}\big(\sup_{s\leq 1}|\delta^2_{cs}| > \eta\big) \to 0$, and then from (3.32) and (3.29) that $\mathbb{P}\big(\sup_{t\leq T}|\epsilon^{2,c}_t| > \eta\big) \to 0$.

It is now clear why we only need to check steps (1) and (2).

Step (1) Convergence of δ^1_c to 0 under the assumption (3.12) and finite-dimensional convergence to 0 of $(\delta^2_{cs})_{s\in[0,1]}$ under the assumptions (3.11).

Actually, as noticed at the the end of Sect. 3.2, we only need to show that the r.v. $\delta^2_c \to 0$ in distribution as $c \to \infty$. For this purpose, we use the Lévy-Khintchine formula (3.23) and get: for all $\theta \in \mathbb{R}$, $\mathrm{Re}(\theta) \geq 0$,

$$-\frac{\log \mathbb{E}[e^{-\theta\delta^1_c}]}{\lambda_c} = \int_{\mathbb{R}} \mathbb{E}\left[(1-e^{-\theta X_\infty/r(c\lambda_c)})1_{(0\leq u\leq\tau)}\right]du = \mathbb{E}\left[\tau(1-e^{-\theta X_\infty/r(c\lambda_c)})\right]. \tag{3.33}$$

The laplace exponent of $\delta^2_c = \sum_{k=1}^{\infty} X^k(\infty)\, 1_{((c-\tau_k)_+\leq T_k\leq c)}$ is calculated similarly:

$$\begin{aligned}-\frac{\log \mathbb{E}[e^{-\theta\delta^2_c}]}{\lambda_c} &= \int_{\mathbb{R}} \mathbb{E}\left[(1-e^{-\theta X_\infty/r(c\lambda_c)})1_{((c-\tau)_+\leq u\leq c)}\right]du \\ &= \mathbb{E}[(c\wedge\tau)(1-e^{-\theta X_\infty/r(c\,\lambda_c)})].\end{aligned} \tag{3.34}$$

We will also use the following: for any positive r.v. Y and any $\eta > 0$, there exists a number $K_\eta > 0$ such that:

$$\mathbb{P}(Y > \eta) \leq K_\eta\, \mathbb{E}[1-e^{-Y}]. \tag{3.35}$$

The latter is true since

$$\begin{aligned}\eta\,\mathbb{E}[1-e^{-Y/\eta}] &= \int_0^\infty e^{-x/\eta}\mathbb{P}(Y > x)\mathrm{d}x \geq \eta\int_0^\eta e^{-x/\eta}\mathbb{P}(Y > \eta)\mathrm{d}x \\ &= \eta^2(1-e^{-1})\mathbb{P}(Y > \eta).\end{aligned}$$

Then, since for all $y \geq 0$, $1-e^{-y} \leq 1\wedge y$, we deduce that the Laplace exponent of δ^i_c satisfies

$$\begin{aligned}0 &\leq -\log\mathbb{E}[e^{-\delta^i_c}] = \lambda_c\mathbb{E}\left[(\tau\, 1_{(i=1)} + \tau\wedge c\, 1_{(i=2)})(1-e^{-X_\infty/r(c\,\lambda_c)})\right] \\ &\leq \rho_c := \lambda_c\mathbb{E}\left[(\tau 1_{(i=1)} + \tau\wedge c\, 1_{(i=2)})\left(1\wedge\frac{X_\infty}{r(c\,\lambda_c)}\right)\right],\end{aligned} \tag{3.36}$$

Finally, (3.35) and (3.36) imply

$$\mathbb{P}(\delta^n_c > \eta) \leq K_\eta\mathbb{E}[1-e^{-\delta^i_c}] \leq K_\eta(1-e^{-\rho_c}) \leq K_\eta\rho_c \tag{3.37}$$

and the claim $\rho_c \longrightarrow 0$, being the conditions (3.11) for $i = 1$ and (3.12) for $i = 2$, is the key of the trick.

Step (2) Tightness of the family $\{(\delta^2_{cs})_{s\in[0,1]},\ c > 0\}$ *under the assumptions (3.10) and (3.11).*

Recall the representation (3.31) where the processes D^c are introduced. A simplified version of lemma 3.32 p.352 [8] insures tightness of the family $\{(\delta^2_{c.}),\ c > 0\}$ if the following holds: both families $\{C^c,\ c > 0\}$ and $\{D^c,\ c > 0\}$ are tight and for all $\eta > 0$

$$\lim_{c\to\infty} \mathbb{P}(\sup_{s\le 1} \Delta C^c_s > \eta) = 0.$$

In proposition 3, we have already shown tightness of the family $\{C^c,\ c > 0\}$ since C^c converges functionally in law. That was due to its Lévy character (i.e. independent and stationary increments). Moreover, because the arrival process N that jumps with size equal to 1, C^c satisfies:

$$\Delta C^c_s = C^c_s - C^c_{s-} = \frac{1}{r(c\,\lambda_c)}\, X^{N_{cs}}(\infty)\, 1_{(\Delta N_{cs}=1)}.$$

Then, using the independence between $(X^k)_{k\in\mathbb{Z}}$ and N, we have for all $\eta > 0$:

$$\begin{aligned}
\mathbb{P}(\sup_{s\le 1} \Delta C^c_s > \eta) &= \mathbb{P}(\exists\, s \le 1\ :\ \Delta C^c_s > \eta)\\
&= \mathbb{P}(\exists\, t \in [0,c]\ :\ \Delta N_t = 1,\ X^{N_t}(\infty) > \eta\, r(c\,\lambda_c))\\
&= \sum_{k=1}^{\infty} \mathbb{P}(\exists\, t \in [0,c]\ :\ T_k = t,\ X^k(\infty) > \eta\, r(c\,\lambda_c))\\
&= \mathbb{P}(X_\infty > \eta\, r(c\,\lambda_c)) \sum_{k=1}^{\infty} \mathbb{P}(\exists\, t \in [0,c]\ :\ T_k = t)\\
&= \mathbb{P}(X_\infty > \eta\, r(c\,\lambda_c))\, \mathbb{P}(\sup_{t\le c} \Delta N_t = 1)\\
&\le \mathbb{P}(X_\infty > \eta\, r(c\,\lambda_c)) \to 0.
\end{aligned}$$

The process D^c is not Lévy but has independent increments. It only remains to show its tightness, and actually we will prove a stronger result: the process D^c also converge functionally in law to the same stable process S as C^c. This is an application of corollary 4.43 p. 440 [8] that says that convergence functionally in law of processes with independent increments is equivalent to convergence of their characteristic functions locally uniformly in time : for each $u \in \mathbb{R}$ and each finite interval $[T, T'] \subset [0, 1]$

$$\mathbb{E}[e^{iuD^c_s}] \longrightarrow \mathbb{E}[e^{iuS_s}], \quad \text{uniformly in } s \in [T, T']. \tag{3.38}$$

Lévy-Khintchine representation (3.23) and then (3.17) imply

$$\log \mathbb{E}[e^{\mathrm{i}uD_s^c}] = \lambda_c \int_{\mathbb{R}} \mathbb{E}[(e^{\mathrm{i}\,u\,X_\infty/r(c\,\lambda_c)} - 1)\,1_{u\leq(cs-\tau)_+}]\,du \;-\; \mathrm{i}\,u\,s\,\frac{d(c\,\lambda_c)}{r(c\,\lambda_c)}$$

$$= \lambda_c\,\mathbb{E}[(cs-\tau)_+\,(e^{\mathrm{i}\,u\,X_\infty/r(c\,\lambda_c)} - 1)] \;-\; \mathrm{i}\,u\,c\,\lambda_c\,s\,\mathbb{E}\left[h\left(\frac{X_\infty}{r(c\,\lambda_c)}\right)\right]$$

$$= \log \mathbb{E}[e^{\mathrm{i}uC_s^c}] \;+\; \lambda_c\,\mathbb{E}[(\tau\wedge cs)\,(1 - e^{\mathrm{i}\,u\,X_\infty/r(c\,\lambda_c)})].$$

Denoting $z_{c,s} = \lambda_c\,\mathbb{E}[\,(\tau\wedge cs)\,(e^{\mathrm{i}\,u\,X_\infty/r(c\,\lambda_c)} - 1)\,]$ and using (3.25) we have

$$|z_{c,s}| \leq (2\vee|u|)\,\lambda_c\,\mathbb{E}\left[\,(\tau\wedge c)\,\left(\frac{X_\infty}{r(c\,\lambda_c)}\wedge 1\right)\right].$$

Using assumption (3.11) or (3.12), we see that $z_{c,s}$ go 0 uniformly in s. Now, write

$$\mathbb{E}[e^{\mathrm{i}uD_s^c}] - \mathbb{E}[e^{\mathrm{i}uS_s}] = \left(\,\mathbb{E}[e^{\mathrm{i}uC_s^c}] - \mathbb{E}[e^{\mathrm{i}uS_s}]\,\right)e^{z_{c,s}} + (e^{z_{c,s}} - 1)\,\mathbb{E}[e^{\mathrm{i}uS_s}]$$

$$\left|\,\mathbb{E}[e^{\mathrm{i}uD_s^c}] - \mathbb{E}[e^{\mathrm{i}uS_s}]\,\right| \leq \left|\,\mathbb{E}[e^{\mathrm{i}uC_s^c}] - \mathbb{E}[e^{\mathrm{i}uS_s}]\,\right|\; e^{|z_{c,s}|} + |z_{c,s}|\,e^{|z_{c,s}|}.$$

Since C^c has independent increments and using again corollary 4.43 p. 440 [8], we get that the quantity $\left|\,\mathbb{E}[e^{\mathrm{i}uD_s^c}] - \mathbb{E}[e^{\mathrm{i}uS_s}]\,\right|$ goes to 0 with the required uniformity. We conclude that (3.38) is true. □

Proof of theorem 5. We have all the ingredients for this proof. Recall $A^c = C^c + \epsilon^{1,c} - \epsilon^{2,c}$ and $B^c = C^c - \epsilon^{2,c}$. The process C^c converges functionally in law to a stable process and the processes $\epsilon^{1,c}$, $\epsilon^{2,c}$ satisfy lemma 4. Assertion a) in theorem 5 has already been proved in proposition 3. Assertions b) and c) are a direct application of lemma 3.31 p.352 [8] (which is kind of Slutzy theorem). The weaker assertion d) is proved using Slutzky theorem. □

Acknowledgements The authors would like to thank the referees for pointing out the obscure points and for their helpful comments, which improved the presentation of this paper. This work was funded by Atlantic Canada Opportunity Agency (ACOA) and through grant from the National Science and Engineering Research Council of Canada (NSERC) to Dr. Vartan Choulakian.

Appendix : Regular Variation

In this appendix, the r.v U has a regularly varying distribution tail with index $-\alpha$, $\alpha > 0$ and could play the rule of the r.v. X_∞ and also of the r.v τ: there exists a deterministic function r such that for each $x > 0$,

$$\lim_{u\to\infty} u\,\mathbb{P}(U > r(u)\,x) = \frac{1}{x^\alpha}. \tag{3.39}$$

Condition (3.39) is equivalent to one of the the definitions: (1) existence of a slowly varying function L (i.e. for all $x > 0, lim_{u\to\infty} L(ux)/L(u) = 1$) such that $\mathbb{P}(U>x) = L(x)x^{-\alpha}$; (2) dilation of the right tail of U gives $\lim_{u\to\infty} \mathbb{P}(U > ux)/\mathbb{P}(U > u) = x^{-\alpha}$. It also implies the existence of moments of any order $0 < \epsilon < \alpha$ for the r.v. U: $\mathbb{E}[U^\epsilon] < \infty$. When $\alpha \in (0,2)$, it is said in the literature that U has a *heavy tail.*

In regular variation theory, the function $r(u)$ is the asymptotic inverse of the function $x^\alpha/L(x)$ and necessarily $r(u)$ is regularly varying with index $1/\alpha$, i.e

$$\lim_{u\to\infty} \frac{L(r(u))}{l(u)^\alpha} = 1 \textit{ and } r(u) = u^{1/\alpha}\, l(u) \tag{3.40}$$

for some slowly varying function l. Furthermore, it is known that *there exits an increasing function $r'(u)$ which is equivalent to $r(u)$ when $u \to \infty$.* This is the reason why we can take an increasing version for the function $r(u)$. A useful result for dominating slowly varying functions is known as Potter's bounds and is given by theorem 1.5.6 [2]: If L is a slowly varying function bounded away from 0 and ∞ on each interval $[A,\infty)$, then for all $\rho, \rho' > 0$, there exits $K(\rho) < 1 < K(\rho')$ such that

$$K(\rho)(x/y)^\rho \le L(x)/L(y) \le K(\rho')(x/y)^{-\rho'}, \; (0 < x \le y). \tag{3.41}$$

Proposition 3.1. *Assume (3.39) and let $u \to \infty$, then*

$$\mathbb{E}\left[\left(\frac{U}{r(u)}\right)^q \wedge 1\right] \overset{+\infty}{\sim} \frac{q}{q-\alpha}\,\frac{1}{u} \qquad \text{for all } q > \alpha \tag{3.42}$$

$$\mathbb{E}\left[\left(\frac{U}{u}\right)^q \wedge 1\right] \overset{+\infty}{\sim} \begin{cases} \dfrac{q}{q-\alpha}\; P(U > u) \text{ for all } q > \alpha \\[2ex] \dfrac{\mathbb{E}[U^q]}{u^q} \qquad\qquad \text{for all } q < \alpha. \end{cases} \tag{3.43}$$

Proof. Equivalence (3.42) is a direct application of (3.43) and (3.39). For (3.43), $q > \alpha$ just write

$$\mathbb{E}\left[\left(\frac{U}{u}\right)^q \wedge 1\right] = \int_0^1 \mathbb{P}\left(\frac{U}{u} > v^{1/q}\right) \mathrm{d}v = q \int_0^\infty \left(\frac{u}{v}\right)^{-q} 1_{(\frac{u}{v}\ge 1)}\, \mathbb{P}(U > v)\, \frac{\mathrm{d}v}{v}.$$

For $q < \alpha$, write $\mathbb{E}[\,(\frac{U}{u})^q \wedge 1\,] = q \displaystyle\int_0^\infty v^{-q}\, 1_{(v\ge 1)}\, \mathbb{P}(U > \frac{u}{v}) \frac{\mathrm{d}v}{v}$. Then apply ARANDELOVIĆ's theorem 4.1.6 [2]. □

Proposition 3.2. *Assume (3.39). Let U a r.v. with regularly varying distribution tail of order $-\alpha < 0$ and Y be an increasing H–self-similar process independent from*

U and $\tau = \inf\{t : Y_t \geq U\}$. Assume that Y_1 admits a density f and $E[Y_1^{-\rho}] < \infty$ if ρ is in the neighborhood of α. Then τ regularly varying distribution tail of order $-\alpha H$ and, more precisely, as $u \to \infty$,

$$\mathbb{P}(\tau > u) \overset{+\infty}{\sim} E[Y_1^{-\alpha}]\,\mathbb{P}(U > u^H). \tag{3.44}$$

Proof. The result is obvious since

$$\mathbb{P}(\tau > u) = \mathbb{P}(Y_u < U) = \mathbb{P}(Y_1\, u^H < U) = \int_{(0,\infty)} \mathbb{P}(Y_1\, u^H < v)\mathbb{P}(U \in \mathrm{d}v)$$
$$= \frac{1}{u^H}\int_0^\infty f\left(\frac{v}{u^H}\right)\mathbb{P}(U > v)\mathrm{d}v = \int_0^\infty \left[\frac{v}{u^H} f\left(\frac{v}{u^H}\right)\right]\mathbb{P}(U > v)\frac{\mathrm{d}v}{v}.$$

Then apply again ARANDELOVIĆ's theorem 4.1.6. □

Appendix : Lévy Processes and Stable Lévy Motions

Let $(\Omega, \mathcal{F}, (\mathcal{F}_t)_{t\geq 0}, \mathbb{P})$ be a stochastic basis. We say that a process X on this basis is a Lévy process, if it has independent and stationary increments. It is entirely characterized by its Lévy-Khintchine exponent: for all $u \in \mathbb{R}$, $\varphi(u) = \log \mathbb{E}[e^{\mathrm{i}\,u\,X_1}] = \mathrm{i}\,u\,d - \frac{b^2 u^2}{2} + \int [e^{\mathrm{i}\,u\,x} - 1 - \mathrm{i}\,u\,g(x)]\,\nu(\mathrm{d}x)$ where $d \in \mathbb{R}, b \geq 0$ and the Lévy measure ν is a positive measure giving no mass to 0, integrating $x^2 \wedge 1$. The truncation function g could be any real bounded function such that $\lim_{x\to 0}(g(x) - x)/x^2 = 0$. A process S is called *stable process* with index $\alpha \in (0, 2]$, if it is a Lévy process and (i) $\alpha = 2$ and $\nu = 0$ (drifted and scaled Brownian motion) or (ii) $b = 0$ and $0 < \alpha < 2$ and $\nu(\mathrm{d}x) = (c_+ 1_(x > 0) + c_- 1_(x < 0))x^{-\alpha-1}\mathrm{d}x)$, $c_+, c_- \geq 0$. A standard totally skewed to the right stable process $(0 < \alpha < 2)$ has $d = c_- = 0, c_+ = 1$ with the truncation function equal to $g(x) = \sin x$. A standard stable subordinator $(0 < \alpha < 1)$ has $d = c_- = 0, c_+ = 1$ with the truncation function equal $g(x) = 0$.

References

1. Billingsley P.: Convergence of Probability Measures, 2nd edn. Wiley, New York (1999)
2. Bingham N.H., Goldie C.M., Teugels J.L.: Regular Variation. Cambridge University Press, Cambridge (1987)
3. Bremaud P.: An insensitivity property of Lindberg's estimate for delayed claims. J. Appl. Probab. **37**, 914–917 (2000)
4. Çağlar M.: A Long-Range Dependent Workload Model for Packet Data Traffic. Math. Oper. Res. **29**(1), 92-105 (2004)

5. Embrecht P., Maejima M.: Selfsimilar Processes. Princeton Univ Press, Princeton and Oxford (2002)
6. Fasen V.: Modeling network traffic by a cluster Poisson input process with heavy and light-tailed file sizes. Queueing Systems: Theory and Applications **66**(4), 313–350 (2010)
7. Fasen V., Samorodnitsky G.: A fluid cluster Poisson input process can look like a fractional Brownian motion even in the slow growth aggregation regime. Adv. in App. Probab. **41**(2), 393–427 (2009)
8. Jacod J., Shiryaev A.N.: Limit Theorems for Stochastic Processes. New York, Springer (1987)
9. Jedidi W.: Stable Processes, Mixing, and Distributional Properties. I. Theory Probab. Appl. **52**(4), 580–593 (2008)
10. Jedidi W.: Stable Processes, Mixing, and Distributional Properties. II. Theory Probab. Appl. **53**(1), 81–105 (2009)
11. Kaj I.: The long-memory infinite source Poisson model: scaling limit and its representation. Technical report, Uppsala, **19** (2003)
12. Kaj I., Taqqu M.S.: Convergence to fractional Brownian motion and to the Telecom process: the integral representation approach. In: In and Out of Equilibrium 2. Sidoravicius V., Vares M.E.: (eds.) Series: Progress in Probability, **60**. pp. 383-427. Birkhäuser (2008)
13. Klüppelberg C., Mikosch T., Schärf A.: Regular variation in the mean and stable limits for Poisson shot noise. Bernoulli **9**, 467–496 (2003)
14. Konstantopoulos T., Lin S.J.: Macroscopic models for long-range dependent network traffic. Queueing Sys. Th. Appl. **28**(1-3), 215–243 (1998)
15. Maulik K., Resnick S., Rootzen H.: Asymptotic independence and a network traffic model. J. of App. Prob. **39**, 671–699 (2002)
16. Maulik K., Resnick, S.: Small and large time scale analysis of a network traffic model. Queueing Systems, **43**(3), 221–250 (2002)
17. Maulik K., Resnick, S.: The self-similar and multifractal nature of a network traffic model. Stoch. Models. **19**(4), 549–577, Dekker (2003)
18. Mikosch T., Resnick, S., Rootzén, H., Stegeman, A.W.: Is network traffic approximated by stable Lévy motion or fractional Brownian motion?. Ann. Appl. Probab. **12**(1), 23–68 (2002)
19. Mikosch T., Samorodnitsky, G.: Scaling limits for cumulative input processes. Math. Oper. Res. **32**, 890-919 (2007)
20. Resnick S., Van Den Berg, E.: Weak Convergence of high-speed network traffic models. J. Appl. Probab. **37**(2), 575-597 (2000)
21. Sato K.I.: Lévy Processes and Infinitely Divisible Distributions. Cambridge University Press, Camrbridge (1999)
22. Whitt W.: Limits for cumulative input processes to queues. Probab. Engrg. Inform. Sci. **14**, 123–150 (2000)
23. Whitt, W.: Stochastic-Process Limits: An Introduction to Stochastic-Process Limits and their Application to Queues. Springer Series in Op. Res. and Fin. Eng. Hardcover (2002)
24. Zolotarev, V.M.: One Dimensional Stable Laws. Trans. Math. Mono. Am. Math. Soc. **31**(2), 424–425 (1986)

Chapter 4
The Lower Classes of the Sub-Fractional Brownian Motion

Charles El-Nouty

Abstract Let $\{B_H(t), t \in \mathbb{R}\}$ be a fractional Brownian motion with Hurst index $0 < H < 1$. Consider the sub-fractional Brownian motion X_H defined as follows :

$$X_H(t) = \frac{B_H(t) + B_H(-t)}{\sqrt{2}}, t \geq 0.$$

We characterize the lower classes of the sup-norm statistic of X_H by an integral test.

4.1 Introduction and Main Results

Let $\{B_H(t), t \in \mathbb{R}\}$ be a fractional Brownian motion (fBm) with Hurst index $0 < H < 1$, i.e. a centered Gaussian process with stationary increments satisfying $B_H(0) = 0$, with probability 1, and $\mathbb{E}(B_H(t))^2 = | t |^{2H}, t \in \mathbb{R}$.

Consider the sub-fractional Brownian motion (sfBm)

$$X_H(s) = \frac{B_H(s) + B_H(-s)}{\sqrt{2}}, s \geq 0,$$

and define its sup-norm statistic

$$Y(t) = \sup_{0 \leq s \leq t} | X_H(s) |, t \geq 0.$$

We have for any $t \geq s$

C. El-Nouty (✉)
UMR 557 Inserm/ U1125 Inra/ Cnam/ Université Paris XIII, SMBH-Université Paris XIII, 74 rue Marcel Cachin, 93017 Bobigny Cedex, FRANCE
e-mail: c.el-nouty@uren.smbh.univ-paris13.fr

M. Zili and D.V. Filatova (eds.), *Stochastic Differential Equations and Processes*, Springer Proceedings in Mathematics 7, DOI 10.1007/978-3-642-22368-6_4,

$$cov\Big(X_H(t)X_H(s)\Big) = s^{2H} + t^{2H} - \frac{1}{2}\Big((s+t)^{2H} + (t-s)^{2H}\Big)$$

and therefore

$$\mathrm{Var} X_H(t) = \left(2 - 2^{2H-1}\right) t^{2H}.$$

The sfBm was introduced by [1]. This process is an interesting one on its own. Indeed, when $H > 1/2$, it arises from occupation time fluctuations of branching particle systems. Roughly speaking, the sfBm has some of the main properties of the fBm, but it has non-stationary increments. Note also that, when $H = 1/2$, X_H is the famous Brownian motion. We refer to [1,2,4,11,12] for further information.

Let us recall some basic properties of the sfBm which we will also use.

Proposition 4.1. *The sfBm is a self-similar process with index H, i.e. $X_H(at)$ and $a^H X_H(t)$ have the same distribution for all $a > 0$.*

Proposition 4.2. *We have for any $t \geq s$*

$$\mathbb{E}\Big(X_H(t) - X_H(s)\Big)^2 = -2^{2H-1}\left(t^{2H} + s^{2H}\right) + (t+s)^{2H} + (t-s)^{2H}.$$

Moreover,

$$L_1\,(t-s)^{2H} \leq \mathbb{E}\left(X_H(t) - X_H(s)\right)^2 \leq L_2\,(t-s)^{2H}, \tag{4.1}$$

where $L_1 = \min\left(1, 2 - 2^{2H-1}\right)$ and $L_2 = \max\left(1, 2 - 2^{2H-1}\right)$.

Proposition 4.3. *The sfBm can be represented as a random integral, i.e.:*

$$X_H(t) = \int_{\mathbb{R}} G(t,s)\,dW(s), \tag{4.2}$$

where $\{W(s), s \in \mathbb{R}\}$ is a Wiener process,

$$G(t,s) = k_{2H}^{-1}\left(\max(t-s,0)^{H-1/2} + \max(-t-s,0)^{H-1/2} - 2\max(-s,0)^{H-1/2}\right)$$

and

$$k_{2H} = \left(2\left(\int_{-\infty}^{0}\left(\left(1-s\right)^{H-1/2} - (-s)^{H-1/2}\right)ds + \frac{1}{2H}\right)\right)^{1/2}.$$

Proposition 4.1 implies that, for all $\epsilon > 0$,

$$\mathbb{P}(Y(t) \leq \epsilon\, t^H) = \mathbb{P}(Y(1) \leq \epsilon) := \phi(\epsilon).$$

The behavior of the small ball function ϕ is given in the following lemma.

Lemma 4.1. *There exists a constant* $K_0, 0 < K_0 \leq 1$, *depending on* H *only, such that for* $0 < \epsilon < 1$

$$\exp\Big(-\frac{1}{K_0\epsilon^{1/H}}\Big) \leq \phi(\epsilon) \leq \exp\Big(-\frac{K_0}{\epsilon^{1/H}}\Big).$$

Let $\{Z(t), t \geq 0\}$ be a stochastic process defined on the basic probability space $(\Omega, \mathscr{A}, \mathbb{P})$. We recall now two definitions of the Lévy classes, stated in [9].

Definition 4.1. The function $f(t), t \geq 0$, belongs to the lower-lower class of the process $Z, (f \in LLC(Z))$, if for almost all $\omega \in \Omega$ there exists $t_0 = t_0(\omega)$ such that $Z(t) \geq f(t)$ for every $t > t_0$.

Definition 4.2. The function $f(t), t \geq 0$, belongs to the lower-upper class of the process $Z, (f \in LUC(Z))$, if for almost all $\omega \in \Omega$ there exists a sequence $0 < t_1 = t_1(\omega) < t_2 = t_2(\omega) < \ldots$ *with* $t_n \to +\infty$, as $n \to +\infty$, such that $Z(t_n) \leq f(t_n), n \in \mathbb{N}^*$.

The study of the lower classes of the fBm was initiated by [10] and extended in [5, 6]. The aim of this paper is to characterize the lower classes of $Y(.)$. Our main result is given in the following theorem.

Theorem 4.1. *Let* $f(t)$ *be a function of* $t \geq 0$ *such that* $f(t) \geq 1$. *Then we have, with probability 1,*

$$f \in LLC(Y)$$

if and only if

$$\frac{f(t)}{t^H} \textit{ is bounded and } \int_0^{+\infty} f(t)^{-1/H}\phi\left(\frac{f(t)}{t^H}\right)dt < +\infty.$$

Let us make some comments on the above theorem. Theorem 4.1 is similar to theorem 1.2 of ([10], p. 193) when the author studied the lower classes of the sup-norm statistic of the fBm and to theorem 1.1 of [7] when it was studied the lower classes of the sup-norm statistic of the Riemann-Liouville process. This is not really surprising. Indeed these three processes are quite close: they have an integral representation of the following type (up to a constant)

$$\int_0^t (t-s)^\alpha dW(s) + \int_{-\infty}^0 h(t,s)dW(s),$$

where α is suitably chosen and $h(t,.) \in L^2(\mathbb{R}_-)$. However, the proof of the necessity part of theorem 4.1 requires a precise expression of the function h. When $h = 0$ (i.e. the Riemann-Liouville process), just few changes in the proofs of [5, 6] were necessary. But, when the function h is more complex (the sfBM case), we have to establish some technical results. This is the flavor of this paper. Finally the sfBm is a second example that the non-stationarity property of the increments has no role in the study in the lower classes, as soon as the methodology introduced by [10] works. The first example was given in [7].

Although theorem 4.1 depends on an unknown function ϕ, it is sharp. Indeed, set

$$f(t) = \lambda \frac{t^H}{(\log\log t)^H}, t \geq 3, \lambda > 0.$$

Combining lemma 4.1 with a careful computation, we get that, if $\lambda < K_0^H$ then $f \in LLC(Y)$, else if $\lambda \geq K_0^{-H}$, then $f \in LUC(Y)$.

In Sect. 4.2, we prove lemma 4.1 and state some basic results on ϕ. The main steps of the sufficiency are given in Sect. 4.3. The proof of the necessity is postponed to Sects. 4.4 and 4.5. Section 4.4 consists in constructing some well-chosen sequences and a suitable set whereas we end the proof of the necessity in Sect. 4.5 by establishing some key small ball estimates. The proofs which are similar to those in [5–7] will consequently be omitted.

In the sequel, there is no loss of generality to assume $H \neq 1/2$.

For notational convenience, we define $a_t = \frac{f(t)}{t^H}$ and $b_t = \phi(a_t) = \phi\left(\frac{f(t)}{t^H}\right)$.

4.2 Proof of Lemma 4.1 and Preliminary Results

Before stating some preliminary results, we have to prove lemma 4.1.

Proof of Lemma 4.1. By using (4.1), we have for any $h > 0$

$$\mathrm{Var}\Big(X_H(t+h) - X_H(t)\Big) = \mathbb{E}\Big(X_H(t+h) - X_H(t)\Big)^2 \leq L_2\, h^{2H} \leq 2\, h^{2H}. \quad (4.3)$$

Moreover, we deduce from (4.2) that

$$\begin{aligned} &\mathrm{Var}\,(X_H(t+h) \mid X_H(s),\ 0 \leq s \leq t) \\ &\geq \mathrm{Var}\Big(k_{2H}^{-1} \int_t^{t+h} (t+h-s)^{H-1/2}\, dW(s)\Big) = k_{2H}^{-2}\, \frac{1}{2H}\, h^{2H}. \quad (4.4) \end{aligned}$$

Since (4.3) and (4.4) hold, we can apply theorem 2.1 of [8]. The proof of lemma 4.1 is therefore complete. □

Set $\psi = -\log\phi$. Thus, ψ is positive and nonincreasing. A straight consequence of lemma 4.1 is given in the following lemma.

Lemma 4.2. *We have for* $0 < \epsilon < 1$

$$\frac{1}{K_1 \epsilon^{1/H}} \leq \psi(\epsilon) \leq \frac{K_1}{\epsilon^{1/H}},$$

where $K_1 \geq 1/K_0$.

Lemma 4.3. *ψ is convex.*

Lemma 4.3 (see [3]) implies the existence of the right derivative ψ' of ψ. Thus, $\psi' \leq 0$ and $\mid \psi' \mid$ is nonincreasing.

Lemma 4.4. *There exists a constant $K_2 \geq \sup\left(2^{1+1/H} K_1, 2(2K_1^2)^H K_1\right)$, such that we have for $0 < \epsilon < 1/K_2$*

$$\frac{1}{K_2\, \epsilon^{1+1/H}} \leq \mid \psi'(\epsilon) \mid \leq \frac{K_2}{\epsilon^{1+1/H}}.$$

Lemma 4.5. *We have for $\epsilon_1 > \epsilon/2$*

$$\exp\left(-K_3 \frac{\mid \epsilon_1 - \epsilon \mid}{\epsilon^{1+1/H}}\right) \leq \frac{\phi(\epsilon_1)}{\phi(\epsilon)} \leq \exp\left(K_3 \frac{\mid \epsilon_1 - \epsilon \mid}{\epsilon^{1+1/H}}\right), \tag{4.5}$$

where $K_3 \geq K_2\, 2^{1+1/H}$.

Lemma 4.6. *There exists $\beta = (H/K_2)^H > 0$, such that for $\epsilon < \beta$, the function $\epsilon^{-1/H}\phi(\epsilon)$ increases.*

4.3 Sufficiency

Suppose here that a_t is bounded and $\int_0^{+\infty} a_t^{-1/H}\, b_t\, \frac{dt}{t} < +\infty$. We want to prove that $f(t) \leq Y(t)$ for t large enough.

Lemma 4.7. $\lim_{t \to +\infty} a_t = 0.$

In order to prove the sufficiency, we need to construct some special sequences $\{t_n, n \geq 1\}, \{u_n, n \geq 1\}$ and $\{v_n, n \geq 1\}$. This will be done by recursion as follows, where L is a parameter depending on H only, such that $L - 2H > 0$. We start with $t_1 = 1$. Having constructed t_n, we set

$$u_{n+1} = t_n\left(1 + a_{t_n}^{1/H}\right),$$

$$v_{n+1} = \inf\{u > t_n, f(u) \geq f(t_n)(1 + L\, a_{t_n}^{1/H})\},$$

and

$$t_{n+1} = \min(u_{n+1}, v_{n+1}).$$

Lemma 4.8. $\lim_{n \to +\infty} t_n = +\infty.$

The key part of the proof of the sufficiency is stated in the following lemma.

Lemma 4.9. *If* $Y(t_n) \geq f(t_n)(1 + L\, a_{t_n}^{1/H})$, *for* $n \geq n_0$, *then* $f(t) \leq Y(t)$ *for* $t \geq t_{n_0}$.

We have also $\mathbb{P}\Big(Y(t_n) < f(t_n)(1 + L\, a_{t_n}^{1/H})\Big) = \phi\Big(a_{t_n}(1 + L\, a_{t_n}^{1/H})\Big)$.

Hence, the proof of the sufficiency will be achieved if we can show that this later series converges. By applying (4.5) with $\epsilon = a_{t_n}$ and $\epsilon_1 = \epsilon(1 + L\,\epsilon^{1/H})$, it suffices to prove that

$$\sum_{n=1}^{\infty} b_{t_n} < +\infty.$$

Lemma 4.10. *(i) If* n *is large enough and* $t_{n+1} = u_{n+1}$, *then we have*

$$b_{t_n} \leq K \int_{t_n}^{t_{n+1}} a_t^{-1/H} b_t \frac{dt}{t},$$

where $K = 2^{1/H} \exp(HK_3)$.

(ii) If n *is large enough and* $t_{n+1} = v_{n+1}$, *then we can choose the parameter* L *depending on* H *only such that*

$$L > 2H \text{ and } b_{t_n} \leq \frac{1}{\lambda} b_{t_{n+1}},$$

where $\lambda = \exp\Big((L - 2H)/2^{1+1/H} K_2\Big) > 1$.

To end the proof of the sufficiency, we consider the set $J = \{n_k \in \mathbb{N}^*, k \geq 1, t_{n_k+1} = u_{n_k+1}\}$. We deduce from the first part of lemma 4.10 that

$$\sum_{n_k \in J} b_{t_{n_k}} \leq K \sum_{n_k \in J} \int_{t_{n_k}}^{t_{n_k+1}} a_t^{-1/H} b_t \frac{dt}{t} \leq K \int_0^{+\infty} a_t^{-1/H} b_t \frac{dt}{t} < +\infty.$$

Let n_{k-1} and n_k be two consecutive terms of J. If there exists an integer n such that $n_{k-1} < n < n_k$, then set $p = n_k - n$. Since $n \in \mathbb{N}^* - J$, we have $t_{n+1} = v_{n+1}$. The second part of lemma 4.10 implies that

$$b_{t_n} \leq \lambda^{-p}\, b_{t_{n_k}}.$$

Thus, we obtain by setting $n_0 = 0$

$$\sum_{n \in \mathbb{N}^* - J} b_{t_n} = \sum_{k=1}^{\infty} \Bigg(\sum_{n_{k-1} < n < n_k} b_{t_n} \Bigg) \leq \frac{K\lambda}{\lambda - 1} \int_0^{\infty} a_t^{-1/H}\, b_t\, \frac{dt}{t} < +\infty.$$

4.4 Necessity 1

Suppose here that, with probability 1, $f(t) \leq Y(t)$ for all t large enough. We want to prove that a_t is bounded and $\int_0^\infty a_t^{-1/H} b_t \frac{dt}{t} < +\infty$.

In the sequel, there is no loss of generality to assume that f is a continuous function of $t \geq 0$.

Lemma 4.11. *a_t is bounded and* $\lim_{t \to +\infty} a_t = 0$.

To prove the necessity, we will show that $f \in LUC(Y)$ when $\int_0^\infty a_t^{-1/H} b_t \frac{dt}{t} = +\infty$ and $\lim_{t \to +\infty} a_t = 0$. The first step consists in constructing a suitable sequence.

Lemma 4.12. *When* $\int_0^\infty a_t^{-1/H} b_t \frac{dt}{t} = +\infty$ *and* $\lim_{t \to +\infty} a_t = 0$, *we can find a sequence* $\{t_n, n \geq 1\}$ *with the three following properties*

$$t_{n+1} \geq t_n(1 + a_{t_n}^{1/H}),$$

$$\textit{For n large enough,} \quad m \geq n \Rightarrow \frac{a_{t_m}}{t_m^H} \leq 2\, \frac{a_{t_n}}{t_n^H},$$

and

$$\sum_{n=1}^{\infty} b_{t_n} = +\infty.$$

To continue the proof of the necessity, we need the following definition.

Definition 4.3. Consider the interval $A_k = [2^k, 2^{k+1}[, \ k \in \mathbb{N}$. If $a_{t_i}^{-1/H} \in A_k, \ i \in \mathbb{N}^*$, then we note $u(i) = k$.

Set $I_k = \{i, u(i) = k\}$ which is finite by lemma 4.11 and $N_k = \exp(K_0\, 2^{k-1})$, where K_0 was defined in lemma 4.1 and depends on H only.

Lemma 4.13. *There exist a constant K_4 depending on H only and a set J with the following properties*

$$\sum_{n \in J} b_{t_n} = +\infty.$$

Given $n \in J, m \in J, n < m$ *and an integer* k *such that* $k \geq \min(u(n), u(m))$ *and* $card(I_k \cap [n, m]) > N_k$, *we have*

$$\frac{t_m}{t_n} \geq \exp\left(\exp\left(K_4\, 2^{\min(u(n),u(m))}\right)\right).$$

4.5 Necessity 2

Consider now the events $E_n = \{Y(t_n) < f(t_n)\}$. We have directly $\mathbb{P}(E_n) = b_{t_n}$, and therefore $\sum_{n \in J} b_{t_n} = +\infty$.

Given $n \in J$, J can be rewritten as follows $J = J' \cup \left(\bigcup_{k\in\mathbb{N}} J_k\right) \cup J"$, where $J' = \{m \in J, t_n \leq t_m \leq 2t_n\}$, $J_k = \{m \in J \cap I_k, t_m > 2t_n, card(I_k \cap [n,m]) \leq N_k\}$ and $J" = J - \left(J' \cup \left(\bigcup_{k\in\mathbb{N}} J_k\right)\right)$.

Lemma 4.14. *$\sum_{m\in J'} \mathbb{P}(E_n \cap E_m) \leq K' b_{t_n}$ and $\sum_{m\in(\cup_k J_k)} \mathbb{P}(E_n \cap E_m) \leq K" \, b_{t_n}$, where K' and $K"$ are numbers.*

The key step of the proof of lemma 4.14 consists in determining a general upper bound of $\mathbb{P}(E_n \cap E_m)$. This is the aim of the following lemma.

Lemma 4.15. *Consider $0 < t < u$, and $\theta, \nu > 0$. Then, we have $\mathbb{P}\Big(\{Y(t) \leq \theta t^H\} \cap \{Y(u) \leq \nu\}\Big)$*

$$\leq \exp(K_5)\, \mathbb{P}\left(Y(t) \leq \theta t^H\right) \exp\left(-\frac{K_5(u-t)}{\nu^{1/H}}\right), \tag{4.6}$$

where K_5 depends on H only.

Proof. Set $F_1 = \{Y(t) \leq \theta\, t^H\}$ and $F_2 = \{Y(u) \leq \nu\}$. We have

$$\mathbb{P}(F_1 \cap F_2) \leq \mathbb{P}\left(F_1 \cap \left\{\sup_{t\leq s\leq u} \mid X_H(s) \mid \leq \nu\right\}\right).$$

Denote by $[x]$ the integer part of a real x. We consider the sequence $z_k, k \in \{0, .., n\}$, where $z_0 = t, z_{k+1} = z_k + \delta$ and $n = [(u-t)/\delta]$. Let G_k be the event defined by

$$G_k = F_1 \cap \left\{\sup_{t\leq s\leq z_k} \mid X_H(s) \mid \leq \nu\right\}.$$

We have $F_1 \cap F_2 \subset G_k$. Moreover, we have $G_{k+1} \subset G_k \cap \{\mid Z \mid \leq 2\nu\}$, where $Z = X_H(z_{k+1}) - X_H(z_k)$.

By (4.2), Z can be rewritten as follows $Z = Z_1 + Z_2$, where

$$Z_1 = k_{2H}^{-1} \int_{z_k}^{z_{k+1}} (z_{k+1} - u)^{H-1/2} dW(u).$$

Note also that Z_1 and Z_2 are independent.

Since $\mathbb{P}\Big(\mid Z_1 + x \mid \leq \nu\Big)$ is maximum at $x = 0$ and Z_1 and G_k are independent, we have

$$\mathbb{P}(G_{k+1}) \leq \mathbb{P}(G_k)\mathbb{P}\Big(\mid Z_1 \mid \leq 2\nu\Big).$$

The integral representation of Z_1 implies that $\mathbb{E}(Z_1) = 0$ and

$$\mathrm{Var} Z_1 = k_{2H}^{-2} \int_{z_k}^{z_{k+1}} (z_{k+1} - u)^{2H-1} du = \frac{1}{2Hk_{2H}^2} \delta^{2H}.$$

So, we have $\mathbb{P}(| Z_1 | \leq 2\nu) = \Phi\left(\frac{2\sqrt{2H}k_{2H}\nu}{\delta^H}\right)$, where Φ denotes the distribution function of the absolute value of a standard Gaussian random variable. Then, we obtain

$$\mathbb{P}(G_{k+1}) \leq \mathbb{P}(G_k)\Phi\left(\frac{2\sqrt{2H}k_{2H}\nu}{\delta^H}\right),$$

and therefore $\mathbb{P}(F_1 \cap F_2) \leq \mathbb{P}(F_1)\Phi\left(\frac{2\sqrt{2H}k_{2H}\nu}{\delta^H}\right)^n$.

Choosing $\delta = \nu^{1/H}$, we get $K_5 = -\log\Phi\left(2\sqrt{2H}\, k_{2H}\right)$. (4.6) is proved. □

To prove lemma 4.14, just apply inequality (4.6) by setting $u = t_m, t = t_n, \theta = a_{t_n}$ and $\nu = f(t_m)$.

Lemma 4.16. *There exists an integer p such that, if $n > \sup_{s\leq p} (\sup I_s)$, then, for $m \in J", m > n$, given $\epsilon > 0$, we have $\mathbb{P}(E_n \cap E_m) \leq (1+\epsilon)\, b_{t_n} b_{t_m}$.*

The proof of lemma 4.16 is based on the two following lemmas. The first one is a general result on Gaussian processes, whereas the second one gives a specific property on some probabilities of the sfBm.

Lemma 4.17. *Let $\{X(t), 0 \leq t \leq 1\}$ be a separable, centered, real-valued Gaussian process such that $X(0) = 0$ with probability 1, and satisfying*

$$\left(\mathbb{E}(X(t+h) - X(t))^2\right)^{1/2} \leq \psi(h) \leq c_\psi\, h^\beta, \beta > 0.$$

Then, we have for $c_\psi^{-1}\delta > 1$

$$\mathbb{P}\left(\sup_{0\leq s\leq 1} | X(s) | \geq \delta\right) \leq \frac{1}{C} \exp\left(-C\,(c_\psi^{-1}\delta)^2\right),$$

where C is a positive constant independent of c_ψ and δ.

Lemma 4.18. *Let α be a real number such that $1/2 < \alpha < 1$. Set $r = \min\left(\frac{1-H}{3}, \frac{(1-\alpha)H}{3}\right)$. Then, we have for $u \geq 2t$*

$$\mathbb{P}\left(Y(t) \leq \theta t^H, Y(u) \leq \nu u^H\right)$$
$$\leq \phi(\theta)\phi(\nu)\exp\left(2\left(\frac{t}{u}\right)^r K_3\left(\frac{1}{\theta^{1+1/H}} + \frac{1}{\nu^{1+1/H}}\right)\right) + \frac{6}{C}\exp\left(-\frac{C}{K^2}\left(\frac{u}{t}\right)^r\right), \tag{4.7}$$

where $K > 0$ depends on H only, K_3 was defined in lemma 4.5 and C in lemma 4.17.

Proof. Set $Q = \mathbb{P}\big(Y(t) \le \theta\, t^H, Y(u) \le \nu\, u^H\big)$.

Set $v = \sqrt{ut}$. If $t = o(u)$ then $t = o(v)$ and $v = o(u)$. $X_H(s)$ can be split as follows

$$X_H(s) = R_1(s) + R_2(s),$$

where $R_1(s) = \int_{|x|\le v} G(s,x)\, dW(x)$.

Note that $R_1(s)$ and $R_2(s)$ are independent. Then, given $\delta > 0$, we have

$$Q \le \phi(\theta + 2\delta)\phi(\nu + 2\delta) \tag{4.8}$$

$$+3\mathbb{P}\Big(\sup_{0\le s\le t} \mid R_2(s) \mid > \delta\, t^H\Big) + 3\mathbb{P}\Big(\sup_{0\le s\le u} \mid R_1(s) \mid > \delta u^H\Big). \tag{4.9}$$

First, we get by (4.5)

$$\phi(\theta + 2\delta)\phi(\nu + 2\delta) \le \phi(\theta)\phi(\nu)\ \exp\left(2\delta K_3\Big(\frac{1}{\theta^{1+1/H}} + \frac{1}{\nu^{1+1/H}}\Big)\right).$$

If we choose $\delta = \left(\frac{t}{u}\right)^r$, then we get the first term of the RHS of (4.7).

Next, we want to obtain an upper bound for

$$\mathbb{P}\Big(\sup_{0\le s\le t} \mid R_2(s) \mid > \delta\, t^H\Big) = \mathbb{P}\Big(\sup_{0\le s\le 1} \mid R_2(s) \mid > \delta\Big).$$

We have

$$R_2(s+h) - R_2(s) = \frac{1}{k_{2H}}\left(\int_{-\infty}^{-v/t} \Big((s+h-x)^{H-1/2} - (s-x)^{H-1/2}\Big)\, dW(x)\right.$$
$$\left. + \int_{-\infty}^{-v/t} \Big((-s-h-x)^{H-1/2} - (-s-x)^{H-1/2}\Big)\, dW(x)\right).$$

Let us introduce the following notation. Set

$$g_1(s,x,h) := (s+h-x)^{H-1/2} - (s-x)^{H-1/2} \tag{4.10}$$

and

$$g_2(s,x,h) := (-s-h-x)^{H-1/2} - (-s-x)^{H-1/2}. \tag{4.11}$$

We can establish

$$\mid g_1(s,x,h) \mid \le \mid H - \frac{1}{2} \mid h \mid -x \mid^{H-3/2} \tag{4.12}$$

and similarily

$$| g_2(s,x,h) | \leq | H - \frac{1}{2} | \; h \; | -s-h-x |^{H-3/2} . \tag{4.13}$$

Hence, since $(a+b)^2 \leq 2(a^2+b^2)$ for any $a,b > 0$, we get by using (4.12) and (4.13)

$$\begin{aligned}
&\mathbb{E}\Big(R_2(s+h) - R_2(s)\Big)^2 \\
&\leq \frac{2}{k_{2H}^2}\left(\int_{-\infty}^{-v/t} | g_1(s,x,h) |^2 \; dx + \int_{-\infty}^{-v/t} | g_2(s,x,h) |^2 \; dx\right) \\
&\leq \frac{2}{k_{2H}^2} \frac{| H-\frac{1}{2} |^2 \; h^2}{2-2H}\left(\left(\frac{v}{t}\right)^{2H-2} + \left(-s-h+\frac{v}{t}\right)^{2H-2}\right) \\
&\leq \frac{2}{k_{2H}^2} \frac{| H-\frac{1}{2} |^2 \; h^2}{2-2H}\left(\left(\frac{v}{t}\right)^{2H-2} + \left(\frac{v}{t}-1\right)^{2H-2}\right).
\end{aligned}$$

Since we suppose $u \geq 2t$ in lemma 4.18, we get $\frac{v}{t} \geq \sqrt{2}$. By choosing $0 < \gamma \leq 1 - \frac{1}{\sqrt{2}}$, we have $\frac{v}{t} - 1 \geq \gamma \frac{v}{t}$. Hence

$$\begin{aligned}
\mathbb{E}\Big(R_2(s+h) - R_2(s)\Big)^2 &\leq \frac{2}{k_{2H}^2} \frac{| H-\frac{1}{2} |^2}{2-2H}\left(1+\gamma^{2H-2}\right)\left(\frac{v}{t}\right)^{2H-2} h^2 \\
&:= K^2 \left(\frac{v}{t}\right)^{2H-2} h^2.
\end{aligned}$$

An application of lemma 4.17 with $\beta = 1$, $c_\psi = K \left(\frac{v}{t}\right)^{H-1}$ and $c_\psi^{-1}\delta > 1$, implies that

$$\mathbb{P}\left(\sup_{0\leq s\leq 1} | R_2(s) | > \delta\right) \leq \frac{1}{C} \exp\Big(-\frac{C}{K^2\left(\frac{v^2}{t^2}\right)^{H-1}} \delta^2\Big).$$

Set $\delta = \left(\frac{t}{u}\right)^r$. Since $v^2 = ut$ and $r \leq \frac{1-H}{3}$, we have

$$\begin{aligned}
\mathbb{P}\left(\sup_{0\leq s\leq 1} | R_2(s) | > \delta\right) &\leq \frac{1}{C} \exp\left(-\frac{C}{K^2\left(\frac{t}{u}\right)^{1-H-2r}}\right) \\
&\leq \frac{1}{C} \exp\Big(-\frac{C}{K^2\left(\frac{t}{u}\right)^{r}}\Big).
\end{aligned} \tag{4.14}$$

Finally, we want to establish a similar result for $R_1(s)$.
First, we remark that

$$\mathbb{P}\left(\sup_{0\le s\le u} \mid R_1(s)\mid > \delta\, u^H\right) = \mathbb{P}\left(\sup_{0\le s\le 1} \mid R_1(s)\mid > \delta\right).$$

To obtain a suitable upper bound for the above probability, we need to establish a technical lemma which we shall prove later.

Lemma 4.19. *Let* $1/2 < \alpha < 1$. *Then there exists a constant* $K > 0$ *such that*

$$\mathbb{E}\Big(R_1(s+h) - R_1(s)\Big)^2 \le K^2 \left(\frac{v}{u}\right)^{2H-2\alpha H} h^{2\alpha H}. \tag{4.15}$$

Combining lemma 4.19 with lemma 4.17 ($\beta{=}\alpha H, c_\psi{=}K\left(\frac{v}{u}\right)^{H-\alpha H}$ and $c_\psi^{-1}\delta >$ 1), we get

$$\mathbb{P}\left(\sup_{0\le s\le 1} \mid R_1(s)\mid > \delta\right) \le \frac{1}{C}\exp\left(-\frac{C}{K^2\left(\frac{v}{u}\right)^{2H-2\alpha H}}\,\delta^2\right).$$

Set $\delta = \left(\frac{t}{u}\right)^r$. Since $v^2 = ut$ and $r \le \frac{(1-\alpha)H}{3}$, we have

$$\begin{aligned}\mathbb{P}\Big(\sup_{0\le s\le 1} \mid R_1(s)\mid > \delta\Big) &\le \frac{1}{C}\exp\left(-\frac{C}{K^2\left(\frac{t}{u}\right)^{H-\alpha H-2r}}\right)\\ &\le \frac{1}{C}\exp\left(-\frac{C}{K^2\left(\frac{t}{u}\right)^{r}}\right).\end{aligned} \tag{4.16}$$

Combining (4.14) and (4.16) with (4.8), we get the last term of the RHS of (4.7) and achieve the proof of lemma 4.18. □

Proof of Lemma 4.19. We deduce from (4.2) and the definition of R_1 that the representation of $R_1(s+h) - R_1(s)$, as a random integral, is given by

$$\begin{aligned}&R_1(s+h) - R_1(s)\\ &= \frac{1}{k_{2H}}\Bigg(\int_{-v/u}^{\min(s+h,v/u)} (s+h-x)^{H-1/2} dW(x)\\ &\quad - \int_{-v/u}^{\min(s,v/u)} (s-x)^{H-1/2}\, dW(x)\end{aligned}$$

$$+\int_{-v/u}^{\max(-s-h,-v/u)} (-s-h-x)^{H-1/2}\, dW(x)$$
$$\left.-\int_{-v/u}^{\max(-s,-v/u)} (-s-x)^{H-1/2}\, dW(x)\right)$$
$$=\frac{1}{k_{2H}}\left(\int_{-v/u}^{\max(-s-h,-v/u)} \Big(g_1(s,x,h)+g_2(s,x,h)\Big)\, dW(x)\right.$$
$$+\int_{\max(-s-h,-v/u)}^{\max(-s,-v/u)} \Big(g_1(s,x,h)-(-s-x)^{H-1/2}\Big)\, dW(x)$$
$$\left.+\int_{\max(-s,-v/u)}^{\min(s,v/u)} g_1(s,x,h)\, dW(x)+\int_{\min(s,v/u)}^{\min(s+h,v/u)} (s+h-x)^{H-1/2}\, dW(x)\right)$$
$$=\frac{1}{k_{2H}}\Big(V_1+V_2+V_3+V_4\Big).$$

Recall that the functions g_1 and g_2 were defined in (4.10) and (4.11). Since

$$\mathbb{E}\Big(R_1(s+h)-R_1(s)\Big)^2=\frac{1}{k_{2H}^2}\sum_{i=1}^{4}\mathbb{E}V_i^2,$$

it suffices to prove (4.15) for any $\mathbb{E}V_i^2,\ 1\le i\le 4$.

Consider $\mathbb{E}V_4^2$ first. Since

$$\mathbb{E}V_4^2=\int_{\min(s,v/u)}^{\min(s+h,v/u)} (s+h-x)^{2H-1}\, dx,$$

we have to investigate the three following cases.

Case 1. $v/u\le s<s+h$. Obviously $\mathbb{E}V_4^2=0$.

Case 2. $s<v/u\le s+h$. Standard computations imply

$$\mathbb{E}V_4^2=\frac{h^{2H}}{2H}\left(1-\left(1-\frac{(v/u)-s}{h}\right)^{2H}\right)$$
$$=\frac{h^{2H}}{2H}\left(\frac{2H\,((v/u)-s)}{h}+o\left(\frac{(v/u)-s}{h}\right)\right).$$

Then there exists a constant $C_1>0$ such that

$$\mathbb{E}V_4^2\le C_1^2\left(\frac{v}{u}-s\right)h^{2H-1}=C_1^2\left(\frac{v}{u}-s\right)h^{2\alpha H}h^{2H-2\alpha H-1}.$$

Recall that $1/2 < \alpha < 1$. Since $H - \alpha H - \frac{1}{2} = \frac{1}{2}(2H - 2\alpha H - 1) < 0$, we get

$$\mathbb{E}V_4^2 \le C_1^2 \left(\frac{v}{u} - s\right)^{2H-2\alpha H} h^{2\alpha H} \le C_1^2 \left(\frac{v}{u}\right)^{2H-2\alpha H} h^{2\alpha H}.$$

Case 3. $s < s + h \le v/u$. Since we have

$$\mathbb{E}V_4^2 = \frac{h^{2H}}{2H} = \frac{h^{2\alpha H}\, h^{2H-2\alpha H}}{2H},$$

there exists a constant $C_2 > 0$ such that

$$\mathbb{E}V_4^2 \le C_2^2 \left(\frac{v}{u} - s\right)^{2H-2\alpha H} h^{2\alpha H} \le C_2^2 \left(\frac{v}{u}\right)^{2H-2\alpha H} h^{2\alpha H}.$$

(4.15) is therefore established for $\mathbb{E}V_4^2$.
Consider $\mathbb{E}V_3^2$ now. We have

$$\mathbb{E}V_3^2 = \int_{\max(-s,-v/u)}^{\min(s,v/u)} g_1(s,x,h)^2\, dx,$$

where the function g_1 was defined in (4.10).

Since $H - \alpha H - \frac{1}{2} = \frac{1}{2}(2H - 2\alpha H - 1) < 0$, $y \to (y - x)^{H-\alpha H-1/2}$ is a decreasing function. Then, we have

$$\begin{aligned} |\, g_1(x,s,h)\,| &\le |\, H - \frac{1}{2}\,|\,(s-x)^{H-\alpha H-1/2}\,\left|\int_s^{s+h} (y-x)^{\alpha H-1} dy\right| \\ &= |\, H - \frac{1}{2}\,|\,(s-x)^{H-\alpha H-1/2}\,\frac{(s+h-x)^{\alpha H} - (s-x)^{\alpha H}}{\alpha H}. \end{aligned}$$

Note that the function $x \to (s+h-x)^{\alpha H} - (s-x)^{\alpha H}$ is positive and increasing. Then, we have $(s+h-x)^{\alpha H} - (s-x)^{\alpha H} \le h^{\alpha H}$. Hence, we have

$$|\, g_1(x,s,h)\,| \le \frac{|\, H - \frac{1}{2}\,|}{\alpha H} (s-x)^{H-\alpha H-1/2}\, h^{\alpha H}. \tag{4.17}$$

Let us investigate the two following cases.
Case 1. $-s \le -v/u$. We have by (4.17)

$$\begin{aligned} \mathbb{E}V_3^2 &= \int_{-v/u}^{v/u} g_1(s,x,h)^2\, dx \\ &\le \left(\frac{H - \frac{1}{2}}{\alpha H}\right)^2 \frac{h^{2\alpha H}}{2H - 2\alpha H} \left(\left(s + v/u\right)^{2H-2\alpha H} - \left(s - v/u\right)^{2H-2\alpha H}\right) \end{aligned}$$

$$\leq \left(\frac{H-\frac{1}{2}}{\alpha H}\right)^2 \frac{h^{2\alpha H}}{2H-2\alpha H} s^{2H-2\alpha H}\left(2(2H-2\alpha H)\frac{v}{us}+o\left(\left(\frac{v}{us}\right)^2\right)\right).$$

Then there exists a constant $C_3 > 0$ such that

$$\mathbb{E}V_3^2 \leq C_3^2\, h^{2\alpha H}\, s^{2H-2\alpha H-1}\,\frac{v}{u} \leq C_3^2\left(\frac{v}{u}\right)^{2H-2\alpha H} h^{2\alpha H}.$$

Case 2. $-v/u < -s$. We have

$$\mathbb{E}V_3^2 \leq \left(\frac{H-\frac{1}{2}}{\alpha H}\right)^2 h^{2\alpha H}\int_{-s}^{s}\left(s-x\right)^{2H-2\alpha H-1}dx$$

$$\leq \left(\frac{H-\frac{1}{2}}{\alpha H}\right)^2 \frac{h^{2\alpha H}}{2H-2\alpha H}(2s)^{2H-2\alpha H}$$

and consequently (4.15).

(4.15) is therefore established for $\mathbb{E}V_3^2$.

Let us turn to $\mathbb{E}V_2^2$. Since $(a-b)^2 \leq 2(a^2+b^2)$ for any $a, b > 0$, we get

$$\mathbb{E}V_2^2 \leq 2\left(\int_{\max(-s-h,-v/u)}^{\max(-s,-v/u)} g_1(s,x,h)^2\, dx + \int_{\max(-s-h,-v/u)}^{\max(-s,-v/u)} (-s-x)^{2H-1}\, dx\right). \tag{4.18}$$

Set $I_1 := \int_{\max(-s-h,-v/u)}^{\max(-s,-v/u)} (-s-x)^{2H-1}\, dx$ and

$$I_2 := \int_{-v/u}^{\max(-s,-v/u)} g_1(s,x,h)^2\, dx$$

$$= \int_{-v/u}^{\max(-s-h,-v/u)} g_1(s,x,h)^2\, dx + \int_{\max(-s-h,-v/u)}^{\max(-s,-v/u)} g_1(s,x,h)^2\, dx.$$

To prove (4.15) for $\mathbb{E}V_2^2$, it suffices by (4.18) to show it for I_1 and I_2.

Consider I_1 first. We have to investigate the two following cases.

Case 1. $-v/u \leq -s-h < -s$. We have

$$I_1 = \int_{-s-h}^{-s} (-s-x)^{2H-1}\, dx = \frac{h^{2H}}{2H} = \frac{h^{2\alpha H}\, h^{2H-2\alpha H}}{2H}$$

$$\leq \frac{1}{2H}\,(v/u-s)^{2H-2\alpha H}\, h^{2\alpha H} \leq \frac{1}{2H}\,(v/u)^{2H-2\alpha H}\, h^{2\alpha H}.$$

Case 2. $-s-h < -v/u$. When $-s < -v/u,\ I_1 = 0$. So assume $-v/u < -s$. We have

$$\begin{aligned} I_1 &= \int_{-v/u}^{-s} (-s-x)^{2H-1}\, dx = \frac{1}{2H}\, (v/u-s)^{2H} \\ &= \frac{1}{2H}\, (v/u-s)^{2\alpha H}\, (v/u-s)^{2H-2\alpha H} \\ &\le \frac{1}{2H}\, h^{2\alpha H}\, (v/u-s)^{2H-2\alpha H} \le K^2\, (v/u)^{2H-2\alpha H}\, h^{2\alpha H}. \end{aligned}$$

Consider I_2 now. We have by (4.17)

$$I_2 \le \frac{(H-1/2)^2}{(\alpha H)^2}\, h^{2\alpha H} \int_{-v/u}^{\max(-s,-v/u)} (s-x)^{2H-2\alpha H-1}\, dx. \tag{4.19}$$

Let us investigate the two following cases.
Case 1. $-s \le -v/u$. Obviously $I_2 = 0$.
Case 2. $-v/u < -s$. We deduce from (4.19)

$$\begin{aligned} I_2 &\le \frac{(H-1/2)^2}{(\alpha H)^2}\, \frac{h^{2\alpha H}}{2H-2\alpha H}\, \left(s+v/u\right)^{2H-2\alpha H} \\ &\le \frac{(H-1/2)^2}{(\alpha H)^2}\, \frac{2^{2H-2\alpha H}}{2H-2\alpha H}\, \left(v/u\right)^{2H-2\alpha H}\, h^{2\alpha H}, \end{aligned}$$

that proves (4.15) for I_2.

Finally let us consider $\mathbb{E}V_1^2$. We have

$$\begin{aligned} \mathbb{E}V_1^2 &\le 2\left(\int_{-v/u}^{\max(-s-h,-v/u)} g_1(s,x,h)^2\, dx + \int_{-v/u}^{\max(-s-h,-v/u)} g_2(s,x,h)^2\, dx\right) \\ &\le 2\left(I_2 + \int_{-v/u}^{\max(-s-h,-v/u)} g_2(s,x,h)^2\, dx\right), \end{aligned}$$

where the function g_2 was defined in (4.11).

To establish (4.15) for $\mathbb{E}V_1^2$, it suffices to show it for

$$\int_{-v/u}^{\max(-s-h,-v/u)} g_2(s,x,h)^2\, dx.$$

When $\max(-s-h,-v/u) = -v/u$, the above integral equals 0. So let us assume $-v/u < -s-h$. We have

$$|\ g_2(s,x,h)\ | = |\ (H-1/2) \int_{-s}^{-s-h} (y-x)^{H-3/2} dy\ |$$

$$= \mid H - 1/2 \mid \mid \int_{-s}^{-s-h} (y-x)^{H-\alpha H-1/2} \, (y-x)^{\alpha H-1} \, dy \mid$$

$$\leq \mid H - 1/2 \mid \; (-s-h-x)^{H-\alpha H-1/2} \, \frac{(-s-x)^{\alpha H} - (-s-h-x)^{\alpha H}}{\alpha H}.$$

Note that the function $x \to (-s-x)^{\alpha H} - (-s-h-x)^{\alpha H}$ is positive and increasing. Then, since $-v/u < x < -s-h$, we get $(-s-x)^{\alpha H} - (-s-h-x)^{\alpha H} \leq h^{\alpha H}$. Hence, we have

$$\mid g_2(x,s,h) \mid \leq \frac{\mid H - \frac{1}{2} \mid}{\alpha H} \, (-s-h-x)^{H-\alpha H-1/2} \, h^{\alpha H}. \tag{4.20}$$

We get by (4.20)

$$\int_{-v/u}^{-s-h} g_2(s,x,h)^2 dx \leq \frac{(H-1/2)^2}{(\alpha H)^2} \, \frac{h^{2\alpha H}}{2H - 2\alpha H} \, (-s-h+v/u)^{2H-2\alpha H}$$
$$\leq K^2 \, (v/u)^{2H-2\alpha H} \, h^{2\alpha H}.$$

The proof of lemma 4.19 is now complete. □

Let us end the proof of theorem 4.1. Combining lemma 4.14 with lemma 4.16, we show that, given $\epsilon > 0$, there exists a real number $K > 0$ and an integer p such that

$$\forall n \in J, \; n \geq p \Rightarrow \sum_{m \in J, m > n} \mathbb{P}(E_n \cap E_m) \; \leq \; b_{t_n} \left(K + (1+\epsilon) \sum_{m \in J, m > n} b_{t_m} \right).$$

Since (4.6) holds, an application of corollary (2.3) of ([10], p. 198) yields

$$\frac{1}{1+2\epsilon} \leq \mathbb{P}\left(\bigcup_{n \in J} E_n \right) = \mathbb{P}\left(\bigcup_{n \in J} \{Y(t_n) \leq f(t_n)\} \right),$$

and consequently $f \in LUC(Y)$. The proof of theorem 4.1 is now complete. □

References

1. Bojdecki, T., Gorostiza, L.G., Talarczyk, A.: Sub-fractional Brownian motion and its relation to occupation times. Statist. Probab. Lett. **69**, 405–419 (2004)
2. Bojdecki, T., Gorostiza, L.G., Talarczyk, A.: Some extensions of fractional Brownian motion and sub-fractional Brownian motion related to particle systems. Electron. Comm. Probab. **12**, 161–172 (2007)
3. Borell, C.: Convex measures on locally convex spaces. Ark. Math. **12**, 239–252 (1974)

4. Dzhaparidze, K., Van Zanten, H.: A series expansion of fractional Brownian motion. Probab. Th. Rel. Fields **103**, 39–55 (2004)
5. El-Nouty, C.: On the lower classes of fractional Brownian motion. Studia Sci. Math. Hungar. **37**, 363–390 (2001)
6. El-Nouty, C.: Lower classes of fractional Brownian motion under Hölder norms. Limit Theorems in Probability and Statistics, Balatonlelle, 1999, (I. Berkes, E. Csáki, M. Csörgő, eds.) János Bolyai Mathematical Society, Budapest (2002)
7. El-Nouty, C.: Lower classes of the Riemann-Liouville process. Bull. Sci. Math. **135**, 113–123 (2011)
8. Monrad, D., Rootzen, H.: Small values of Gaussian processes and functional laws of the iterated logarithm. Probab. Th. Rel. Fields **101**, 173–192 (1995)
9. Révész, P.: Random walk in random and non-random environments. World Scientific Publishing Co., Teaneck, NJ (1990)
10. Talagrand, M.: Lower classes of fractional Brownian motion. J. Theoret. Probab. **9**, 191–213 (1996)
11. Tudor, C.: Some properties of the sub-fractional Brownian motion. Stochastics **79**, 431–448 (2007)
12. Tudor, C.: On the Wiener integral with respect to a sub-fractional Brownian motion on an interval. J. Math. Anal. Appl. **351**, 456–468 (2009)

Chapter 5
On the Bounded Variation of the Flow of Stochastic Differential Equation

Mohamed Erraoui and Youssef Ouknine

Abstract We consider a stochastic differential equation, driven by a Brownian motion, with non Lipschitz coefficients. We consider the class **BV** which is larger than Sobolev space and got a sufficient condition for a solution of stochastic differential equation to belong to the class **BV**. As a consequence we prove that the corresponding flow is, almost surely, almost every where derivable with respect to initial data. The result is a partial extension of the result of N. Bouleau and F. Hirsch on the derivability, with respect to the initial data, of the solution of a stochastic differential equation with Lipschitz coefficients.

5.1 Introduction

Let us consider the 1-dimensional stochastic differential equation

$$X(t) = X_0 + \int_0^t \sigma(X(s))dB(s) + \int_0^t b(X(s))ds, \tag{5.1}$$

where σ and b are $\mathbb{R}$-valued measurables functions, $\{B(t),\ t \geq 0\}$ is a 1-dimensional Brownian motion on the filtered space $(\Omega, \mathscr{F}, (\mathscr{F}_t)_{t\leq 0}, P)$ and X_0 is the initial condition independent of B. We assume that:

M. Erraoui (✉) · Y. Ouknine
Faculté des Sciences Semlalia, Département de Mathématiques, Université Cadi Ayyad
BP 2390, Marrakech, Maroc
e-mail: erraoui@ucam.ac.ma; ouknine@ucam.ac.ma

M. Zili and D.V. Filatova (eds.), *Stochastic Differential Equations and Processes*, Springer Proceedings in Mathematics 7, DOI 10.1007/978-3-642-22368-6_5,

A.1 σ and b are continuous on $\mathbb{R}$ with linear growth condition

$$|\sigma(x)| + |b(x)| \leq L(1 + |x|),$$

for every $x \in \mathbb{R}$, where L is a positive constant.

A.2 The equation (5.1) has pathwise uniqueness in the sense that if whenever (X, B) and (X', B') are two solutions defined on the same filtered space with $B = B'$ and $X_0 = X'_0$ a.s., then X and X' are indistinguishable.

In the sequel we will abbreviate "stochastic differential equation" to SDE. Furthermore, we use the notation $X_x(\cdot)$ if we impose the condition $X_0 = x$ a.s. on the solution. It is well-known that assumption (A.1) ensures the existence of weak solution $\{X_x(t),\ t \geq 0\}$, for every $x \in \mathbb{R}$, which becomes strong under assumption (A.2), cf. [6].

For $x \leq y$ fixed, we define stopping time,

$$S = \inf\{t > 0 : X_x(t) > X_y(t)\}.$$

It is well known that on the set $[S < +\infty]$ the process $\{\tilde{B}(t) = B(S+t) - B(S),\ t \geq 0\}$ is again a Brownian motion. Moreover the processes $\{X_x(S+t)\,, t \geq 0\}$ and $\{X_y(S+t)\,, t \geq 0\}$ satisfy the following SDEs:

$$X_x(S+t) = X_x(S) + \int_0^t \sigma(X_x(S+s))d\tilde{B}(s) + \int_0^t b(X_x(S+s))ds,$$

and

$$X_y(S+t) = X_y(S) + \int_0^t \sigma(X_y(S+s))d\tilde{B}(s) + \int_0^t b(X_y(S+s))ds.$$

Since $X_x(S) = X_y(S)$ a.s. on $[S < +\infty]$ then thanks to pathwise uniqueness, we have

$$P[X_x(S+t)1_{[S<+\infty]} = X_y(S+t)1_{[S<+\infty]},\ \forall\, t \geq 0] = 1.$$

It follows that

$$P[X_x(t) \leq X_y(t),\ \forall\, t \geq 0] = 1.$$

We conclude that P-almost all w, for any $t \geq 0$ the Borel function $x \mapsto X_x(t)(w)$ is increasing and consequently is differentiable a.e. with respect to Lebesgue measure in the sense of classical differentiation. It should be noted that this result is obtained without assuming the Lipschitz condition on the coefficients. The main tool used is the comparison theorem which is no longer valid in higher dimensions. A question arises: What remains of the above conclusion for SDE in $\mathbb{R}^n$?

The main result of this note provides a partial answer to this question. First, it is important to recall the differentiability result for SDE in $\mathbb{R}^n$ with Lipschitz coefficients due to Bouleau and Hirsch [1]. Precisely they have shown that the corresponding flow is, almost surely, almost everywhere derivable with respect to the initial data for any time. Their proof is based on the fact that the solution lives in a subspace, which is a Dirichlet space, of the Sobolev space $H^1_{loc}(\mathbb{R}^n)$. Unfortunately when the diffusion coefficient is not necessary Lipschitz the solutions would not belong to Sobolev space in general. Indeed we give sufficient conditions for solutions of stochastic differential equations to belong a larger class which is merely the class of functions of bounded variation which we denote by **BV**. We note that this idea was recently used by Kusuoka in [5] to show the existence of density for SDE with non Lipschitz coefficients. Finally we use the relation between differentiability and the class **BV** to prove differentiability of the solutions with respect to the initial data.

5.2 Class BV

For $d \geq 1$, we denote by $(\Omega, \mathscr{F}, (\mathscr{F}_t)_{t\geq 0}, P, (B_t)_{t\geq 0})$ the standard Brownian motion in $\mathbb{R}^d$ starting from 0. That is $\Omega = C_0([0, +\infty), \mathbb{R}^d)$ is the Banach space of continuous functions, null at time 0, equipped with the Borel σ-field $\mathscr{F}$. The canonical process $B = \{B_t : t \geq 0\}$ is defined by $B_t(\omega) = \omega(t)$, the probability measure $\mathbb{P}$ on $(\Omega, \mathscr{F})$ is the Wiener measure and $\mathscr{F}_t$ is the σ-field $\sigma\{B_s, 0 \leq s \leq t\}$. Let $\mathbf{h}$ be a fixed continuous positive function on $\mathbb{R}^n$ such that

$$\int_{\mathbb{R}^n} \mathbf{h}(\mathbf{x})d\mathbf{x} = 1 \text{ and } \int_{\mathbb{R}^n} |\mathbf{x}|^2\, \mathbf{h}(\mathbf{x})d\mathbf{x} < +\infty.$$

Let $\hat{\Omega}$ be the product space $\mathbb{R}^n \times \Omega$, which is also a Fréchet space. We endow $\hat{\Omega}$ with the measure $(\hat{\mathscr{F}}, \hat{P})$ where $\hat{\mathscr{F}}$ denotes the Borel σ-field on $\hat{\Omega}$ and $\hat{P}$ denotes the product measure $\mathbf{h}(\mathbf{x})\, d\mathbf{x} \times P$. For simplicity we choose $\mathbf{h}$ the form of

$$\mathbf{h}(\mathbf{x}) = \mathbf{h}(x_1, \ldots, x_n) = \frac{1}{(\sqrt{2\pi})^n} e^{-\sum_{i=1}^n x_i^2/2} = \prod_{i=1}^n h(x_i)$$

where $h(x_i) = \dfrac{1}{\sqrt{2\pi}} e^{-x_i^2/2}$.

Now we give the definition of class **BV** as follows

Definition 5.1. We define $\mathbf{BV}(\mathbb{R}^n \times \Omega)$ the total set of Borel functions F on $(\hat{\Omega}, \hat{\mathscr{F}}, \hat{P})$ such that there exists a Borel function $\hat{F}$ on $(\hat{\Omega}, \hat{\mathscr{F}}, \hat{P})$ satisfying that:

(i) $F = \hat{F}$ $\hat{P}$-a.s.
(ii) $\forall\, w \in \Omega$, $\mathbf{x} \to \hat{F}(\mathbf{x}, w)$ is a function of bounded variation on each compact in $\mathbb{R}^n$.

Now we give a criterion that a Borel function belongs to the class $V(\mathbb{R}^n \times \Omega)$.

Theorem 5.1. *Let $p > 1$ and assume that there exists a sequence $\{F_j : j \in \mathbb{N}\}$ in $L^p(\hat{\Omega}, \hat{\mathscr{F}}, \hat{P})$ such that:*

B.1 F_j *converges to* F *almost surely.*
B.2 $\{F_j : j \in \mathbb{N}\}$ *are uniformly bounded in* $L^p(\hat{\Omega}, \hat{\mathscr{F}}, \hat{P})$.
B.3 *For all* $(\mathbf{x}, w) \in \hat{\Omega}$, $i \in \{1, \ldots, n\}$ *and* $j \in \mathbb{N}$, $t \to F_j(\mathbf{x}+te_i, w)$ *is absolutely continuous with respect to the one-dimensional Lebesgue measure on any finite interval.*
B.4 $\left\{\nabla_i F_j := \liminf_{t\to 0} \frac{F_j(\mathbf{x}+te_i, w) - F_j(\mathbf{x}, w)}{t} \quad j \in \mathbb{N}\right\}$ *are uniformly bounded in* $L^1(\hat{\Omega}, \hat{\mathscr{F}}, \hat{P})$.

Then $F \in \mathbf{BV}(\mathbb{R}^n \times \Omega)$.

Proof. We will give the proof in several steps using the same approach and keeping the same notations as in [5].

First we identify $\mathbf{x}$ as $(x_1, \tilde{x})$ and $\hat{P}$ as $h(x_1)dx_1 \times \tilde{h}(\tilde{x})d\tilde{x} \times P$ where $\tilde{h}(\tilde{x}) = \prod_{i=2}^{n} h(x_i)$. Since the family $\{F_j : j \in \mathbb{N}\}$ is uniformly bounded in $L^p(\hat{\Omega}, \hat{\mathscr{F}}, \hat{P})$, with $p > 1$, then it is uniformly integrable. Moreover it satisfies the condition (B.1) then by Vitali's theorem $F \in L^1(\hat{\Omega}, \hat{\mathscr{F}}, \hat{P})$ and F_j converges to F in $L^1(\hat{\Omega}, \hat{\mathscr{F}}, \hat{P})$. Let $M > 0$ and define a function $\phi \in C^\infty(\mathbb{R})$ such that

$$(i)\ 0 \le \phi \le 1,\ (ii)\ 0 \le \phi' \le 1,\ (iii)\ \phi(y) = \begin{cases} 1\ if\ |y| \le M \\ 0\ if\ |y| \ge M+1. \end{cases} \tag{5.2}$$

For $x_1 \in \mathbb{R}$ and $m \in \mathbb{N}^*$, we set

$$\psi(x_1, \tilde{x}, w) = F(x_1, \tilde{x}, w)\phi(x_1),$$

and

$$\Psi_m(x_1, \tilde{x}, w) = 2^m \int_0^{1/2^m} \psi(x_1 + u, \tilde{x}, w)du.$$

It is easy to see that:

- $x_1 \mapsto \Psi_m(x_1, \tilde{x}, w)$ is absolutely continuous and has its support in $[-(M+2), M+2]$.
- There exists a constant $C > 0$, which depend only on M, such that $\sup_m \hat{\mathbb{E}}\,|\Psi_m| \le C\,\hat{\mathbb{E}}\,|\psi|$.
- $\Psi_m(x_1, \tilde{x}, w) = 2^m \int_{x_1}^{x_1+1/2^m} \psi(u, \tilde{x}, w)\,du$, $x_1 \mapsto \Psi_m(x_1, \tilde{x}, w)$ is differentiable a.e and

$$\nabla_1 \Psi_m(x_1, \tilde{x}, w) = 2^m[\psi(x_1 + 1/2^m, \tilde{x}, w) - \psi(x_1, \tilde{x}, w)] \quad dx_1\ a.e.$$

The first step is to show that $\int_{\mathbb{R}} \mid \nabla_1 \Psi_m(x_1, \tilde{x}, w) \mid dx_1$ is increasing in m. Indeed

$$\begin{aligned}
\int_{\mathbb{R}} \mid \nabla_1 \Psi_m(x_1, \tilde{x}, w) \mid dx_1 &= 2^m \int_{\mathbb{R}} |\psi(x_1 + 1/2^m, \tilde{x}, w) - \psi(x_1, \tilde{x}, w)| \, dx_1 \\
&\leq 2^m \int_{\mathbb{R}} \left|\psi(x_1 + 1/2^m, \tilde{x}, w) - \psi(x_1 + 1/2^{m+1}, \tilde{x}, w)\right| dx_1 \\
&\quad + 2^m \int_{\mathbb{R}} \left|\psi(x_1 + 1/2^{m+1}, \tilde{x}, w) - \psi(x_1, \tilde{x}, w)\right| dx_1 \\
&= 2^{m+1} \int_{\mathbb{R}} \left|\psi(x_1 + 1/2^{m+1}, \tilde{x}, w) - \psi(x_1, \tilde{x}, w)\right| dx_1 \\
&= \int_{\mathbb{R}} \mid \nabla_1 F^{m+1}(x_1 + u, \tilde{x}, w) \mid dx_1.
\end{aligned}$$

The second step is to prove that for $(\tilde{h}(\tilde{x})d\tilde{x} \times P)$-almost all $(\tilde{x}, w)$

$$\sup_m \int_{\mathbb{R}} \mid \nabla_1 \Psi_m(x_1, \tilde{x}, w) \mid dx_1 < +\infty. \tag{5.3}$$

To this end we need the following estimate

$$\hat{\mathbb{E}} \left|\psi(x_1 + t, \tilde{x}, w) - \psi(x_1 + s, \tilde{x}, w)\right| \leq C_M \left|t - s\right|, \tag{5.4}$$

where C_M is a positive constant which depend on M. In what follows we establish the proof of (5.4).

For $x_1 \in \mathbb{R}$ and $j \in \mathbb{N}$, we set

$$\psi_j(x_1, \tilde{x}, w) = F_j(x_1, \tilde{x}, w)\phi(x_1).$$

Let us remark that ψ_j has support in $[-(M+1), M+1]$ and converge to ψ in $L^1(\hat{\Omega}, \hat{\mathscr{F}}, \hat{P})$ which follows from the convergence of F_j to F. Moreover, as a consequence of assertions (i) and (ii) of condition (5.2) we have, for all $x_1 \in [-(M+1), M+1]$, that

$$|\nabla_1 \psi_j(x_1, \tilde{x}, w)| \leq |F_j(x_1, \tilde{x}, w)| + |\nabla_1 F_j(x_1, \tilde{x}, w)|. \tag{5.5}$$

Then, for $t, s \in \mathbb{R}$, we obtain

$$\begin{aligned}
&\hat{\mathbb{E}} \left|\psi(x_1 + t, \tilde{x}, w) - \psi(x_1 + s, \tilde{x}, w)\right| \\
&\quad = \lim_j \hat{\mathbb{E}} \left|\psi_j(x_1 + t, \tilde{x}, w) - \psi_j(x_1 + s, \tilde{x}, w)\right| \\
&\quad = \frac{1}{\sqrt{2\pi}} \lim_j \mathbb{E} \int_{\mathbb{R}^{n-1}} \int_{\mathbb{R}} \left|\psi_j(x_1 + t, \tilde{x}, w) - \psi_j(x_1 + s, \tilde{x}, w)\right| e^{-x_1^2/2} dx_1 \tilde{h}(\tilde{x}) d\tilde{x}
\end{aligned}$$

$$\leq \frac{1}{\sqrt{2\pi}} \liminf_j \mathbb{E} \int_{\mathbb{R}^{n-1}} \int_{\mathbb{R}} \left| \int_s^t \nabla_1 \psi_j(x_1 + u, \tilde{x}, w) du \right| dx_1 \tilde{h}(\tilde{x}) d\tilde{x}$$

$$\leq \frac{1}{\sqrt{2\pi}} \liminf_j \mathbb{E} \int_s^t \int_{\mathbb{R}^{n-1}} \int_{\mathbb{R}} |\nabla_1 \psi_j(x_1 + u, \tilde{x}, w)| \, dx_1 \tilde{h}(\tilde{x}) d\tilde{x} du$$

$$\leq \frac{1}{\sqrt{2\pi}} |t - s| e^{(M+2)^2/2} \sup_j \hat{\mathbb{E}}(|F_j(x_1, \tilde{x}, w)| + |\nabla_1 F_j(x_1, \tilde{x}, w)|),$$

where the last inequality is a consequence of (5.5). Now using (B.2) and (B.4) we obtain

$$\hat{\mathbb{E}} |\psi(x_1 + t, \tilde{x}, w) - \psi(x_1 + s, \tilde{x}, w)| \leq C_M |t - s|,$$

where $C_M = \frac{1}{\sqrt{2\pi}} e^{(M+2)^2/2} \sup_j \hat{\mathbb{E}}(|F_j(x_1, \tilde{x}, w)| + |\nabla_1 F_j(x_1, \tilde{x}, w)|)$.
Now we are able to make out (5.3). By monotone convergence theorem we have

$$\mathbb{E} \int_{\mathbb{R}^{n-1}} \sup_m \int_{\mathbb{R}} |\nabla_1 \Psi_m(x_1, \tilde{x}, w)| \, dx_1 \tilde{h}(\tilde{x}) d\tilde{x}$$
$$= \sup_m \mathbb{E} \int_{\mathbb{R}^{n-1}} \int_{\mathbb{R}} |\nabla_1 \Psi_m(x_1, \tilde{x}, w)| \, dx_1 \tilde{h}(\tilde{x}) d\tilde{x}$$

Since Ψ_m has support in $[-(M+2), M+2]$ then

$$\mathbb{E} \int_{\mathbb{R}^{n-1}} \sup_m \int_{\mathbb{R}} |\nabla_1 \Psi_m(x_1, \tilde{x}, w)| \, dx_1 \tilde{h}(\tilde{x}) d\tilde{x} \leq \sqrt{2\pi} \, e^{(M+2)^2/2} \, \hat{\mathbb{E}} |\nabla_1 \Psi_m(x_1, \tilde{x}, w)|$$

$$\leq \sqrt{2\pi} \, e^{(M+2)^2/2} \sup_m 2^m \, \hat{\mathbb{E}} |\psi(x_1 + 1/2^m, \tilde{x}, w) - \psi(x_1, \tilde{x}, w)|.$$

Now using (5.4) we obtain

$$\mathbb{E} \int_{\mathbb{R}^{n-1}} \sup_m \int_{\mathbb{R}} |\nabla_1 \Psi_m(x_1, \tilde{x}, w)| \, dx_1 \tilde{h}(\tilde{x}) d\tilde{x} \leq \sqrt{2\pi} \, e^{(M+2)^2/2} C_M. \tag{5.6}$$

Therefore, for $(\tilde{h}(\tilde{x}) d\tilde{x} \times P)$-almost all $(\tilde{x}, w)$,

$$\sup_m \int_{\mathbb{R}} |\nabla_1 \Psi_m(x_1, \tilde{x}, w)| \, dx_1 < +\infty.$$

On the other hand, by the definition of Ψ_m, for all $(\tilde{x}, w)$ there exists a function $\bar{F}_1(\cdot, \tilde{x}, w)$ so that

$$\lim_m \Psi_m(x_1, \tilde{x}, w) = \bar{F}_1(x_1, \tilde{x}, w) \phi(x_1)$$

$$\bar{F}_1(., \tilde{x}, w) = F(., \tilde{x}, w), \quad dx_1 - a.e.$$

Hence, by Corollary 5.3.4 of [7], we have $\bar{F}_1(.,\tilde{x},w)\phi(.)$ is a function of bounded variation on $\mathbb{R}$ for $(\tilde{h}(\tilde{x})d\tilde{x}\times P)$-almost all $(\tilde{x},w)$. So, for $(\tilde{h}(\tilde{x})d\tilde{x}\times P)$-almost all $(\tilde{x},w)$, and for all $M>0$, $\bar{F}_1(.,\tilde{x},w)$ is a function of bounded variation on $[-M,M]$ such that

$$V_{[-M,M]}\bar{F}_1(.,\tilde{x},w)\leq \sup_m \int_{\mathbb{R}} |\nabla_1 \Psi_m(x_1,\tilde{x},w)|\,dx_1,$$

where $V_{[-M,M]}(f)$ denotes the total variation of the function f on $[-M,M]$. It follows from (5.6) that

$$\mathbb{E}\int_{\mathbb{R}^{n-1}} V_{[-M,M]}\bar{F}_1(.,\tilde{x},w)\tilde{h}(\tilde{x})d\tilde{x}\leq \mathbb{E}\int_{\mathbb{R}^{n-1}}\sup_m\int_{\mathbb{R}}|\nabla_1\Psi_m(x_1,\tilde{x},w)|\,dx_1\tilde{h}(\tilde{x})d\tilde{x}$$
$$\leq \sqrt{2\pi}\;e^{(M+2)^2/2}C_M.$$

Thus for P-almost all w and for each rectangular cell $\mathscr{R}$ of the form $\prod_{i=2}^{n}[-M,M]$ we have

$$\int_{\mathscr{R}} V_{[-M,M]}\bar{F}_1(.,\tilde{x},w)d\tilde{x}<+\infty. \tag{5.7}$$

Now for $2\leq i\leq n$, we write $\mathbf{x}$ as $(x_i,\tilde{x})$ where $\tilde{x}\in\mathbb{R}^{n-1}$. Then using the same procedure as that used above we obtain that there exists a function $\bar{F}_i(\cdot,\tilde{x},w)$ so that

$$\bar{F}_i(.,\tilde{x},w)=F(.,\tilde{x},w),\;\; dx_i-a.e.$$

Moreover for P-almost all w and for each rectangular cell $\mathscr{R}$ in $\mathbb{R}^{n-1}$ we have

$$\int_{\mathscr{R}} V_{[-M,M]}\bar{F}_i(x_i,\tilde{x},w)d\tilde{x}<+\infty. \tag{5.8}$$

Therefore, we conclude by Theorem 5.3.5 of [7] that $F\in\mathbf{BV}(\mathbb{R}^n\times\Omega)$.

Let F is in $\mathbf{BV}(\mathbb{R}^n\times\Omega)$ and $\hat{F}$ its associated according to the above definition. Then for each $1\leq i\leq n$ the function $t\to\hat{F}(\mathbf{x}+te_i,w)$ is a function of bounded variation on any finite interval. Consequently it is differentiable a.e and for all $w\in\Omega$

$$\nabla_i\hat{F}(\mathbf{x},w)=\liminf_{t\to 0}\frac{\hat{F}(\mathbf{x}+te_i,w)-\hat{F}(\mathbf{x},w)}{t}.$$

We can prove as in Bouleau-Hirsch [2] that $\nabla_i F$ is well defined $\hat{P}$-a.s. and this definition does not depend on the representative $\hat{F}$, up to $\hat{P}$-a.s. equality. This is given in the following proposition

Proposition 5.1. *If $F\in\mathbf{BV}(\mathbb{R}^n\times\Omega)$ then for P-almost all $w\in\Omega$ and for all $i\in\{1,\ldots\ldots,n\}$ we have:*

(i) $t \to F(\mathbf{x} + te_i, w)$ *is a function of bounded variation on any finite interval.*

(ii) $\frac{\partial}{\partial x_i} F(\mathbf{x}, w) = \nabla_i \hat{F}(\mathbf{x}, w)\ d\mathbf{x}$*-a.e.*

5.3 Applications to Stochastic Differential Equations

Now we consider if the corresponding flow of the following stochastic differential equation whose coefficients are not Lipschitz continuous

$$\begin{cases} dX_{\mathbf{x}}^k(t) = \sum_{j=1}^d \sigma_j^k(t, X_{\mathbf{x}}^k(t))dB^j(t) + b^k(t, X_{\mathbf{x}}(t))dt \quad k = 1, 2, \ldots, n, \\ X_{\mathbf{x}}(0) = \mathbf{x}, \end{cases} \tag{5.9}$$

is almost surely almost everywhere derivable or not. Let

$$\sigma = \left(\sigma_j^k\right)_{k=1,\ldots,n, j=1,\ldots,d} \in C_b([0, +\infty) \times \mathbb{R} \to \mathbb{R}^n \otimes \mathbb{R}^d),$$

$$b = \left(b^k\right)_{k=1,\ldots,n} \in C_b([0, +\infty) \times \mathbb{R}^n \to \mathbb{R}^n),$$

We assume the following

H.1 $\max_k \left|b^k(t, \mathbf{x}) - b^k(t, \mathbf{x}')\right| \leq K\left|\mathbf{x} - \mathbf{x}'\right|_{\mathbb{R}^n}$, for all $t \geq 0$, $\mathbf{x}$, $\mathbf{x}' \in \mathbb{R}^n$.

H.2 The n-dimensional stochastic differential equation (5.9) has pathwise uniqueness.

It is well-known that, under assumption **H.1**, $\nabla_i b^k$ exists for each $i, k \in \{1, \cdots, n\}$ almost everywhere and satisfies

$$\left|(\nabla_i b^k)(t, \mathbf{x})\right| \leq K, \quad \forall t \geq 0. \tag{5.10}$$

Let $\hat{B}_t(\mathbf{x}, w) = B_t(w)$, $t \in [0, +\infty)$ and $\hat{\mathscr{F}}_t$ the least σ-field containing the $\hat{P}$-negligible sets of $\hat{\mathscr{F}}$ for which all $\hat{B}_s$, $0 \leq s \leq t$, are measurable.

Then $\left(\hat{\Omega}, \hat{\mathscr{F}}, (\hat{\mathscr{F}}_t)_{t\in[0,+\infty)}, \hat{P}, (\hat{B}_t)_{t\in[0,+\infty)}\right)$ is a Brownian motion in $\mathbb{R}^d$ starting from 0.

Let us consider the following SDE

$$\begin{cases} d\hat{X}^k(t) = \sum_{j=1}^d \sigma_j^k(t, \hat{X}^k(t))d\hat{B}^j(t) + b^k(t, \hat{X}(t))dt \quad k = 1, 2, \ldots, n, \\ \hat{X}(0) = \mathbf{x}, \end{cases} \tag{5.11}$$

Now, we state a result which is due to Bouleau and Hirsch. It highlights the relationship between the solutions of SDE (5.9) and (5.11).

Proposition 5.2. *There exists a $\hat{P}$-negligible set $\hat{N}$ such that*

$$\forall(\mathbf{x}, w) \in \hat{N}^c \ \ \forall t \geq 0 \ \ \hat{X}(t)(\mathbf{x}, w) = X_{\mathbf{x}}(t)(w)$$

For the first, we will show a lemma which makes the most important role in this paper.

Lemma 5.1. *Let $f = (f_j)_{j=1,2,\ldots,d}$ be an $\mathbb{R}^d$ – valued measurable function on $[0, +\infty) \times \hat{\Omega}$, and g be a measurable function on $[0, +\infty) \times \hat{\Omega}$. We assume that a 1-dimensional $\mathscr{F}_t$-adapted continuous process $\hat{Y} = (\hat{Y}(t))$ on $(\hat{\Omega}, \hat{\mathscr{F}}, \hat{P})$ satisfies the stochastic differential equation*

$$\hat{Y}(t) = y + \sum_{j=1}^{d} \int_0^t f_j(s, y, w)\hat{Y}(s) d\hat{B}_s^j + \int_0^t g(s, y, w) ds, \tag{5.12}$$

where $y \in \mathbb{R}$. Moreover, we assume that

$$\max_j \sup_{t,y,w} \left| f_j(t, y, w) \right| < \infty, \tag{5.13}$$

and there exists a constant L satisfying that

$$|g(t, y, w)| \leq L \left| \hat{Y}(t) \right|, \quad \textit{for all}\, (t, (y, w)) \in [0, +\infty) \times \hat{\Omega}. \tag{5.14}$$

Then, for each $T \geq 0$ there exists a constant C which depends on only T, L such that

$$\sup_{t \in [0,T]} \hat{\mathbb{E}} \left| \hat{Y}(t) \right| \leq C.$$

Proof. For $(t, (y, w)) \in [0, +\infty) \times \hat{\Omega}$, define:

$$\hat{g}(t, y, w) = \begin{cases} \dfrac{g(t, y, w)}{\hat{Y}(t)} & \textit{if } \hat{Y}(t) \neq 0, \\ 0 & \textit{if } \hat{Y}(t) = 0. \end{cases}$$

Then it follows from (5.14) that

$$|\hat{g}(s, y, w)| \leq L, \quad \textit{for all}\, (t, (y, w)) \in [0, +\infty) \times \hat{\Omega}.$$

Moreover it is easy to see that $\hat{Y}$ satisfies the following Linear stochastic differential equation

$$\hat{Y}(t) = y + \sum_{j=1}^{d} \int_0^t f_j(s, y, w)\hat{Y}(s) d\hat{B}_s^j + \int_0^t \hat{g}(s, y, w)\hat{Y}(s) ds. \tag{5.15}$$

Thus we have

$$\hat{Y}(t) = y \exp\left(\sum_{j=1}^{d} \int_0^t f_j(s,y,w)d\hat{B}_s^j + \int_0^t \hat{g}(s,y,w)ds \right.$$
$$\left. - \frac{1}{2}\sum_{j=1}^{d} \int_0^t f_j(s,y,w)^2 ds \right). \tag{5.16}$$

Since $\hat{g}$ is bounded by L we deduce the following estimate

$$|\hat{Y}(t)| \leq |y|e^{LT} \exp\left(\sum_{j=1}^{d} \int_0^t f_j(s,y,w)d\hat{B}_s^j - \frac{1}{2}\sum_{j=1}^{d} \int_0^t f_j(s,y,w)^2 ds\right)$$

for all $t \in [0,T]$.

Using condition (5.13), yields that the exponential

$$\exp\left(\sum_{j=1}^{d} \int_0^t f_j(s,y,w)d\hat{B}_s^j - \frac{1}{2}\sum_{j=1}^{d} \int_0^t f_j(s,y,w)^2 ds\right)$$

is a $(\hat{\mathscr{F}}_t - \hat{P})$-martingale. Therefore, for all $t \in [0,T]$, we obtain

$$\hat{\mathbb{E}}\left|\hat{Y}(t)\right| \leq |y|e^{LT}.$$

This establishes the lemma.

The next lemma is a version of the above lemma about derivative of a stochastic differential equation.

Lemma 5.2. *We assume that*

$$\sigma = \left(\sigma_j^k\right)_{k=1,\ldots,n,j=1,\ldots,d} \in C_b([0,+\infty) \times \mathbb{R} \to \mathbb{R}^n \otimes \mathbb{R}^d),$$

$\sigma_j^k(t,\cdot) \in C^\infty(\mathbb{R})$, $t \in [0,+\infty)$ *and* $\partial_x \sigma_j^k(\cdot,\cdot)$ *is bounded on compact subsets of* $[0,+\infty) \times \mathbb{R}$,

$$b = \left(b^k\right)_{k=1,\ldots,n} \in C_b([0,+\infty) \times \mathbb{R}^n \to \mathbb{R}^n),$$
$$b^k(t,\cdot) \in C^\infty(\mathbb{R}^n \to \mathbb{R}^n)),\ t \in [0,+\infty),$$

and assumptions ***(H.1)***–***(H.2)*** *hold. Then, for each* $T \geq 0$, *the solution* $\hat{X}$ *on* $(\hat{\Omega}, \hat{\mathscr{F}}, \hat{P})$ *of the stochastic differential equation (5.11) satisfies*

$$\sup_{t\in[0,T]} \hat{\mathbb{E}} \left| \nabla_i \hat{X}^k(t) \right| \le C \qquad \textit{for all}\, k = 1, 2, \ldots, n. \tag{5.17}$$

Proof. A consequence of the continuity of the coefficients the SDE (5.11) has a weak solution. Under assumption (**H.2**) this solution becomes strong. Furthermore for each $T \ge 0$ we have

$$\max_k \sup_{t\in[0,T]} \hat{\mathbb{E}} |\hat{X}^k(t)|^2 < +\infty,$$

which implies that the sequence of stopping times

$$S_n := \inf\{t \in [0, +\infty) : |\hat{X}(t)| > n\} \quad (\inf(\emptyset) = +\infty)$$

converges to $+\infty$. Since σ and b are sufficiently smooth it follows from [K] that for each $k \in \{1, \ldots, n\}$

$$\nabla_i \hat{X}^k(t) = \delta_{ik} + \sum_{j=1}^{d} \int_0^t \partial_x \sigma_j^k(s, \hat{X}^k(s)) \nabla_i \hat{X}^k(s) d\hat{B}^j(s)$$
$$+ \int_0^t (\nabla_i b^k)(s, \hat{X}(s)) \nabla_i \hat{X}^k(s) ds,$$

where δ_{ik} denotes the Kronecker's delta. By the Lipschitz condition of b, we have

$$\left| (\nabla_i b^k)(t, \hat{X}(t)) \right| \le K, \quad \forall t \ge 0.$$

Moreover if we set

$$f_j(s,x,w) = \partial_x \sigma_j^k(s, \hat{X}^k(s \wedge S_n)),\ g(s,x,w) = (\nabla_i b^k)(s, \hat{X}(s \wedge S_n)) \nabla_i \hat{X}^k(s \wedge S_n)$$

then it is simple to see that the conditions (5.13) and (5.14) are satisfied. We then get that $\nabla_i X_x^k(s \wedge S_n)$ is solution of SDE (5.12). Hence, we obtain (5.17) for $\nabla_i X_x^k(s \wedge S_n)$ from the previous lemma. Letting n tend to $+\infty$, we arrive at the desired result.

Now we will show a sufficient condition for solutions of stochastic differential equations to belong to the class $\mathbf{BV}(\mathbb{R}^n \times \Omega)$. The advantage is that we assume only bounded on the diffusion coefficient σ.

Theorem 5.2. *Assume that (**H.1**)–(**H.2**) hold. Then, the solution $\hat{X}^k(t)$ is in $\mathbf{BV}(\mathbb{R}^n \times \Omega)$ for all $t \in [0, +\infty)$ and each $k = 1, 2, \ldots, n$.*

Proof. The regularization procedure enables the existence of a sequences $\{\sigma_m; m \in \mathbb{N}\}$ and $\{b_m; m \in \mathbb{N}\}$ such that

$$\sigma_m = \left(\sigma^k_{j,m}\right)_{k=1,\dots,n,j=1,\dots,d} \in C_b([0,+\infty)\times\mathbb{R}\to\mathbb{R}^n\otimes\mathbb{R}^d),$$

$$\sigma^k_{j,m}(t,\cdot)\in C^\infty(\mathbb{R}),\ t\in[0,+\infty),$$

$$b_m = \left(b^k_m\right)_{k=1,\dots,n} \in C_b([0,+\infty)\times\mathbb{R}^n\to\mathbb{R}^n),$$

$$b^k_m(t,\cdot)\in C^\infty(\mathbb{R}^n\to\mathbb{R}^n)),\ t\in[0,+\infty),$$

$$\sup_m \left|(\nabla_i b^k_n)(t,\mathbf{x})\right| \le K,\ \ \forall\, t\ge 0,\ \mathbf{x}\in\mathbb{R}^n.$$

$$\lim_m \sup_{t\in[0,T]}\sup_{x\in\mathscr{K}} |\sigma_m(t,x)-\sigma(t,x)|_{\mathbb{R}^n\otimes\mathbb{R}^d} + |b_m(t,x)-b(t,x)|_{\mathbb{R}^n} = 0$$

for each compact $\mathscr{K}$ in $\mathbb{R}^n$. Let $\{Z^n_{\mathbf{x}}; n\in\mathbb{N}\}$ be the strong solutions of the SDE (5.9) with coefficients σ_n and b_n and the Brownian motion B. Then, by a standard method of stochastic differential equations and (5.17) of Lemma 5.2 we have for all $i\in\{1,\cdots,n\}$ and $t\in[0,+\infty)$

$$\sup_m \mathbb{E}\,|Z^m_{\mathbf{x}}(t)|^2_{\mathbb{R}^n} < \infty \text{ and } \sup_m \mathbb{E}\,|\nabla_i (Z^m_{\mathbf{x}}(t)|_{\mathbb{R}^n} < \infty. \tag{5.18}$$

On the other hand we have from [3] that

$$\lim_{n\to+\infty}\ \sup_{x\in\mathscr{K}} \mathbb{E}\,(\max_{t\in[0,T]} |Z^n_{\mathbf{x}}(t) - X_{\mathbf{x}}(t)|^2_{\mathbb{R}^n}) = 0. \tag{5.19}$$

Let $M>0$ and Φ be a fixed function in $C^\infty(\mathbb{R}^n)$ satisfying

$$\Phi(\mathbf{x}) = 1 \text{ if } \mathbf{x}\in\prod_{i=1}^n[-M,M],$$

$$\Phi(\mathbf{x}) = 0 \text{ if } \mathbf{x}\notin\prod_{i=1}^n[-M+1,M+1],$$

$$|\nabla_i\Phi(\mathbf{x})| \le C \text{ (constant) for all } \mathbf{x}\in\mathbb{R}^n.$$

Let $\left\{\hat{Z}^m; m\in\mathbb{N}\right\}$ be the strong solution of the SDE (5.11) with coefficients σ_m and b_m and the Brownian motion $\hat{B}$. It follows from (5.18–5.19) and Proposition 5.2 that

$$\sup_m \hat{\mathbb{E}}\left|\hat{Z}^m(t)\,\Phi(\mathbf{x})\right|^2_{\mathbb{R}^n} < \infty \text{ and } \sup_m \hat{\mathbb{E}}\left|\nabla_i(\hat{Z}^m(t)\,\Phi(\mathbf{x}))\right|_{\mathbb{R}^n} < \infty, \tag{5.20}$$

and

$$\lim_{m\to+\infty} \hat{\mathbb{E}}\,(\max_{t\in[0,T]} |\hat{Z}^m(t)\Phi(\mathbf{x}) - \hat{X}(t)\Phi(\mathbf{x})|^2_{\mathbb{R}^n}) = 0. \tag{5.21}$$

It is easy to see that (5.20) means that the family $\{\hat{Z}^m \Phi, m \in \mathbb{N}\}$ satisfies condition (**B.2**) and (**B.4**) of Theorem 7.1. Furthermore, for each $m \in \mathbb{N}$, $\hat{Z}^m \Phi$ satisfies the smoothness property (**B.3**), cf. Kunita [4]. It follows from (5.21) that there exists a subsequence of $\{\hat{Z}^m \Phi, m \in \mathbb{N}\}$ which converges $\hat{P}$-almost surely to $\hat{X}\Phi$. For simplicity, we also denote the subsequence by $\{\hat{Z}^m \Phi, m \in \mathbb{N}\}$ again. Thus, we can use Theorem 2.2, and we have $\hat{X}(t)\Phi(\mathbf{x})$ is in $\mathbf{BV}(\mathbb{R}^n \times \Omega)$ and consequently $\hat{X}(t) \in \mathbf{BV}(\mathbb{R}^n \times \Omega)$ for all $t \in [0, +\infty)$.

Now as a consequence of the foregoing we obtain the following result

Theorem 5.3. *Assume that (**H.1**)–(**H.2**) hold. Then, for almost all w the flow $\mathbf{x} \mapsto X_{\mathbf{x}}(t, w)$ is a function of bounded variation on each compact in $\mathbb{R}^n$.*

Acknowledgements The authors wish to thank the anonymous referee for the suggestion of a simple proof of Lemma 1. This work was supported by the Academy Hassan II of Sciences and Technology

References

1. Bouleau, N. Hirsch, F.: On the derivability with respect to the initial data of the solution of a stochastic differential equation with lipschitz coefficients. In: Séminaire de Théorie du Potential Paris, **9**, 39–57. Springer, Berlin (1989)
2. Bouleau, N. Hirsch, F.: Propriétés d'absolue continuité dans les espaces de Dirichlet et application aux équations différentielles stochastiques. In: Séminaire de Probabilités, XX, 131–161. Springer, Berlin (1986)
3. Kaneko, H., Nakao, S.: A note on approximation for stochastic differential equations. In: Séminaire de Probabilités, XXII, 155–162. Springer, Berlin (1988)
4. Kunita, H.: Stochastic differential equations and stochastic flows of diffeomorphisms. In: École d'été de probabilités de Saint-Flour, XII-1982, 143–303. Springer, Berlin (1984)
5. Kusuoka, S.: Existence of densities of solutions of stochastic differential equations by Malliavin calculus. J. Funct. Anal. **258**, 758–784 (2010)
6. Revuz, D. Yor, M.: Continuous martingales and Brownian motion. Springer-Verlag, Berlin (1999)
7. Ziemer William, P.: Weakly differentiable functions. Sobolev spaces and functions of bounded variation. Springer-Verlag, New York (1989)

Chapter 6
Stochastic Volatility and Multifractional Brownian Motion

Antoine Ayache and Qidi Peng

Abstract In order to make stochastic volatility models more realistic some authors (see for example Comte and Renault J. Econom. 73:101–150, 1996; Comte and Renault Math. Financ. 8:291–323, 1998; Gloter, A., Hoffmann, M.: Stochastic volatility and fractional Brownian motion Stoch. Proc. Appl. 113:143–172, 2004; Rosenbaum Stoch. Proc. Appl. 118:1434–1462, 2008) have proposed to replace the Brownian motion governing the volatility by a more flexible stochastic process. This is why, we introduce multifractional stochastic volatility models; their main advantage is that they allow to account variations with respect to time of volatility local roughness. Our definition of multifractional stochastic volatility models is inspired by that of the fractional stochastic volatility models previously introduced by Gloter and Hoffmann (Gloter, A., Hoffmann, M.: Stochastic volatility and fractional Brownian motion, Stoch. Proc. 502 Appl. 113:143–172, 2004). The main goal of our article is to extend to these new models some theoretical results concerning statistical inference which were obtained in (Gloter, A., Hoffmann, M.: Stochastic volatility and fractional Brownian motion, Stoch. Proc. Appl. 113:143–172, 2004). More precisely, assuming that the functional parameter $H(\cdot)$ of multifractional Brownian motion is known, we construct, in a general framework, an estimator of integrated functional of the volatility, and we derive from it, in the linear case, an estimator of a parameter θ related to the volatility.

A. Ayache (✉)
U.M.R. CNRS 8524, Laboratory Paul Painlevé, University Lille 1, 59655 Villeneuve d'Ascq Cedex, France
e-mail: Antoine.Ayache@math.univ-lille1.fr

Q. Peng
U.M.R. CNRS 8524, Laboratory Paul Painlevé, University Lille 1, 59655 Villeneuve d'Ascq Cedex, France
e-mail: Qidi.Peng@math.univ-lille1.fr

M. Zili and D.V. Filatova (eds.), *Stochastic Differential Equations and Processes*, Springer Proceedings in Mathematics 7, DOI 10.1007/978-3-642-22368-6_6,

6.1 Introduction and Motivation

Stochastic volatility models are extensions of the well-known Black and Scholes model. Hull and White [22] and other authors in mathematical finance (see for instance [30] and [25]) introduced them in the eighties in order to account the volatility effects of exogenous arrivals of information. Our article is inspired from a work of Gloter and Hoffmann [21, 22] which concerns statistical inference in a parametric stochastic volatility model driven by a fractional Brownian motion (fBm for short). Namely, the model considered in [21,22] can be expressed as:

$$\begin{cases} Z(t) = z_0 + \int_0^t \sigma(s)\,\mathrm{d}W(s) \\ \sigma(s) = \sigma_0 + \Phi(\theta, B_\alpha(s)), \end{cases} \tag{6.1}$$

where:

- $Z(t)$ denotes logarithm of the price of the underlying asset, the original price z_0 is supposed to be deterministic and known.
- $\{W(s)\}_{s\in[0,1]}$ denotes a standard Brownian motion (Bm for short).
- $\{\sigma(s)\}_{s\in[0,1]}$ denotes the volatility process (σ_0 is real-valued and known); the deterministic function $x \mapsto \Phi(\theta, x)$, through which it is defined, is known up to a real-valued parameter θ. For the sake of convenience, one sets for every $x \in \mathbb{R}$,

$$f(x) := (\sigma_0 + \Phi(\theta, x))^2 \tag{6.2}$$

and throughout this article one assumes that the function f belongs to the set $C^2_{pol}(\mathbb{R})$. Observe that for every integer $l \geq 0$, $C^l_{pol}(\mathbb{R})$ denotes the vector space of l-times continuously differentiable functions over the real line, which slowly increase at infinity as well as their derivative of any order, more formally,

$$C^l_{pol}(\mathbb{R}) := \Big\{h \in C^l(\mathbb{R}) : \exists c, K > 0\,, \forall\, x \in \mathbb{R}, \ \sum_{k=0}^{l} |h^{(k)}(x)| \leq c\left(1 + |x|^K\right)\Big\}. \tag{6.3}$$

- $\{B_\alpha(s)\}_{s\in[0,1]}$ denotes a fractional Brownian motion (fBm for short) with Hurst parameter α (see e.g. [15,16,29]), which is assumed to be independent on the Bm $\{W(s)\}_{s\in[0,1]}$; one makes the latter independence assumption for the stochastic integral $\int_0^t \sigma(s)\,\mathrm{d}W(s)$ to be well-defined. Note that the idea of replacing the Bm governing the volatility by a fBm is due to Comte and Renault (see [13,14]), who have proposed to do so in order to account some long memory effects.

In order to clearly explain the main goal of our article, we need to briefly present some of the main results obtained in [21, 22]. In the latter articles, it is assumed that one observes a discretized trajectory of the process $\{Z(t)\}_{t\in[0,1]}$, namely the high frequency data $Z(j/n)$, $j = 0, \ldots, n$. Also, it is assumed that the fBm B_α governing the volatility is hidden; however one knows the value of its Hurst parameter α, moreover $\alpha \in (1/2, 1)$. Though, the hypothesis that the Hurst parameter is known may seem to be restrictive from a practical point of view, it has already been made by other authors (see for example [35]), in some settings more or less related to that of the model (6.1). Under additional technical assumptions, we will not give here for the sake of simplicity, Gloter and Hoffmann [21, 22] have obtained the following results (1) and (2):

1. By using the notion of generalized quadratic variation, one can construct estimators of integrated functional of the volatility of the form: $\int_0^1 f'(B_\alpha(s))^2 h(\sigma^2(s))\,\mathrm{d}s$, where f' is the derivative of f and $h \in C^1_{pol}(\mathbb{R})$ is arbitrary and fixed. Note that the problem of the estimation of such quantities is of some importance in its own right, since more or less similar integrals appear in some option pricing formulas (see for instance [22]).
2. Thanks to the result (1), it is possible to build a minimax optimal estimator of the unknown parameter θ. Also, it is worth noticing that, it has been shown in [21, 22] that the minimax rate of convergence for estimating θ, in the setting of the model (6.1), is not the usual rate $n^{-1/2}$ but the slower rate $n^{-1/(4\alpha+2)}$, which deteriorates when the Hurst parameter α increases; basically, the reason for this unusual phenomenon is that the volatility is hidden and the Brownian motion W makes the approximation of the volatility more noisy.

Let us now present the main motivation behind the introduction of multifractional stochastic volatility models. To this end, first we need to introduce the notion of pointwise Hölder exponent. Let $\{X(s)\}_{s\in[0,1]}$ be a stochastic process whose trajectories are with probability 1, continuous and nowhere differentiable functions (this is the case of fBm and of multifractional Brownian motion which will soon be introduced), $\rho_X(t)$ the pointwise Hölder exponent of the process $\{X(s)\}_{s\in[0,1]}$ at an arbitrary time t, is defined as,

$$\rho_X(t) = \sup\left\{\rho \in [0,1] : \limsup_{\tau\to 0} \frac{X(t+\tau) - X(t)}{|\tau|^\rho} = 0\right\}.$$

The quantity of $\rho_X(t)$ provides a measure of $\{X(s)\}_{s\in[0,1]}$ roughness (i.e. of the maximum of the fluctuations amplitudes of $\{X(s)\}_{s\in[0,1]}$) in a neighborhood of t; the smaller $\rho_X(t)$ is the rougher (i.e. the more fluctuating) is $\{X(s)\}_{s\in[0,1]}$ in t neighborhood. In [1] numerical evidences have shown that for a better understanding of stock price dynamics, it is important to analyze volatility local roughness. With this respect, fractional stochastic volatility model has a serious limitation: its volatility local roughness cannot evolve with time; more precisely, when $\Phi(\theta,\cdot)$ is a continuously differentiable function with a nowhere vanishing derivative, then one has, almost surely, at any time t, $\rho_\sigma(t) = \alpha$, where α is the constant Hurst parameter of the fBm $\{B_\alpha(s)\}_{s\in[0,1]}$ and $\rho_\sigma(t)$ the pointwise Hölder exponent at t, of the volatility process $\{\sigma(s)\}_{s\in[0,1]}$ defined in (6.1). The latter limitation is due to the fact that the local roughness of $\{B_\alpha(s)\}_{s\in[0,1]}$ itself cannot change from time to time, namely one has almost surely, for all t, $\rho_{B_\alpha}(t) = \alpha$ (see e.g. [23]). In order to overcome this drawback, we propose to replace in (6.1), the fBm $\{B_\alpha(s)\}_{s\in[0,1]}$ by a multifractional Brownian motion (mBm for short), denoted in all the sequel by $\{X(s)\}_{s\in[0,1]}$, which is independent on $\{W(s)\}_{s\in[0,1]}$. Thus, we obtain a new stochastic volatility model we call multifractional stochastic volatility model. Its precise definition is the following:

$$\begin{cases} Z(t) = z_0 + \int_0^t \sigma(s)\,\mathrm{d}W(s) \\ \sigma(s) = \sigma_0 + \Phi(\theta, X(s)), \end{cases} \tag{6.4}$$

where z_0, σ_0, θ, $\Phi(\theta,\cdot)$ and $\{W(s)\}_{s\in[0,1]}$ satisfy the same assumptions as before; note that the stochastic integral $\int_0^t \sigma(s)\,\mathrm{d}W(s)$ is well-defined since the mBm $\{X(s)\}_{s\in[0,1]}$ is assumed to be independent on the Bm $\{W(s)\}_{s\in[0,1]}$. In order to clearly explain the reason why multifractional stochastic volatility model allows to overcome the limitation of fractional volatility model we have already pointed out we need to make some brief recalls concerning mBm. The latter non stationary increments centered Gaussian process was introduced independently in [8] and [26], in order to avoid some drawbacks coming from the fact that the Hurst parameter of fBm cannot evolve with time. The mBm $\{X(s)\}_{s\in[0,1]}$ can be obtained by substituting to the constant Hurst parameter α in the harmonizable representation of fBm:

$$B_\alpha(s) = \int_{\mathbb{R}} \frac{e^{is\xi}-1}{|\xi|^{\alpha+1/2}}\,\mathrm{d}\widehat{B}(\xi), \tag{6.5}$$

a function $H(\cdot)$ depending continuously on time and with values in $(0,1)$. The process $\{X(s)\}_{s\in[0,1]}$ can therefore be expressed as,

$$X(s) := \int_{\mathbb{R}} \frac{e^{is\xi}-1}{|\xi|^{H(s)+1/2}}\,\mathrm{d}\widehat{B}(\xi). \tag{6.6}$$

Throughout our article not only we assume that $H(\cdot)$ is continuous but also that it is a C^2-function, actually we need to impose this condition in order to be able to estimate the correlations between the generalized increments of local averages of mBm (see Proposition 6.3.6). Moreover, to obtain Part (ii) of Lemma 6.3.2, we need to assume that $H(\cdot)$ is with values in $(1/2, 1)$.

For the sake of clarity, notice that $\mathrm{d}\widehat{B}$ is defined as the unique complex-valued stochastic Wiener measure which satisfies for all $f \in L^2(\mathbb{R})$,

$$\int_{\mathbb{R}} f(s)\,\mathrm{d}B(s) = \int_{\mathbb{R}} \widehat{f}(\xi)\,\mathrm{d}\widehat{B}(\xi), \tag{6.7}$$

where $\{B(s)\}_{s\in\mathbb{R}}$ denotes a real-valued Wiener process and $\widehat{f}$ the Fourier transform of f. Observe that it follows from (6.7) that, one has up to a negligible deterministic smooth real-valued multiplicative function (see [12, 34]), for all $s \in [0,1]$,

$$\int_{\mathbb{R}} \left\{ |s+x|^{H(s)-1/2} - |x|^{H(s)-1/2} \right\} \mathrm{d}B(x) = \int_{\mathbb{R}} \frac{e^{is\xi}-1}{|\xi|^{H(s)+1/2}} \mathrm{d}\widehat{B}(\xi),$$

which implies that the process $\{X(s)\}_{s\in[0,1]}$ is real-valued.

Since several years, there is an increasing interest in the study of mBm and related processes (see for instance [2–7, 9–11, 17–19, 31–34]). The usefulness of such processes as models in financial frame has been emphasized by several authors (see for example [9–11, 23, 24]). Generally speaking, mBm offers a larger spectrum of applications than fBm, mainly because its local roughness can be prescribed via its functional parameter $H(\cdot)$ and thus is allowed to change with time; more precisely, one has almost surely, for all t, $\rho_X(t) = H(t)$, where $\rho_X(t)$ denotes the pointwise Hölder exponent at t of the mBm $\{X(s)\}_{s\in[0,1]}$. It is worth noticing that the latter result, in turn, implies that in the model (6.4) the volatility local roughness can evolve with time, namely when $\Phi(\theta,\cdot)$ is a continuously differentiable function with a nowhere vanishing derivative, then one has, almost surely, at any time t, $\rho_\sigma(t) = H(t)$, where $\rho_\sigma(t)$ is the pointwise Hölder exponent at t, of the volatility process $\{\sigma(s)\}_{s\in[0,1]}$ defined in (6.4).

Having given the main motivation behind multifractional stochastic volatility models, let us now clearly explain the goal of our article. Our aim is to study to which extent it is possible to extend to the setting of these new models Gloter and Hoffmann results (1) and (2) stated above. Basically, we use some techniques which are reminiscent to those in [21, 22]; however new difficulties appear in our multifractional setting. These new difficulties are essentially due to the fact that local properties of mBm change from one time to another.

Throughout our article we assume that the functional parameter $H(\cdot)$ of the mBm $\{X(s)\}_{s\in[0,1]}$ is known. We show that the result (1) can be stated in a more general form and can be extended to multifractional stochastic volatility models. The challenging problem of extending the result (2) to these models remains open; the major difficulty in it, consists in precisely determining the minimax rate of convergence for estimating θ. Yet, in the linear case, that is for a model of the form:

$$\begin{cases} Z(t) = z_0 + \int_0^t \sigma(s)\,\mathrm{d}W(s) \\ \sigma(s) = \sigma_0 + \theta X(s), \end{cases} \tag{6.8}$$

assuming that there exists $t_0 \in (0,1)$ such that $H(t_0) = \min_{t\in[0,1]} H(t)$, we give a partial solution to this problem; namely, we show that by localizing Gloter and Hoffmann estimator in a well-chosen neighborhood of t_0, it is possible to obtain an estimator of θ^2 whose rate of convergence can be bounded in probability by $n^{-1/(4H(t_0)+2)}\big(\log n\big)^{1/4}$.

6.2 Statement of the Main Results

Let us consider an integrated functional of the volatility of the form:

$$\int_0^1 \left(f'(X(s))\right)^2 h(Y(s))\,\mathrm{d}s, \tag{6.9}$$

where, $\{X(s)\}_{s\in[0,1]}$ denotes the mBm, $\{Y(s)\}_{s\in[0,1]}$ is the process defined as

$$Y(s) := f(X(s)) := \sigma^2(s) := \left(\sigma_0 + \Phi(\theta, X(s))\right)^2, \tag{6.10}$$

and h an arbitrary function of $C^1_{pol}(\mathbb{R})$. An important difficulty in the problem of the nonparametric estimation of the integral (6.9) comes from the fact that the process $\{Y(s)\}_{s\in[0,1]}$ is hidden; as we have mentioned before, we only observe the sample $(Z(0), Z(1/n), \ldots, Z(1))$, where $\{Z(t)\}_{t\in[0,1]}$ is the process defined in (6.4). Let us first explain how to overcome this difficulty. $\overline{Y}_{i,N}$, $i = 0, \ldots, N-1$, the local average values of the process $\{Y(s)\}_{s\in[0,1]}$ over a grid $\{0, 1/N, \ldots, 1\}$, $N \geq 1$ being an arbitrary integer, are defined, for all $i = 0, \ldots, N-1$, as

$$\overline{Y}_{i,N} := N \int_{\frac{i}{N}}^{\frac{i+1}{N}} Y(s)\,\mathrm{d}s. \tag{6.11}$$

Let us now assume that N is a well chosen integer depending on n (this choice will be made more precise in the statements of Theorems 6.2.2 and 6.2.3 given below), such an N is denoted by N_n; moreover, we set

$$m_n := [n/N_n] \text{ and for every } i = 0, \ldots, N_n, j_i := [in/N_n], \tag{6.12}$$

with the convention that $[\cdot]$ is the integer part function. The key idea to overcome the difficulty, we have already pointed out, consists in using the fact that for n big enough, $\overline{Y}_{i,N_n}$ can be approximated by

$$\widehat{Y}_{i,N_n,n} := N_n \sum_{k=0}^{j_{i+1}-j_i-1} \left(Z\left(\frac{j_i+k+1}{n}\right) - Z\left(\frac{j_i+k}{n}\right)\right)^2. \tag{6.13}$$

The rigorous proof of the latter approximation result relies on Itô formula, it is given in [22] page 157 in the case of fBm and it can be easily extended to the case of mBm. This is why we will only give here a short heuristic proof. Using (6.13), (6.4), the fact that n is big enough, and (6.10), one has

$$\begin{aligned}
\widehat{Y}_{i,N_n,n} &= N_n \sum_{k=0}^{j_{i+1}-j_i-1} \left(\int_{\frac{j_i+k}{n}}^{\frac{j_i+k+1}{n}} \sigma(s)\, \mathrm{d}W(s) \right)^2 \\
&\approx N_n \sum_{k=0}^{j_{i+1}-j_i-1} \left(\int_{\frac{j_i+k}{n}}^{\frac{j_i+k+1}{n}} \mathrm{d}W(s) \right)^2 \sigma^2 \left(\frac{j_i+k}{n} \right) \\
&= N_n \sum_{k=0}^{j_{i+1}-j_i-1} \left(W\left(\frac{j_i+k+1}{n} \right) - W\left(\frac{j_i+k}{n} \right) \right)^2 Y\left(\frac{j_i+k}{n} \right) \\
&\approx N_n \left(n^{-1} \sum_{k=0}^{j_{i+1}-j_i-1} Y\left(\frac{j_i+k}{n} \right) \right),
\end{aligned} \tag{6.14}$$

where the latter approximation follows from the fact that $\left(W\left(\frac{j_i+k+1}{n}\right) - W\left(\frac{j_i+k}{n}\right)\right)^2$, $k = 0, \ldots, j_{i+1} - j_i - 1$ are i.i.d random variables whose expectation equals n^{-1}. Then noticing that $n^{-1} \sum_{k=0}^{j_{i+1}-j_i-1} Y\left(\frac{j_i+k}{n}\right)$ is a Riemann sum which, in view of (6.12), converges to the integral $\int_{\frac{i}{N_n}}^{\frac{i+1}{N_n}} Y(s)\,\mathrm{d}s$; it follows from (6.11) and (6.14) that

$$\widehat{Y}_{i,N_n,n} \approx \overline{Y}_{i,N_n}.$$

The main goal of Sect. 6.3 is to construct estimators of the integrated functional of the volatility

$$V(h; \mu_N, \nu_N) := \frac{1}{\nu_N - \mu_N} \int_{\mu_N}^{\nu_N} \left(f'(X(s))\right)^2 h(Y(s))\,\mathrm{d}s, \tag{6.15}$$

where $(\mu_N)_N$ and $(\nu_N)_N$ are two arbitrary sequences satisfying: for every N, $0 \leq \mu_N < \nu_N \leq 1$ and $\lim_{N \to +\infty} N(\nu_N - \mu_N) = +\infty$.

Observe that when we take for every N, $\mu_N = 0$ and $\nu_N = 1$, then the integral in (6.15) is equal to the integral in (6.9).

In order to be able to state the main two results of Sect. 6.3, one needs to introduce some additional notations. Throughout this article one denotes by $a = (a_0, \ldots, a_p)$ a finite sequence of $p + 1$ arbitrary fixed real numbers whose $M(a)$ first moments vanish i.e. one has

$$\sum_{k=0}^{p} k^l a_k = 0, \text{ for all } l = 0, \ldots, M(a) - 1$$

and one always assumes that $M(a) \geq 3$ (observe that one has necessarily $p > M(a)$). For each integer $N \geq p + 1$ and any $i = 0, \ldots, N - p - 1$, $\Delta_a \overline{Y}_{i,N}$ is the generalized increment of local average values of Y, defined as

$$\Delta_a \overline{Y}_{i,N} := \sum_{k=0}^{p} a_k \overline{Y}_{i+k,N} \tag{6.16}$$

and $\Delta_a \overline{X}_{i,N}$ is the generalized increment of local average values of mBm X, defined as

$$\Delta_a \overline{X}_{i,N} := \sum_{k=0}^{p} a_k \overline{X}_{i+k,N}, \tag{6.17}$$

where

$$\overline{X}_{i,N} := N \int_{\frac{i}{N}}^{\frac{i+1}{N}} X(s)\,\mathrm{d}s. \tag{6.18}$$

At last, for each integer n big enough and any $i = 0, \ldots, N_n - p - 1$, one denotes by $\Delta_a \widehat{Y}_{i,N_n,n}$ the generalized increment defined as

$$\Delta_a \widehat{Y}_{i,N_n,n} := \sum_{k=0}^{p} a_k \widehat{Y}_{i+k,N_n,n}. \tag{6.19}$$

One is now in position to state the two main results of Sect. 6.3. The following theorem provides an estimator of the integrated functional of the volatility $V(h;\mu_N,\nu_N)$ starting from $\overline{Y}_{i,N}$, $\mu_N \le i/N \le \nu_N$, the local average values of the process $\{Y(s)\}_{s\in[0,1]}$ over the grid $\{0, 1/N, \ldots, 1\} \cap [\mu_N, \nu_N]$. It also provides an upper bound of the rate of convergence.

Theorem 6.2.1. *For every integer $N \ge p+1$ and for every function $h \in C^1_{pol}(\mathbb{R})$ one sets*

$$\overline{V}(h;\mu_N,\nu_N) := \frac{1}{N(\nu_N - \mu_N)} \sum_{i\in\mathscr{I}(\mu_N,\nu_N)} \frac{(\Delta_a \overline{Y}_{i,N})^2}{C(i/N) N^{-2H(i/N)}} h(\overline{Y}_{i,N}), \tag{6.20}$$

where:

- *$\mathscr{I}(\mu_N,\nu_N)$ denotes the set of indices,*

$$\mathscr{I}(\mu_N,\nu_N) = \big\{ i \in \{0,\ldots,N-p-1\} : \mu_N \le i/N \le \nu_N \big\}; \tag{6.21}$$

- *For all $s \in [0,1]$,*

$$C(s) := \int_{\mathbb{R}} \frac{|e^{i\eta}-1|^2 |\sum_{k=0}^{p} a_k e^{ik\eta}|^2}{|\eta|^{2H(s)+3}}\,\mathrm{d}\eta. \tag{6.22}$$

Then there exists a constant $c > 0$, such that one has for each integer $N \ge p+1$,

$$\mathbb{E}\Big\{ \big| \overline{V}(h;\mu_N,\nu_N) - V(h;\mu_N,\nu_N) \big| \Big\} \le c\Big(N(\nu_N - \mu_N) \Big)^{-1/2}. \tag{6.23}$$

Recall that the integrated functional of the volatility $V(h;\mu_N,\nu_N)$ has been defined in (6.15).

In view of the previous theorem, in order to construct an estimator of $V(h;\mu_{N_n},\nu_{N_n})$ starting from the observed data $Z(j/n)$, $j=0,\ldots,n$, a natural idea consists in replacing in (6.20), the $\overline{Y}_{i,N_n}$'s by their approximations $\widehat{Y}_{i,N_n,n}$. However (this has already been noticed in [21, 22] in the case where X is the fBm, $\mu_{N_n}=0$ and $\nu_{N_n}=1$),

$$\frac{1}{N_n(\nu_{N_n}-\mu_{N_n})}\sum_{i\in\mathscr{I}(\mu_{N_n},\nu_{N_n})}\frac{(\Delta_a\widehat{Y}_{i,N_n,n})^2}{C(i/N_n)N_n^{-2H(i/N_n)}}h(\widehat{Y}_{i,N_n,n})-V(h;\mu_{N_N},\nu_{N_n})$$

does not converge to zero in the $L^1(\Omega)$ norm; one needs therefore to add the correction term:

$$-\frac{1}{N_n(\nu_{N_n}-\mu_{N_n})}\sum_{i\in\mathscr{I}(\mu_{N_n},\nu_{N_n})}\frac{2\|a\|_2^2(\widehat{Y}_{i,N_n,n})^2}{C(i/N_n)N_n^{-2H(i/N_n)}m_n}h(\widehat{Y}_{i,N_n,n}),$$

where $\|a\|_2=\sqrt{\sum_{k=0}^p a_k^2}$ denotes the Euclidian norm of a. More precisely, the following theorem holds.

Theorem 6.2.2. *For every integer n big enough and $h\in C^1_{pol}(\mathbb{R})$, one sets*

$$\widehat{V}(h;\mu_{N_n},\nu_{N_n}):=\frac{1}{N_n(\nu_{N_n}-\mu_{N_n})}\sum_{i\in\mathscr{I}(\mu_{N_n},\nu_{N_n})}\left(\frac{(\Delta_a\widehat{Y}_{i,N_n,n})^2}{C(i/N_n)N_n^{-2H(i/N_n)}}-\frac{2\|a\|_2^2(\widehat{Y}_{i,N_n,n})^2}{C(i/N_n)N_n^{-2H(i/N_n)}m_n}\right)h(\widehat{Y}_{i,N_n,n}),\tag{6.24}$$

where m_n is as in (6.12). Then assuming that

$$\sup_n m_n^{-1}N_n^{2\max_{s\in[\mu_{N_n},\nu_{N_n}]}H(s)}<+\infty,\tag{6.25}$$

it follows that there exists a constant $c>0$, such that one has for all n big enough,

$$\mathbb{E}\left\{\left|\widehat{V}(h;\mu_{N_n},\nu_{N_n})-V(h;\mu_{N_n},\nu_{N_n})\right|\right\}\le c\Big(N_n(\nu_{N_n}-\mu_{N_n})\Big)^{-1/2}.\tag{6.26}$$

Remark 6.2.1. When the mBm X is a fBm with Hurst parameter $\alpha\in(1/2,1)$, one can take in Theorems 6.2.1 and 6.2.2, $H(\cdot)=\alpha$, $\mu_{N_n}=0$ and $\nu_{N_n}=1$; then one recovers Theorem 3 in [22] and Proposition 1 in [20].

Let us now turn to Sect. 6.4. The goal of this section is to construct an estimator of θ^2 in the setting of a linear stochastic volatility model driven by a mBm, that is a model of the type (6.8); and also to give an evaluation of the rate of convergence of this estimator in terms of $\min_{t\in[0,1]} H(t)$. Notice that, in Sect. 6.4, we assume that $\theta \neq 0$ and there exists $t_0 \in (0,1)$ such that

$$H(t_0) = \min_{t\in[0,1]} H(t). \tag{6.27}$$

In order to be able to state the main result of this section, we need to introduce some additional notations. For n big enough, we set,

$$\mathscr{E}_{N_n}^{min}(t_0) = t_0 - \frac{1}{\sqrt{\log(N_n)}}, \tag{6.28}$$

$$\mathscr{E}_{N_n}^{max}(t_0) = t_0 + \frac{1}{\sqrt{\log(N_n)}} \tag{6.29}$$

and

$$\begin{aligned} \mathscr{V}_{N_n,t_0} &:= \mathscr{I}\left(\mathscr{E}_{N_n}^{min}(t_0), \mathscr{E}_{N_n}^{max}(t_0)\right) \\ &= \left\{ i \in \{0,\dots,N_n-p-1\} \,:\, \left|t_0 - \frac{i}{N_n}\right| \le \frac{1}{\sqrt{\log N_n}} \right\}, \end{aligned} \tag{6.30}$$

where $\mathscr{I}$ has been introduced in (6.21).

Let $(a_n)_n$ and $(b_n)_n$ be two arbitrary sequences of positive real numbers. The notation $a_n \asymp b_n$ means there exist two constants $0 < c_1 \le c_2$, such that for all n, one has $c_1 a_n \le b_n \le c_2 a_n$.

We are now in position to state the main result of Sect. 6.4.

Theorem 6.2.3. *Consider a linear stochastic volatility model driven by a mBm. For n big enough, let,*

$$\widehat{\theta}^2_{n,t_0} = \frac{\widehat{V}(1; \mathscr{E}_{N_n}^{min}(t_0), \mathscr{E}_{N_n}^{max}(t_0))}{4\big(2(\log(N_n))^{-1/2} N_n\big)^{-1} \sum_{i\in\mathscr{V}_{N_n,t_0}} \widehat{Y}_{i,N_n,n}}, \tag{6.31}$$

where $\widehat{V}$ has been introduced in (6.24). Assume that

$$N_n \asymp n^{1/(2H(t_0)+1)}. \tag{6.32}$$

Then the sequence of random variables

$$\left(n^{1/(4H(t_0)+2)} (\log n)^{-1/4} (\widehat{\theta}^2_{n,t_0} - \theta^2)\right)_n,$$

is bounded in probability i.e. one has

$$\lim_{\lambda \to +\infty} \limsup_{n \to +\infty} \mathbb{P}\Big\{ n^{1/(4H(t_0)+2)} (\log n)^{-1/4} |\widehat{\theta}^2_{n,t_0} - \theta^2| > \lambda \Big\} = 0. \tag{6.33}$$

Remark 6.2.2. Theorem 6.2.3 is an extension of Proposition 2 in [20].

In fact, Theorem 6.2.3 is a straightforward consequence of the following result.

Theorem 6.2.4. *Consider a linear stochastic volatility model driven by a mBm. For n big enough, we set,*

$$\widehat{\theta}^2_n(\mu_{N_n}, \nu_{N_n}) = \frac{\widehat{V}(1; \mu_{N_n}, \nu_{N_n})}{4(N_n(\nu_{N_n} - \mu_{N_n}))^{-1} \sum_{i \in \mathscr{J}(\mu_{N_n}, \nu_{N_n})} \widehat{Y}_{i,N_n,n}}. \tag{6.34}$$

Assume that $(\mu_{N_n})_n$ *and* $(\nu_{N_n})_n$ *are two convergent sequences and also that*

$$N_n \asymp n^{1/(2\max_{s \in [\mu_{N_n}, \nu_{N_n}]} H(s)+1)}. \tag{6.35}$$

Then the sequence of random variables

$$\Big(n^{1/(4\max_{s \in [\mu_{N_n}, \nu_{N_n}]} H(s)+2)} (\nu_{N_n} - \mu_{N_n})^{1/2} \Big(\widehat{\theta}^2_n(\mu_{N_n}, \nu_{N_n}) - \theta^2 \Big) \Big)_n,$$

is bounded in probability.

6.3 Estimation of Integrated Functionals of the Volatility

6.3.1 *Proof of Theorem 6.2.1 When Y is the mBm*

The goal of this subsection is to show that Theorem 6.2.1 holds in the particular case where the process Y (see (6.10)) is the mBm X itself i.e. the function f is equal to the identity. Namely, we will prove the following theorem.

Theorem 6.3.1. *For every integer* $N \geq p+1$ *and every function* $h \in C^1_{pol}(\mathbb{R})$ *one sets*

$$Q(h; \mu_N, \nu_N) := \frac{1}{\nu_N - \mu_N} \int_{\mu_N}^{\nu_N} h(X(s))\, ds \tag{6.36}$$

and

$$\overline{Q}(h; \mu_N, \nu_N) := \frac{1}{N(\nu_N - \mu_N)} \sum_{i \in \mathscr{J}(\mu_N, \nu_N)} \frac{(\Delta_a \overline{X}_{i,N})^2}{C(i/N) N^{-2H(i/N)}} h(\overline{X}_{i,N}). \tag{6.37}$$

Then there is a constant $c > 0$*, such that one has for each* $N \geq p+1$*,*

$$\mathbb{E}\left\{\left|\overline{Q}(h;\mu_N,\nu_N)-Q(h;\mu_N,\nu_N)\right|\right\}\le c\Big(N(\nu_N-\mu_N)\Big)^{-1/2}. \tag{6.38}$$

Remark 6.3.1. This theorem generalizes Proposition 1 in [22].

Let us explain the main intuitive ideas which lead to the estimator $\overline{Q}(h;\mu_N,\nu_N)$.

- First one approximates the integral $(\nu_N-\mu_N)^{-1}\int_{\mu_N}^{\nu_N} h(X(s))\,\mathrm{d}s$ by the Riemann sum
$$(N(\nu_N-\mu_N))^{-1}\sum_{i\in\mathscr{I}(\mu_N,\nu_N)} h(X(i/N));$$
- Then one approximates the latter quantity by
$$(N(\nu_N-\mu_N))^{-1}\sum_{i\in\mathscr{I}(\mu_N,\nu_N)}\frac{(\Delta_a\overline{X}_{i,N})^2}{\mathrm{Var}(\Delta_a\overline{X}_{i,N})}h(X(i/N));$$
- Finally, one approximates the latter quantity by $\overline{Q}(h;\mu_N,\nu_N)$ since $X(i/N)\simeq\overline{X}_{i,N}$ and $\mathrm{Var}(\Delta_a\overline{X}_{i,N})\simeq C(i/N)N^{-2H(i/N)}$.

Upper bounds of the L^1-norms of the successive approximation errors are given in the following lemma.

Lemma 6.3.2. *Let $h\in C^1_{pol}(\mathbb{R})$, then there exist four constants $c_1,c_2,c_3,c_4>0$, such that the following inequalities hold for every $N\ge p+1$.*

(i)

$$E_1:=\mathbb{E}\left\{\left|\frac{1}{N(\nu_N-\mu_N)}\sum_{i\in\mathscr{I}(\mu_N,\nu_N)}h(X(i/N))-\frac{1}{\nu_N-\mu_N}\int_{\mu_N}^{\nu_N}h(X(s))\,\mathrm{d}s\right|\right\}$$
$$\le c_1\Big(N(\nu_N-\mu_N)\Big)^{-1/2}; \tag{6.39}$$

(ii)

$$E_2:=\mathbb{E}\left\{\left|\frac{1}{N(\nu_N-\mu_N)}\sum_{i\in\mathscr{I}(\mu_N,\nu_N)}\left\{\frac{(\Delta_a\overline{X}_{i,N})^2}{Var(\Delta_a\overline{X}_{i,N})}-1\right\}h(X(i/N))\right|\right\}$$
$$\le c_2\Big(N(\nu_N-\mu_N)\Big)^{-1/2}; \tag{6.40}$$

(iii)

$$E_3:=\mathbb{E}\left\{\left|\frac{1}{N(\nu_N-\mu_N)}\sum_{i\in\mathscr{I}(\mu_N,\nu_N)}\frac{(\Delta_a\overline{X}_{i,N})^2}{Var(\Delta_a\overline{X}_{i,N})}\Big(h(\overline{X}_{i,N})-h(X(i/N))\Big)\right|\right\}$$
$$\le c_3\Big(N(\nu_N-\mu_N)\Big)^{-1/2}; \tag{6.41}$$

(iv)

$$E_4 := \mathbb{E}\left\{\left|\overline{Q}(h;\mu_N,\nu_N) - \widetilde{Q}(h;\mu_N,\nu_N)\right|\right\} \le c_4\Big(N(\nu_N-\mu_N)\Big)^{-1/2}, \quad (6.42)$$

where

$$\widetilde{Q}(h;\mu_N,\nu_N) := \frac{1}{N(\nu_N-\mu_N)} \sum_{i\in\mathscr{J}(\mu_N,\nu_N)} \frac{(\Delta_a \overline{X}_{i,N})^2}{Var(\Delta_a\overline{X}_{i,N})} h(\overline{X}_{i,N}). \quad (6.43)$$

Proof of Theorem 6.3.1: This theorem is a straightforward consequence of Lemma 6.3.2 and the triangle inequality. Indeed, one has that

$$\mathbb{E}\left\{\left|\widetilde{Q}(h;\mu_N,\nu_N) - Q(h;\mu_N,\nu_N)\right|\right\} \le E_1+E_2+E_3+E_4 \le c\Big(N(\nu_N-\mu_N)\Big)^{-1/2}.$$

where $c>0$ is a constant. □

Parts (i) and (iii) of Lemma 6.3.2 are not very difficult to obtain, this is why we will not give their proofs. Part (iv) of the latter lemma is a consequence of the following lemma proved in [29].

Lemma 6.3.3. *There is a constant $c>0$ such that for every $i\in\{0,\dots,N-p-1\}$ one has,*

$$\left|Var(\Delta_a\overline{X}_{i,N}) - C(i/N)N^{-2H(i/N)}\right| \le c\log(N)N^{-1-2H(i/N)}. \quad (6.44)$$

Let us now focus on the proof of Part (ii) of Lemma 6.3.2; this proof relies on some techniques which are more or less similar to those in [21, 22]. First we need to give some preliminary results. The following lemma is a classical result on centered 2-dimensional Gaussian vectors, this is why we have omitted its proof.

Lemma 6.3.4. *Let (Z,Z') be a centered 2-dimensional Gaussian vector and let us assume that the variances of Z and Z' are both equal to the same quantity denoted by v. Then, one has,*

$$\mathbb{E}\left\{(Z^2-v)(Z'^2-v)\right\} = 2\left(Cov(Z,Z')\right)^2. \quad (6.45)$$

Lemma 6.3.5. *For every $N\ge p+1$, let $\rho_N:[\mu_N,\nu_N]\to\mathbb{R}$ be an arbitrary deterministic bounded function and let $\Sigma_N(\rho_N)$ be the quantity defined as,*

$$\Sigma_N(\rho_N) = \sum_{j\in\mathscr{J}(\mu_N,\nu_N)} \left\{\frac{(\Delta_a\overline{X}_{j,N})^2}{Var(\Delta_a\overline{X}_{j,N})} - 1\right\}\rho_N(j/N), \quad (6.46)$$

where $\mathscr{J}(\mu_N, \nu_N)$ is the set introduced in (6.21). Then, there is a constant $c > 0$, non depending on N and ρ_N, such that the inequality,

$$\mathbb{E}\left\{\left(\Sigma_N(\rho_N)\right)^2\right\} \le c\|\rho_N\|_\infty^2 N(\nu_N - \mu_N),$$

where $\|\rho_N\|_\infty := \sup_{x\in[\mu_N,\nu_N]} |\rho_N(x)|$ holds for every $N \ge p+1$.

In order to prove Lemma 6.3.5, we need the following technical proposition which concerns the estimation of the correlation between the generalized increments $\Delta_a \overline{X}_{j,N}$ and $\Delta_a \overline{X}_{j',N}$; we refer to [27] for its proof.

Proposition 6.3.6. *Assume that $H(\cdot)$ is a C^2-function and $M(a) \ge 3$ (recall that $M(a)$ is the number of the vanishing moments of the sequence a). Then there is a constant $c > 0$ such that one has for any integer $N \ge p+1$ big enough and each $j, j' \in \{0, \dots, N-p-1\}$,*

$$\left|Corr\left\{\Delta_a \overline{X}_{j,N}, \Delta_a \overline{X}_{j',N}\right\}\right| \le c\left(\frac{1}{1+|j-j'|}\right), \tag{6.47}$$

where

$$Corr\left\{\Delta_a \overline{X}_{j,N}, \Delta_a \overline{X}_{j',N}\right\} := \mathbb{E}\left\{\frac{\Delta_a \overline{X}_{j,N}}{\sqrt{Var\left\{\Delta_a \overline{X}_{j,N}\right\}}}\frac{\Delta_a \overline{X}_{j',N}}{\sqrt{Var\left\{\Delta_a \overline{X}_{j',N}\right\}}}\right\}.$$

Proof of Lemma 6.3.5: One clearly has that

$$\mathbb{E}\left\{(\Sigma_N(\rho_N))^2\right\} = \sum_{j,j'\in\mathscr{J}(\mu_N,\nu_N)} \rho_N(j/N)\rho_N(j'/N)\mathbb{E}$$
$$\left\{\left(\frac{(\Delta_a \overline{X}_{j,N})^2}{\mathrm{Var}(\Delta_a \overline{X}_{j,N})} - 1\right)\left(\frac{(\Delta_a \overline{X}_{j',N})^2}{\mathrm{Var}(\Delta_a \overline{X}_{j',N})} - 1\right)\right\}.$$

Next it follows from Lemma 6.3.4 and Proposition 6.3.6 that

$$\begin{aligned}
\mathbb{E}\left\{(\Sigma_N(\rho_N))^2\right\} &= 2\sum_{j,j'\in\mathscr{J}(\mu_N,\nu_N)} \rho_N(j/N)\rho_N(j'/N)\left(\mathrm{Corr}(\Delta_a \overline{X}_{j,N}, \Delta_a \overline{X}_{j',N})\right)^2 \\
&\le 2c_1\|\rho_N\|_\infty^2 \sum_{j,j'\in\mathscr{J}(\mu_N,\nu_N)} \left(\frac{1}{1+|j-j'|}\right)^2 \\
&\le 4c_1\|\rho_N\|_\infty^2 \sum_{j\in\mathscr{J}(\mu_N,\nu_N)} \sum_{l=-\infty}^{\infty} \left(\frac{1}{1+|l|}\right)^2 \\
&\le c_2\|\rho_N\|_\infty^2 N(\nu_N - \mu_N),
\end{aligned} \tag{6.48}$$

where c_1 and c_2 are two constants which do not depend on N. □

Lemma 6.3.7. *For every $N \geq p+1$, let $\rho_N : [\mu_N, \nu_N] \to \mathbb{R}$ be an arbitrary bounded deterministic function that vanishes outside a dyadic interval of the form $[k2^{-j_0}\mathscr{L}_N + \mu_N, k'2^{-j_0}\mathscr{L}_N + \mu_N]$, where $\mathscr{L}_N := \nu_N - \mu_N$ and where the integers j_0, k and k' are arbitrary and satisfy $j_0 \geq 0$ and $0 \leq k < k' \leq 2^{j_0}$. Then there exists a constant $c > 0$ which does not depend on N, k, k' and j_0, such that for all integers $N \geq p+1$, and j_1 satisfying $2^{j_0} \leq 2^{j_1} \leq N\mathscr{L}_N < 2^{j_1+1}$, one has*

$$\mathbb{E}\left\{(\Sigma_N(\rho_N))^2\right\} \leq c\|\rho_N\|_\infty^2 (k'-k)2^{j_1-j_0}. \tag{6.49}$$

Proof of Lemma 6.3.7. Let $\mathscr{I}(k,k',j_0,N)$ be the set of indices defined as,

$$\begin{aligned}
&\mathscr{I}(k,k',j_0,N)\\
&\quad = \Big\{ i \in \mathscr{J}(\mu_N,\nu_N) : i \in \big[(k2^{-j_0}\mathscr{L}_N + \mu_N)N, (k'2^{-j_0}\mathscr{L}_N + \mu_N)N\big]\Big\}.
\end{aligned}$$

One has,

$$\mathrm{Card}\big(\mathscr{I}(k,k',j_0,N)\big) \leq N(k'-k)2^{-j_0}\mathscr{L}_N + 1 \leq 4(k'-k)2^{j_1-j_0}, \tag{6.50}$$

where the last inequality follows from the fact that $N\mathscr{L}_N \leq 2^{j_1+1}$. Using the method which allowed us to obtain (6.48) and replacing $\mathscr{J}(\mu_N,\nu_N)$ by $\mathscr{I}(k,k',j_0,N)$, one can show that,

$$\begin{aligned}
&\mathbb{E}\left\{(\Sigma_N(\rho_N))^2\right\}\\
&\quad = 2 \sum_{j,j' \in \mathscr{I}(k,k',j_0,N)} \rho_N(j/N)\rho_N(j'/N)\left(\mathrm{Corr}(\Delta_a \overline{X}_{j,N}, \Delta_a \overline{X}_{j',N})\right)^2\\
&\quad \leq c\|\rho_N\|_\infty^2 (k'-k)2^{j_1-j_0},
\end{aligned} \tag{6.51}$$

where $c > 0$ is a constant which does not depend on k, k', j_0, N. □

We are now in position to prove Part (ii) of Lemma 6.3.2.

Proof of Part (ii) of Lemma 6.3.2: Let $N \geq p+1$ be fixed. Observe that, with probability 1, the function $t \mapsto h(X(t))$ belongs to $C\big([\mu_N,\nu_N]\big)$, the space of continuous functions over $[\mu_N,\nu_N]$. By expanding it in the Schauder basis of this space, one obtains that

$$h(X(t)) = \lambda_0\phi_0(t) + \lambda_1\phi_1(t) + \sum_{j=0}^{\infty}\sum_{k=1}^{2^j} \lambda_{j,k}\phi_{j,k}(t), \tag{6.52}$$

where $\lambda_0 = h(X(\mu_N))$, $\lambda_1 = h(X(\nu_N))$, $\phi_0(t) = (\nu_N - t)\mathscr{L}_N^{-1}$, $\phi_1(t) = (t - \mu_N)\mathscr{L}_N^{-1}$, with $\mathscr{L}_N = \nu_N - \mu_N$,

$$\lambda_{j,k} = 2^{-\frac{j}{2}}\Big\{2h\Big(X\Big(\frac{2k-1}{2^{j+1}}\mathscr{L}_N+\mu_N\Big)\Big) - h\Big(X\Big(\frac{2k}{2^{j+1}}\mathscr{L}_N+\mu_N\Big)\Big)$$
$$-h\Big(X\Big(\frac{2k-2}{2^{j+1}}\mathscr{L}_N+\mu_N\Big)\Big)\Big\},$$

and

$$\phi_{j,k}(t) = 2^{3j/2}\mathscr{L}_N^{-1}\int_{\mu_N}^t \Big(\mathbb{1}_{\big[\frac{(2k-2)\mathscr{L}_N}{2^{j+1}}+\mu_N,\frac{(2k-1)\mathscr{L}_N}{2^{j+1}}+\mu_N\big]}(s)$$
$$-\mathbb{1}_{\big[\frac{(2k-1)\mathscr{L}_N}{2^{j+1}}+\mu_N,\frac{2k\mathscr{L}_N}{2^{j+1}}+\mu_N\big]}(s)\Big)\,ds. \tag{6.53}$$

Observe that the series in (6.52) is, with probability 1, uniformly convergent in $[\mu_N, \nu_N]$. Now let us show that there is a constant $c > 0$ such that

$$\mathbb{E}\{\lambda_0^2 + \lambda_1^2\} \le c, \tag{6.54}$$

and

$$\mathbb{E}\{\lambda_{j,k}^2\} \le c2^{-j(1+2\min_{s\in[\mu_N,\nu_N]} H(s))}, \text{ for every } j,k. \tag{6.55}$$

By using the fact that $h \in C^1_{pol}(\mathbb{R})$ as well as the fact that all the moments of the random variable $\|X\|_\infty := \sup_{s\in[0,1]}|X(s)|$ are finite, one gets,

$$\mathbb{E}|\lambda_0|^2 = \mathbb{E}\Big(h(X(\mu_N))\Big)^2 \le \mathbb{E}\Big(c(1+\|X\|_\infty)^K\Big)^2 < +\infty. \tag{6.56}$$

Similarly, one can show that

$$\mathbb{E}|\lambda_1|^2 < +\infty. \tag{6.57}$$

Combining (6.56) and (6.57), one obtains (6.54). Let us now show that (6.55) holds. Using, the expression of $\lambda_{j,k}$, the inequality $(a+b)^2 \le 2a^2 + 2b^2$ for all $a, b \in \mathbb{R}$ and the triangle inequality one gets,

$$\mathbb{E}|\lambda_{j,k}|^2$$
$$= \mathbb{E}\Bigg|2^{-\frac{j}{2}}\Bigg\{2h\Bigg(X\Bigg(\frac{2k-1}{2^{j+1}}\mathscr{L}_N+\mu_N\Bigg)\Bigg) - h\Bigg(X\Bigg(\frac{2k}{2^{j+1}}\mathscr{L}_N+\mu_N\Bigg)\Bigg)$$
$$-h\Bigg(X\Bigg(\frac{2k-2}{2^{j+1}}\mathscr{L}_N+\mu_N\Bigg)\Bigg)\Bigg\}\Bigg|^2$$

$$\leq 2^{-j}\Bigg\{2\mathbb{E}\left|h\left(X\left(\frac{2k-1}{2^{j+1}}\mathscr{L}_N+\mu_N\right)\right)-h\left(X\left(\frac{2k}{2^{j+1}}\mathscr{L}_N+\mu_N\right)\right)\right|^2$$

$$+2\mathbb{E}\left|h\left(X\left(\frac{2k-1}{2^{j+1}}\mathscr{L}_N+\mu_N\right)\right)-h\left(X\left(\frac{2k-2}{2^{j+1}}\mathscr{L}_N+\mu_N\right)\right)\right|^2\Bigg\}. \quad (6.58)$$

Thus, in view of (6.58), for proving (6.55), it suffices to show that there is a constant $c > 0$ (which does not depend on N, j and k) such that one has,

$$\mathbb{E}\left|h\left(X\left(\frac{2k-1}{2^{j+1}}\mathscr{L}_N+\mu_N\right)\right)-h\left(X\left(\frac{2k}{2^{j+1}}\mathscr{L}_N+\mu_N\right)\right)\right|^2 \leq c2^{-2j\min_{s\in[\mu_N,\nu_N]}H(s)} \quad (6.59)$$

and

$$\mathbb{E}\left|h\left(X\left(\frac{2k-1}{2^{j+1}}\mathscr{L}_N+\mu_N\right)\right)-h\left(X\left(\frac{2k-2}{2^{j+1}}\mathscr{L}_N+\mu_N\right)\right)\right|^2 \leq c2^{-2j\min_{s\in[\mu_N,\nu_N]}H(s)} \quad (6.60)$$

We will only prove that (6.59) holds since (6.60) can be obtained in the same way. By using the fact that $h \in C^1_{pol}(\mathbb{R})$, the Mean Value Theorem and Cauchy-Schwarz inequality, one has

$$\mathbb{E}\left|h\left(X\left(\frac{2k-1}{2^{j+1}}\mathscr{L}_N+\mu_N\right)\right)-h\left(X\left(\frac{2k}{2^{j+1}}\mathscr{L}_N+\mu_N\right)\right)\right|^2$$

$$\leq \mathbb{E}\left(\left(\sup_{s\in[-\|X\|_\infty,\|X\|_\infty]}|h'(s)|\right)\left|X\left(\frac{2k-1}{2^{j+1}}\mathscr{L}_N+\mu_N\right)\right.\right.$$

$$\left.\left.-X\left(\frac{2k}{2^{j+1}}\mathscr{L}_N+\mu_N\right)\right|\right)^2$$

$$\leq \left(\mathbb{E}\left(c(1+\|X\|_\infty)^K\right)^4\right)^{1/2}\left(\mathbb{E}\left|X\left(\frac{2k-1}{2^{j+1}}\mathscr{L}_N+\mu_N\right)\right.\right.$$

$$\left.\left.-X\left(\frac{2k}{2^{j+1}}\mathscr{L}_N+\mu_N\right)\right|^4\right)^{1/2}. \quad (6.61)$$

On the other hand, standard computations (see e.g. [6]) allow to show that, there is a constant $c > 0$ (non depending on N) such that for all $t, t' \in [\mu_N, \nu_N]$, one has

$$\mathbb{E}\,|X(t)-X(t')|^2 \leq c|t-t'|^{2\min_{s\in[\mu_N,\nu_N]}H(s)}.$$

Then the latter inequality, the equivalence of the Gaussian moments and (6.61) implies that (6.59) holds; recall that (6.60) can be obtained in the same way. Next combining (6.58) with (6.59) and (6.60) one gets (6.55).

Now observe that (6.52) and (6.46), entail that

$$\sum_{i \in \mathscr{J}(\mu_N,\nu_N)} \left\{ \frac{(\Delta_a \overline{X}_{i,N})^2}{\operatorname{Var}(\Delta_a \overline{X}_{i,N})} - 1 \right\} h(X(i/N))$$

$$= \lambda_0 \Sigma_N(\phi_0) + \lambda_1 \Sigma_N(\phi_1) + \sum_{j=0}^{\infty} \sum_{k=1}^{2^j} \lambda_{j,k} \Sigma_N(\phi_{j,k}). \tag{6.62}$$

Also observe that by using the triangle inequality, Cauchy-Schwarz inequality, Lemma 6.3.5, (6.54) and the fact that $\|\phi_0\|_\infty = \|\phi_1\|_\infty = 1$, one gets that there is a constant $c > 0$ such that for all N,

$$\begin{aligned}
\mathbb{E}\left|\lambda_0 \Sigma_N(\phi_0) + \lambda_1 \Sigma_N(\phi_1)\right| &\le \mathbb{E}\left|\lambda_0 \Sigma_N(\phi_0)\right| + \mathbb{E}\left|\lambda_1 \Sigma_N(\phi_1)\right| \\
&\le \left(\mathbb{E}\left|\lambda_0\right|^2\right)^{1/2} \left(\mathbb{E}\left|\Sigma_N(\phi_0)\right|^2\right)^{1/2} \\
&\quad + \left(\mathbb{E}\left|\lambda_1\right|^2\right)^{1/2} \left(\mathbb{E}\left|\Sigma_N(\phi_1)\right|^2\right)^{1/2} \\
&\le c\big(N(\nu_N - \mu_N)\big)^{1/2}.
\end{aligned} \tag{6.63}$$

Thus in view of (6.62) and (6.63), in order to finish our proof, it remains to show that there exists a constant $c > 0$ such that for all $N \ge p + 1$, one has,

$$\mathbb{E}\left|\sum_{j=0}^{\infty} \sum_{k=1}^{2^j} \lambda_{j,k} \Sigma_N(\phi_{j,k})\right| \le c(N(\nu_N - \mu_N))^{1/2}. \tag{6.64}$$

Let $j_1 \ge 1$ be the unique integer such that

$$2^{j_1} \le N \mathscr{L}_N < 2^{j_1+1}.$$

It follows from the triangle inequality, Cauchy-Schwarz inequality, (6.55), the fact that for all j, k,

$$\operatorname{supp} \phi_{j,k} \subseteq \left[\frac{k-1}{2^j} \mathscr{L}_N + \mu_N, \frac{k}{2^j} \mathscr{L}_N + \mu_N\right], \tag{6.65}$$

and Lemma 6.3.7 in which one takes $j_0 = j$, that

$$
\begin{aligned}
&\mathbb{E}\Big|\sum_{j=0}^{j_1}\sum_{k=1}^{2^j}\lambda_{j,k}\Sigma_N(\phi_{j,k})\Big| \\
&\le \sum_{j=0}^{j_1}\sum_{k=1}^{2^j}\Big(\mathbb{E}\,|\lambda_{j,k}|^2\Big)^{1/2}\Big(\mathbb{E}\,|\Sigma_N(\phi_{j,k})|^2\Big)^{1/2} \\
&\le c\sum_{j=0}^{j_1}\sum_{k=1}^{2^j}2^{-j(1+2\min_{s\in[\mu_N,\nu_N]}H(s))/2}\Big(\|\phi_{j,k}\|_\infty^2(k-(k-1))2^{j_1-j}\Big)^{1/2} \\
&= c\sum_{j=0}^{j_1}\sum_{k=1}^{2^j}2^{-j(1+\min_{s\in[\mu_N,\nu_N]}H(s))+j_1/2}\|\phi_{j,k}\|_\infty,
\end{aligned}
\tag{6.66}
$$

where $c > 0$ is a constant. Then using (6.66), the fact that $\|\phi_{j,k}\|_\infty \le 2^{j/2}$, the inequalities $\min_{s\in[\mu_N,\nu_N]} H(s) \ge \min_{s\in[0,1]} H(s) > 1/2$ and $2^{j_1} \le N\mathscr{L}_N$, one gets that

$$
\begin{aligned}
\mathbb{E}\Big|\sum_{j=0}^{j_1}\sum_{k=1}^{2^j}\lambda_{j,k}\Sigma_N(\phi_{j,k})\Big| &\le c\sum_{j=0}^{j_1}\sum_{k=1}^{2^j}2^{-j(1/2+\min_{s\in[\mu_N,\nu_N]}H(s))+j_1/2} \\
&\le c2^{j_1/2}\sum_{j=0}^{+\infty}2^{j(1/2-\min_{s\in[\mu_N,\nu_N]}H(s))} \\
&\le c_1(N(\nu_N-\mu_N))^{1/2},
\end{aligned}
\tag{6.67}
$$

where $c_1 = c\sum_{j=0}^{+\infty}2^{j(1/2-\min_{s\in[0,1]}H(s))} < +\infty$. Let us now show that there is a constant $c_2 > 0$ non depending on j_1 and N, such that

$$
\mathbb{E}\Big|\sum_{j=j_1+1}^{\infty}\sum_{k=1}^{2^j}\lambda_{j,k}\Sigma_N(\phi_{j,k})\Big| \le c_2(N(\nu_N-\mu_N))^{1/2}. \tag{6.68}
$$

First observe that, for every fixed (j,k), (6.65) and the inequalities $2^{-j} \le 2^{-j_1-1} < (N\mathscr{L}_N)^{-1}$ imply that there is at most one index $i \in \mathscr{I}(\mu_N,\nu_N)$ such that $\phi_{j,k}(i/N) \ne 0$. Therefore, in view of (6.46), one has

$$
\mathbb{E}\,|\Sigma_N(\phi_{j,k})|^2 \le \|\phi_{j,k}\|_\infty^2\mathbb{E}\left(\frac{(\Delta_a\overline{X}_{i,N})^2}{\mathrm{Var}(\Delta_a\overline{X}_{i,N})}-1\right)^2. \tag{6.69}
$$

Then noticing that $\|\phi_{j,k}\|_\infty \le 2^{j/2}$ and that $\mathbb{E}\left(\frac{(\Delta_a \overline{X}_{i,N})^2}{\mathrm{Var}(\Delta_a \overline{X}_{i,N})} - 1\right)^2 = \mathbb{E}\left(Z^2 - 1\right)^2$, where Z is a standard Gaussian random variable. It follows that

$$\mathbb{E}\,|\Sigma_N(\phi_{j,k})|^2 \le c2^j, \tag{6.70}$$

where $c > 0$ is a constant non depending on N, j and k. Now, for every fixed $j \ge j_1 + 1$, let us denote by $\mathscr{K}_{j,N}$ the set of indices $k \in \{0, \ldots, 2^j\}$ defined as

$$\mathscr{K}_{j,N} = \Big\{k \in \{1, \ldots, 2^j\} : \exists i \in \mathscr{I}(\mu_N, \nu_N) \quad \text{such that} \quad \phi_{j,k}(i/N) \ne 0\Big\}.$$

Observe that when $k \notin \mathscr{K}_{j,N}$ then for every $i \in \mathscr{I}(\mu_N, \nu_N)$, one has $\phi_{j,k}(i/N) = 0$ and as a consequence,

$$\Sigma_N(\phi_{j,k}) = 0. \tag{6.71}$$

On the other hand, by using (6.65) and the fact that for all k and k' satisfying $k \ne k'$, one has

$$\left(\frac{k-1}{2^j}\mathscr{L}_N + \mu_N, \frac{k}{2^j}\mathscr{L}_N + \mu_N\right) \cap \left(\frac{k'-1}{2^j}\mathscr{L}_N + \mu_N, \frac{k'}{2^j}\mathscr{L}_N + \mu_N\right) = \emptyset,$$

it follows that

$$\mathrm{Card}(\mathscr{K}_{j,N}) \le \mathrm{Card}\,\mathscr{I}(\mu_N, \nu_N) \le 2N(\nu_N - \mu_N). \tag{6.72}$$

Next it follows from, (6.71), the triangle inequality, Cauchy-Schwarz inequality, (6.55), (6.70), (6.72), and the inequalities $\min_{s\in[0,1]} H(s) > 1/2$, $2^{-j_1} < 2(N\mathscr{L}_N)^{-1}$, that

$$\begin{aligned}
&\mathbb{E}\Big|\sum_{j=j_1+1}^{+\infty}\sum_{k=1}^{2^j}\lambda_{j,k}\Sigma_N(\phi_{j,k})\Big| \\
&\le \sum_{j=j_1+1}^{+\infty}\sum_{k\in\mathscr{K}_{j,N}}\left(\mathbb{E}\,|\lambda_{j,k}|^2\right)^{1/2}\left(\mathbb{E}\,|\Sigma_N(\phi_{j,k})|^2\right)^{1/2} \\
&\le c\sum_{j=j_1+1}^{+\infty}\sum_{k\in\mathscr{K}_{j,N}} 2^{-j(1+2\min_{s\in[\mu_N,\nu_N]} H(s))/2}2^{j/2} \\
&\le 2cN(\nu_N - \mu_N)\sum_{j=j_1}^{+\infty} 2^{-j\min_{s\in[0,1]} H(s)} \\
&= 2cN(\nu_N - \mu_N)\frac{2^{-j_1\min_{s\in[0,1]} H(s)}}{1 - 2^{-\min_{s\in[0,1]} H(s)}}
\end{aligned}$$

$$\le c'\big(N\mathscr{L}_N\big)^{1-\min_{s\in[0,1]} H(s)}$$
$$\le c'(N(\nu_N-\mu_N))^{1/2}, \tag{6.73}$$

where the constant $c' = 4c\big(1-2^{-\min_{s\in[0,1]} H(s)}\big)^{-1}$ and thus we obtain (6.68). Moreover combining (6.68) with (6.67) one gets (6.64).

Finally, Part (ii) of Lemma 6.3.2 results from (6.62), (6.63) and (6.64). □

6.3.2 Proof of Theorem 6.2.1 When Y is Arbitrary

We need the following lemmas.

Lemma 6.3.8. *For all $N \ge p+1$ and $j \in \{0,\dots,N-p-1\}$, one sets*

$$e_{j,N} = \Delta_a \overline{Y}_{j,N} - f'(\overline{X}_{j,N})\Delta_a \overline{X}_{j,N} \text{ and } e'_{j,N} = h(\overline{Y}_{j,N}) - h(f(\overline{X}_{j,N})).$$

Then for all real $k \ge 1$ there exists a constant $c = c(k) > 0$, such that the inequalities,

$$\mathbb{E}\{|e_{j,N}|^k\} \le cN^{-2H(j/N)k} \text{ and } \mathbb{E}\{|e'_{j,N}|^k\} \le cN^{-H(j/N)k},$$

hold for every $N \ge p+1$ and $j \in \{0,\dots,N-p-1\}$.

Proof of Lemma 6.3.8: The first inequality can be obtained similarly to Relation (34) in [22], by using a second order Taylor expansion for f. The second inequality follows from the regularity properties of f and h and its proof is very close to that of Relation (35) in [22]. □

Lemma 6.3.9. *For all $N \ge p+1$, one sets*

$$e_N^{(1)} = \frac{1}{N(\nu_N-\mu_N)} \sum_{j\in\mathscr{J}(\mu_N,\nu_N)} \frac{\Big(e_{j,N}^2 + 2\Delta_a\overline{X}_{j,N} f'(\overline{X}_{j,N})e_{j,N}\Big)}{C(j/N)N^{-2H(j/N)}} h(\overline{Y}_{j,N})$$

and

$$e_N^{(2)} = \frac{1}{N(\nu_N-\mu_N)} \sum_{j\in\mathscr{J}(\mu_N,\nu_N)} \frac{\Big(\Delta_a\overline{X}_{j,N} f'(\overline{X}_{j,N})\Big)^2}{C(j/N)N^{-2H(j/N)}} e'_{j,N}.$$

Then there is a constant $c > 0$, such that the inequality

$$\mathbb{E}\left\{|e_N^{(1)}|\right\} + \mathbb{E}\left\{|e_N^{(2)}|\right\} \le cN^{-1/2},$$

holds for every $N \ge p+1$.

Proof of Lemma 6.3.9. The lemma can be obtained by using Lemmas 6.3.8, 6.3.3 and standard computations. □

Lemma 6.3.10. *For every function $h \in C^1_{pol}(\mathbb{R})$ one has*

$$V(h; \mu_N, \nu_N) = Q((f')^2 \times h \circ f; \mu_N, \nu_N).$$

Moreover, for each $N \geq p+1$ one has

$$\overline{V}(h; \mu_N, \nu_N) = \overline{Q}((f')^2 \times h \circ f; \mu_N, \nu_N) + e_N^{(1)} + e_N^{(2)}.$$

Proof of Lemma 6.3.10: The lemma can be obtained by using standard computations. □

We are now in position to prove Theorem 6.2.1.

Proof of Theorem 6.2.1: We use Lemmas 6.3.8, 6.3.9 and 6.3.10 as well as Theorem 6.3.1 and we follow the same lines as in the proof of Theorem 3 in [22]. □

6.3.3 *Proof of Theorem 6.2.2*

Theorem 6.2.2 is a straightforward consequence of Theorem 6.2.1 and the following proposition which allows to control the $L^1(\Omega)$ norm of the error one makes when one replaces the estimator $\overline{V}(h; \mu_{N_n}, \nu_{N_n})$ by the estimator $\widehat{V}(h; \mu_{N_n}, \nu_{N_n})$.

Proposition 6.3.11. *For any n big enough one sets*

$$v(N_n, m_n) = (N_n^{-1/2} + m_n^{-1/2})\left(1 + m_n^{-1} N_n^{2\max_{s\in[\mu_{N_n}, \nu_{N_n}]} H(s)}\right).$$

Recall that m_n has been defined in (6.12). Let us assume that (6.25) is satisfied. Then there exists a constant $c > 0$ such that for any n big enough, one has

$$\mathbb{E}\left\{\left|\widehat{V}(h; \mu_{N_n}, \nu_{N_n}) - \overline{V}(h; \mu_{N_n}, \nu_{N_n})\right|\right\} \leq c v(N_n, m_n) = \mathcal{O}(N_n^{-1/2}) \qquad (6.74)$$

The proposition can be obtained similarly to Proposition 2 in [22]. □

6.4 Estimation of the Unknown Parameter in the Linear Case

The goal of this section is to prove Theorem 6.2.4. One needs the following two lemmas. We will not give the proof of the first lemma since it is not difficult.

Lemma 6.4.1. *Assume that* $(\mu_{N_n})_n$ *and* $(\nu_{N_n})_n$ *are two convergent sequences. Then, when n goes to infinity, the random variable* $T_{N_n} := (\nu_{N_n} - \mu_{N_n})^{-1} \int_{\mu_{N_n}}^{\nu_{N_n}} Y(s)\,\mathrm{d}s$ *almost surely converges to an almost surely strictly positive random variable* T.

Lemma 6.4.2. *Assume that (6.25) holds. Then the sequence*

$$\left(\frac{1}{N_n(\nu_{N_n} - \mu_{N_n})} \sum_{i \in \mathscr{I}(\mu_{N_n}, \nu_{N_n})} \widehat{Y}_{i,N_n,n} - \frac{1}{\nu_{N_n} - \mu_{N_n}} \int_{\mu_{N_n}}^{\nu_{N_n}} Y(s)\,\mathrm{d}s\right)_n$$

converges to 0 *in* $L^1(\Omega, \mathscr{F}, \mathbb{P})$ *with the rate* $(N_n(\nu_{N_n} - \mu_{N_n}))^{-1/2}$ *when n goes to infinity. Note that in this lemma we do not necessarily suppose that* Φ *is of the form* $\Phi(x, \theta) = \theta x$.

Proof of Lemma 6.4.2: It follows from (6.11) that for any n big enough one has

$$\begin{aligned} &\mathbb{E}\left|(N_n(\nu_{N_n} - \mu_{N_n}))^{-1} \sum_{i \in \mathscr{I}(\mu_{N_n}, \nu_{N_n})} \widehat{Y}_{i,N_n,n} - (\nu_{N_n} - \mu_{N_n})^{-1} \int_{\mu_{N_n}}^{\nu_{N_n}} Y(s)\,\mathrm{d}s\right| \\ &\leq M_n + \mathscr{O}\big((N_n(\nu_{N_n} - \mu_{N_n}))^{-1}\big), \end{aligned} \tag{6.75}$$

where

$$M_n := \frac{1}{N_n(\nu_{N_n} - \mu_{N_n})} \sum_{i \in \mathscr{I}(\mu_{N_n}, \nu_{N_n})} \mathbb{E}\left|\widehat{Y}_{i,N_n,n} - \overline{Y}_{i,N_n}\right|. \tag{6.76}$$

Next Cauchy-Schwarz inequality, a result similar to Part (ii) of Lemma 7 in [20], (6.25), the inequalities $0 \leq \nu_{N_n} - \mu_{N_n} \leq 1$ and the fact that $H(\cdot)$ is with values in $(1/2, 1)$, imply that for all n big enough,

$$\begin{aligned} M_n &\leq \frac{1}{N_n(\nu_{N_n} - \mu_{N_n})} \sum_{i \in \mathscr{I}(\mu_{N_n}, \nu_{N_n})} \sqrt{\mathbb{E}\left(|\widehat{Y}_{i,N_n,n} - \overline{Y}_{i,N_n}|^2\right)} \\ &= \mathscr{O}(m_n^{-1/2}) = \mathscr{O}\big(N_n^{-\max_{s \in [\mu_{N_n}, \nu_{N_n}]} H(s)}\big) = \mathscr{O}\big((N_n(\nu_{N_n} - \mu_{N_n}))^{-1/2}\big). \end{aligned} \tag{6.77}$$

Finally putting together (6.75), (6.76) and (6.77), one obtains the lemma. □

We are now in position to prove Theorem 6.2.4.

Proof of Theorem 6.2.4: Let us set

$$T_{N_n} = (\nu_{N_n} - \mu_{N_n})^{-1} \int_{\mu_{N_n}}^{\nu_{N_n}} Y(s)\,\mathrm{d}s, \tag{6.78}$$

$$\overline{T}_n = (N_n(\nu_{N_n} - \mu_{N_n}))^{-1} \sum_{i \in \mathscr{I}(\mu_{N_n}, \nu_{N_n})} \widehat{Y}_{i,N_n,n}, \tag{6.79}$$

$$u_n = \widehat{V}(1; \mu_{N_n}, \nu_{N_n}) - V(1; \mu_{N_n}, \nu_{N_n}), \tag{6.80}$$

and

$$v_n = \overline{T}_n - T_{N_n}. \tag{6.81}$$

Moreover, observe that (6.2), (6.15), the fact that $\Phi(x,\theta) = \theta x$, and (6.78) imply that

$$V(1;\mu_{N_n},\nu_{N_n}) = 4\theta^2(\nu_{N_n} - \mu_{N_n})^{-1}\int_{\mu_{N_n}}^{\nu_{N_n}} Y(s)\,\mathrm{d}s = 4\theta^2 T_{N_n}. \tag{6.82}$$

Then it follows from (6.34), (6.78), (6.79), (6.80), (6.81) and (6.82) that

$$\widehat{\theta}_n^2(\mu_{N_n},\nu_{N_n}) - \theta^2 = \frac{u_n - 4\theta^2 v_n}{4T_{N_n} + 4v_n}. \tag{6.83}$$

Therefore, one has for any real $\lambda > 1$ and any integer n big enough

$$\begin{aligned}
&\mathbb{P}\Big((N_n(\nu_{N_n} - \mu_{N_n}))^{1/2}|\widehat{\theta}_n^2(\mu_{N_n},\nu_{N_n})) - \theta| > \lambda\Big)\\
&\le \mathbb{P}\bigg(\Big\{\frac{(N_n(\nu_{N_n} - \mu_{N_n}))^{1/2}|u_n - 4\theta^2 v_n|}{4T_{N_n} + 4v_n} > \lambda\Big\} \cap \Big\{T_{N_n} \ge \lambda^{-1/2}\Big\}\\
&\quad \cap \Big\{|v_n| \le 4^{-1}\lambda^{-1/2}\Big\}\bigg)\\
&\quad + \mathbb{P}\Big(T_{N_n} < \lambda^{-1/2}\Big) + \mathbb{P}\Big(|v_n| > 4^{-1}\lambda^{-1/2}\Big)\\
&\le \mathbb{P}\Big((N_n(\nu_{N_n} - \mu_{N_n}))^{1/2}|u_n - 4\theta^2 v_n| > 3\lambda^{3/2}\Big)\\
&\quad + \mathbb{P}\Big(T_{N_n} < \lambda^{-1/2}\Big) + \mathbb{P}\Big(|v_n| > 4^{-1}\lambda^{-1/2}\Big).
\end{aligned}$$

Next the latter inequality, (6.78), (6.79), (6.80), (6.81), Theorem 6.2.2, Lemma 6.4.2 and Markov inequality, imply that there is a constant $c > 0$ such that for all real $\lambda > 1$, one has

$$\begin{aligned}
&\limsup_{n\to+\infty} \mathbb{P}\Big((N_n(\nu_{N_n} - \mu_{N_n}))^{1/2}\Big|\widehat{\theta}_n^2(\mu_{N_n},\nu_{N_n})) - \theta^2\Big| > \lambda\Big)\\
&\le 3^{-1}c\lambda^{-3/2} + \mathbb{P}\Big(T < \lambda^{-1/2}\Big),
\end{aligned} \tag{6.84}$$

where $T := \lim_{n\to+\infty} T_{N_n}$. Thus, it follows from (6.35), (6.84), (6.78) and Lemma 6.4.1, that Theorem 6.2.4 holds. □

6.5 Histograms of the Estimated Values

We have tested our estimator of θ^2 on simulated data and our numerical results are summarized in the following two histograms:

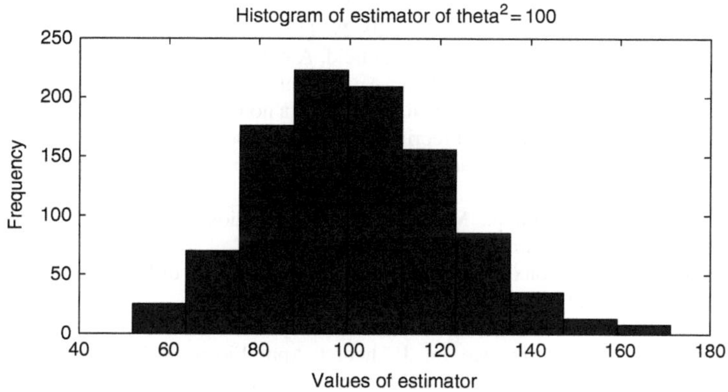

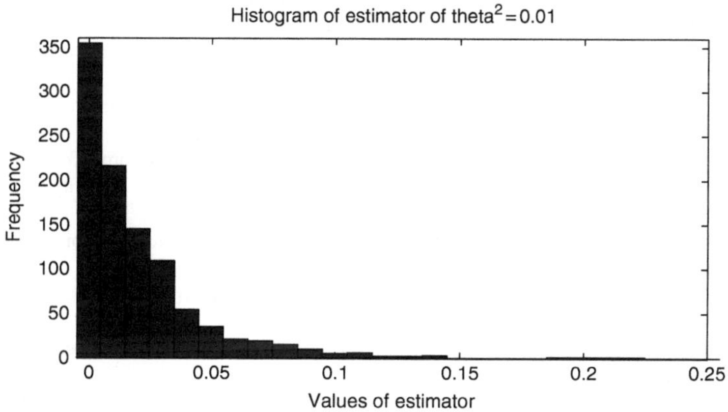

To obtain the first histogram, we have proceeded as follows: we assume that $\theta = 10$ and $H(s) = (s - 0.5)^2 + 0.6$ for all $s \in [0, 1]$, then we simulate 1000 discretized trajectories of the process $\{Z(t)\}_{t\in[0,1]}$, finally we apply our estimator to each trajectory which gives us 1000 estimations of θ^2. The second histogram has been obtained by using a similar method; $H(\cdot)$ is defined in the same way, yet in this case we assume that $\theta = 0.1$.

Acknowledgements The authors are thankful to the anonymous referee for valuable comments which have helped them to improve the earlier version of their paper.

References

1. Alfi, V., Cocceti, F., Petri, A., Pietronero, L.: Roughness and finite size effect in the NYSE stock-price fluctuations. Eur. Phys. J. B. **55**, 135–142 (2007)
2. Ayache, A., Cohen, S., Lévy Véhel, J.: The covariance structure of multifractional Brownian motion, with application to long range dependence. Proceedings IEEE ICASSP. **6**, 3810–3813 (2000)

3. Ayache, A.: Du mouvement brownien fractionnaire au mouvement brownien multifractionnaire. Technique et science informatiques **20–29**, 1133–1152 (2001)
4. Ayache, A.: The generalized multifractional field: A nice tool for the study of the Generalized Multifractional Brownian Motion. J. Four. Anal. Appl. **8**, 581–601 (2002)
5. Ayache, A., El-Nouty, C.: The small ball behavior of a non stationary increments process: The multifractional Brownian motion. Preprint CMLA. **08**, (2004)
6. Ayache, A., Taqqu, M.S.: Multifractional processes with random exponent. Publ. Mat. **49**(2), 459–486 (2005)
7. Ayache, A., Jaffard, S., Taqqu, M.S.: Wavelet construction of generalized multifractional processes. Rev. Mat. Iberoam. **23**(1), 327–370 (2007)
8. Benassi, A., Jaffard, S., Roux, D.: Gaussian processes and pseudodifferential elliptic operators. Rev. Mat. Iberoam. **13**(1), 19–81 (1997)
9. Bianchi, S.: Pathwise identification of the memory function of multifractional Brownian motion with application to finance. Int. J. Theoret. Appl. Financ. **8**(2), 255–281 (2005)
10. Bianchi, S., Pianese, A.: Multifractional properties of stock indices decomposed by filtering their pointwise hölder regularity. Int. J. Theoret. Appl. Financ. **11**(6), 567–595 (2008)
11. Bianchi, S., Pianese, A., Pantanella, A.: Modeling Stock Prices by Multifractional Brownian Motion: An Improved Estimation of the Pointwise Regularity. Quantitative Finance, Forthcoming. Available at SSRN: http://ssrn.com/abstract=1888288 (2011)
12. Cohen, S.: From self-similarity to local self-similarity: the estimation problem, Fractal in Engineering, ed. by Lévy Véhel and C. Tricot. Springer, 3–16 (1999)
13. Comte, F., Renault, E.: Long memory continuous time models. J. Econom. **73**, 101–150 (1996)
14. Comte, F., Renault, E.: Long memory in continuous-time stochastic volatility models. Math. Financ. **8**, 291–323 (1998)
15. Doukhan, P., Oppenheim, G., Taqqu, M.S.: Theory and applications of long-Range dependence. Birkhäuser, Boston (2003)
16. Embrechts, P., Maejima, M.: Selfsimilar Processes. Princeton University Press, Princeton and Oxford (2002)
17. Falconer, K.: Tangent fields and the local structure of random fields. J. Theoret. Probab. **15**, 731–750 (2002)
18. Falconer, K.: The local structure of random processes, J. London Math. Soc. **67**, 657–672 (2003)
19. Falconer, K., Le Guével, R., Lévy Véhel, J.: Localisable moving average stable and multistable processes. Stochastic Models. **25**, 648–672 (2009)
20. Gloter, A., Hoffmann, M.: Stochastic volatility and fractional Brownian motion. Prépublication du laboratoire de Probabilités & Modèles Aléatoires des Universités Paris 6 & Paris 7, 746 (2002)
21. Gloter, A., Hoffmann, M.: Stochastic volatility and fractional Brownian motion. Stoch. Proc. Appl. **113**, 143–172 (2004)
22. Hull, J., White, A.: The pricing of options on assets with stochastic volatilities. J. Financ. **3**, 281–300 (1987)
23. Mandelbrot, B.B.: Fractales, hasard et finance. Paris, Flammarion (1997)
24. Mandelbrot, B.B.: Fractals and scaling in finance: Discontinuity, concentration, risk. New York, Springer (1997)
25. Melino, A., Turnbull, S.M.: Pricing foreign currency options with stochastic volatility. J. Econom. **45**, 239–265 (1990)
26. Peltier, R.F., Lévy Véhel, J.: Multifractional Brownian motion: Definition and preliminary results. Rapport de recherche de l'INRIA, *2645* (1995)
27. Peng, Q.: Statistical inference for hidden multifractional processes in a setting of stochastic volatility models. PhD Thesis, University Lille 1 (2011)
28. Rosenbaum, M.: Estimation of the volatility persitence in a discretely observed diffusion model. Stoch. Proc. Appl. **118**, 1434–1462 (2008)
29. Samorodnitsky, G., Taqqu, M.S.: Stable non-Gaussian random processes. Chapman and Hall, New York, London (1994)

30. Scott, L.: Option pricing when the variance changes randomly: Estimation and an application. J. Finan. Quantit. Anal. **22**, 419–438 (1987)
31. Surgailis, D.: Nonhomogeneous fractional integration and multifractional processes. Stoch. Proc. Appl. **118**, 171–198 (2008)
32. Stoev, S., Taqqu, M.S.: Path properties of the linear multifractional stable motion. Fractals **13**(2), 157–178 (2005)
33. Stoev, S., Taqqu, M.S.: Stochastic properties of the linear multifractional stable motion. Adv. Appl. Probab. **36**(4), 1085–1115 (2004)
34. Stoev, S., Taqqu, M.S.: How rich is the class of multifractional Brownian motions? Stoch. Process Appl. **116**, 200–221 (2006)
35. Tudor, C., Viens, F.: Statistical aspects of the fractional stochastic calculus. Ann. Statist. **35**, 1183–1212 (2007)

Chapter 7
Two-Sided Estimates for Distribution Densities in Models with Jumps

Archil Gulisashvili and Josep Vives*

Abstract The present paper is devoted to applications of mathematical analysis to the study of distribution densities arising in stochastic stock price models. We consider uncorrelated Stein-Stein, Heston, and Hull-White models and their perturbations by compound Poisson processes with jump amplitudes distributed according to a double exponential law. Similar perturbations of the Black-Scholes model were studied by S. Kou. For perturbed models, we obtain two-sided estimates for the stock price distribution density and compare the rate of decay of this density in the original and the perturbed model. It is shown that if the value of the parameter, characterizing the rate of decay of the right tail of the double exponential law, is small, then the stock price density in the perturbed model decays slower than the density in the original model. On the other hand, if the value of this parameter is large, then there are no significant changes in the behavior of the stock price distribution density.

7.1 Introduction

In this paper, we use methods of mathematical analysis to find two-sided estimates for the distribution density of the stock price in stochastic models with jumps. We consider perturbations of the Hull-White, the Stein-Stein, and the Heston models

* The research of Josep Vives is partially supported by grant MTM2009-07203

A. Gulisashvili (✉)
Department of Mathematics, Ohio University, Athens, OH 45701, USA
e-mail: gulisash@ohio.edu

J. Vives
Departament de Probabilitat, Lògica i Estadística, Universitat de Barcelona, Gran Via 585, 08007-Barcelona (Catalunya), Spain
e-mail: josep.vives@ub.edu

M. Zili and D.V. Filatova (eds.), *Stochastic Differential Equations and Processes*, Springer Proceedings in Mathematics 7, DOI 10.1007/978-3-642-22368-6_7,

by compound Poisson processes. In these models, the volatility processes are a geometric Brownian motion, the absolute value of an Ornstein-Uhlenbeck process, and a Cox-Ingersoll-Ross process, respectively. For more information on stochastic volatility models, see [5] and [7].

A stock price model with stochastic volatility is called uncorrelated if standard Brownian motions driving the stock price equation and the volatility equation are independent. In [8, 10], and [11], sharp asymptotic formulas were found for the distribution density of the stock price in uncorrelated Hull-White, Stein-Stein, and Heston models. Various applications of these formulas were given in [9] and [12]. The results obtained in [10] and [11] will be used in the present paper.

It is known that the stock price distribution density in an uncorrelated stochastic volatility model possesses a certain structural symmetry (see formula (7.14) below). This implies a similar symmetry in the Black-Scholes implied volatility, which does not explain the volatility skew observed in practice. To improve the performance of an uncorrelated model, one can either assume that the stock price process and the volatility process are correlated, or add a jump component to the stock price equation or to the volatility equation. The stock price distribution in the resulting model fits the empirical stock price distribution better than in the uncorrelated case. However, passing to a correlated model or adding a jump component may sometimes lead to similar effects or may have different consequences (see e.g. [1] and [2]). Examples of stock price models with jumps can be found in [3, 17], and [18]. We refer the reader to [4] for more information about stock price models with jumps. An interesting discussion of the effect of adding jumps to the Heston model in contained in [16].

An important jump-diffusion model was introduced and studied by Kou in [17] and by Kou and Wang in [18]. This model can be described as a perturbation of the Black-Scholes model by a compound Poisson process with double-exponential law for the jump amplitudes. In the present paper, we consider similar perturbations of stochastic volatility models. Our main goal is to determine whether significant changes may occur in the tail behavior of the stock price distribution after such a perturbation. We show that the answer depends on the relations between the parameters defining the original model and the characteristics of the jump process. For instance, no significant changes occur in the behavior of the distribution density of the stock price in a perturbed Heston or Stein-Stein model if the value of the parameter characterizing the right tail of the double exponential law is large. On the other hand, if this value is small, then the distribution density of the stock price in the perturbed model decreases slower than in the original model. For the Hull-White model, there are no significant changes in the tail behavior of the stock price density, since this density decays extremely slowly.

We will next briefly overview the structure of the present paper. In Sect. 7.2, we describe classical stochastic volatility models and their perturbations by a compound Poisson process. In Sect. 7.3 we formulate the main results of the paper and discuss what follows from them. Finally, in Sect. 7.4, we prove the theorems formulated in Sect. 7.3.

7.2 Preliminaries

In the present paper, we consider perturbations of uncorrelated Stein-Stein, Heston, and Hull-White models by compound Poisson processes. Our goal is to determine whether the behavior of the stock price distribution density in the original models changes after such a perturbation.

The stock price process X and the volatility process Y in the Stein-Stein model satisfy the following system of stochastic differential equations:

$$\begin{cases} dX_t = \mu X_t dt + |Y_t| X_t dW_t \\ dY_t = q(m - Y_t)\, dt + \sigma dZ_t. \end{cases} \tag{7.1}$$

This model was introduced and studied in [20]. The process Y, solving the second equation in (7.1), is called an Ornstein-Uhlenbeck process. We assume that $\mu \in \mathbb{R}$, $q \geq 0$, $m \geq 0$, and $\sigma > 0$.

The Heston model was developed in [14]. In this model, the processes X and Y satisfy

$$\begin{cases} dX_t = \mu X_t dt + \sqrt{Y_t} X_t dW_t \\ dY_t = q(m - Y_t)\, dt + c\sqrt{Y_t} dZ_t, \end{cases} \tag{7.2}$$

where $\mu \in \mathbb{R}$, $q \geq 0$, $m \geq 0$, and $c > 0$. The volatility equation in (7.2) is uniquely solvable in the strong sense, and the solution Y is a non-negative stochastic process. This process is called a Cox-Ingersoll-Ross process.

The stock price process X and the volatility process Y in the Hull-White model are determined from the following system of stochastic differential equations:

$$\begin{cases} dX_t = \mu X_t dt + Y_t X_t dW_t \\ dY_t = \nu Y_t dt + \xi Y_t dZ_t. \end{cases} \tag{7.3}$$

In (7.3), $\mu \in \mathbb{R}$, $\nu \in \mathbb{R}$, and $\xi > 0$. The Hull-White model was introduced in [15]. The volatility process in this model is a geometric Brownian motion.

It will be assumed throughout the paper that standard Brownian motions W and Z in (7.1), (7.2), and (7.3) are independent. The initial conditions for the processes X and Y will be denoted by x_0 and y_0, respectively.

We will next discuss perturbations of the models defined above by a compound Poisson process with jump amplitudes distributed according to a double exponential law. Perturbations of the Black-Scholes model by such jump processes were studied by Kou in [17] and by Kou and Wang in [18]. Some of the methods developed in [17] will be used in the present paper.

Let N be a standard Poisson process with intensity $\lambda > 0$, and consider a compound Poisson process defined by

$$J_t = \sum_{i=1}^{N_t} (V_i - 1), \quad t \geq 0, \tag{7.4}$$

where V_i are positive identically distributed random variables such that the distribution density f of $U_i = \log V_i$ is double exponential, that is,

$$f(u) = p\eta_1 e^{-\eta_1 u} 1\!\!1_{\{u \geq 0\}} + q\eta_2 e^{\eta_2 u} 1\!\!1_{\{u<0\}}, \tag{7.5}$$

where $\eta_1 > 1$, $\eta_2 > 0$, and p and q are positive numbers such that $p + q = 1$. Recall that condition $\eta_1 > 1$ is necessary and sufficient to guarantee that J has finite expectation.

Consider the following stochastic volatility models with jumps:

$$\begin{cases} d\widetilde{X}_t = \mu\widetilde{X}_{t-}dt + |Y_t|\,\widetilde{X}_{t-}dW_t + \widetilde{X}_{t-}dJ_t \\ dY_t = q\,(m - Y_t)\,dt + \sigma dZ_t \end{cases} \tag{7.6}$$

(the perturbed Stein-Stein model),

$$\begin{cases} d\widetilde{X}_t = \mu\widetilde{X}_{t-}dt + \sqrt{Y_t}\widetilde{X}_{t-}dW_t + \widetilde{X}_{t-}dJ_t \\ dY_t = q\,(m - Y_t)\,dt + c\sqrt{Y_t}dZ_t \end{cases} \tag{7.7}$$

(the perturbed Heston model), and

$$\begin{cases} d\widetilde{X}_t = \mu\widetilde{X}_{t-}dt + Y_t\widetilde{X}_{t-}dW_t + \widetilde{X}_{t-}dJ_t \\ dY_t = \nu Y_t dt + \xi Y_t dZ_t \end{cases} \tag{7.8}$$

(the perturbed Hull-White model). It is assumed in (7.6), (7.7), and (7.8) that the compound Poisson process J is independent of standard Brownian motions W and Z.

We will next formulate several results of Gulisashvili and Stein.

Let D_t be the distribution density of the stock price price X_t for a fixed $t > 0$. For the uncorrelated Heston model, there exist constants $A_1 > 0$, $A_2 > 0$, and $A_3 > 2$ such that

$$D_t(x) = A_1(\log x)^{-\frac{3}{4}+\frac{qm}{c^2}} e^{A_2\sqrt{\log x}} x^{-A_3}\left(1 + O\left((\log x)^{-\frac{1}{4}}\right)\right) \tag{7.9}$$

as $x \to \infty$ (see [11]). For the uncorrelated Stein-Stein model, there exist constants $B_1 > 0$, $B_2 > 0$, and $B_3 > 2$ such that

$$D_t(x) = B_1(\log x)^{-\frac{1}{2}} e^{B_2\sqrt{\log x}} x^{-B_3}\left(1 + O\left((\log x)^{-\frac{1}{4}}\right)\right) \tag{7.10}$$

as $x \to \infty$ (see [11]). Finally, in the case of the uncorrelated Hull-White model, there exist constants $b_1 > 0$, b_2 and b_3 such that following formula holds (see [10] and also Theorem 4.1 in [11]):

$$D_t(x) = b_1 x^{-2} (\log x)^{\frac{b_2-1}{2}} (\log\log x)^{b_3}$$

$$\exp\left\{-\frac{1}{2t\xi^2}\left(\log\left[\frac{1}{y_0}\sqrt{\frac{2\log x}{t}}\right] + \frac{1}{2}\log\log\left[\frac{1}{y_0}\sqrt{\frac{2\log x}{t}}\right]\right)^2\right\}$$

$$\left(1 + O\left((\log\log x)^{-\frac{1}{2}}\right)\right) \tag{7.11}$$

as $x \to \infty$. The constants in formulas (7.9), (7.10), and (7.11) depend on t and on the model parameters. Explicit expressions for these constants can be found in [10] and [11]. The constants A_3 and B_3, appearing in (7.9) and (7.10), describe the rate of the power-type decay of the stock price distribution density in the Heston and the Stein-Stein model, respectively. The explicit formulas for these constants are as follows:

$$A_3 = \frac{3}{2} + \frac{\sqrt{8C+t}}{2\sqrt{t}} \quad \text{with} \quad C = \frac{t}{2c^2}\left(q^2 + \frac{4}{t^2}r^2_{\frac{qt}{2}}\right), \tag{7.12}$$

and

$$B_3 = \frac{3}{2} + \frac{\sqrt{8G+t}}{2\sqrt{t}} \quad \text{with} \quad G = \frac{t}{2\sigma^2}\left(q^2 + \frac{1}{t^2}r^2_{qt}\right). \tag{7.13}$$

In (7.12) and (7.13), r_s denotes the smallest positive root of the entire function

$$z \mapsto z\cos z + s\sin z.$$

Formulas (7.12) and (7.13) can be found in [11].

It is known that the distribution density D_t in uncorrelated stochastic volatility models satisfies the following symmetry condition:

$$\left(\frac{x_0 e^{\mu t}}{x}\right)^3 D_t\left(\frac{(x_0 e^{\mu t})^2}{x}\right) = D_t(x), \quad x > 0, \tag{7.14}$$

(see Sect. 2 in [11]). This condition shows that the asymptotic behavior of the stock price distribution density near zero is completely determined by its behavior near infinity.

7.3 Main Results

The following theorems concern the tail behavior of the stock price distribution density in perturbed uncorrelated Stein-Stein, Heston, and Hull-White models:

Theorem 7.1. *Let $\varepsilon > 0$. Then there exist $c_1 > 0$, $c_2 > 0$, and $x_1 > 1$ such that the following estimates hold for the distribution density $\widetilde{D}_t$ of the stock price $\widetilde{X}_t$ in the* *perturbed Heston model:*

$$c_1\left(\frac{1}{x^{A_3}}+\frac{1}{x^{1+\eta_1}}\right)\le \widetilde{D}_t(x)\le c_2\left(\frac{1}{x^{A_3-\varepsilon}}+\frac{1}{x^{1+\eta_1-\varepsilon}}\right) \tag{7.15}$$

for all $x > x_1$. The constants c_2 and x_1 in formula (7.15) depend on ε, while the constant c_1 does not. The constant A_3 in (7.15) is given by (7.12). It depends on t and the model parameters.

Theorem 7.2. *Let $\varepsilon > 0$. Then there exist $c_3 > 0$, $c_4 > 0$, and $x_2 \in (0, 1)$ such that the following estimates hold for the distribution density $\widetilde{D}_t$ of the stock price $\widetilde{X}_t$ in the perturbed Heston model:*

$$c_3\left(x^{A_3-3}+x^{\eta_2-1}\right)\le \widetilde{D}_t(x)\le c_4\left(x^{A_3-3-\varepsilon}+x^{\eta_2-1-\varepsilon}\right) \tag{7.16}$$

for all $0 < x < x_2$. Here A_3 is the same as in Theorem 7.1. The constants c_4 and x_2 in (7.16) depend on ε, while the constant c_3 does not.

Theorem 7.3. *Let $\varepsilon > 0$. Then there exist $c_5 > 0$, $c_6 > 0$, and $x_3 > 1$ such that the following estimates hold for the distribution density $\widetilde{D}_t$ of the stock price $\widetilde{X}_t$ in the perturbed Stein-Stein model:*

$$c_5\left(\frac{1}{x^{B_3}}+\frac{1}{x^{1+\eta_1}}\right)\le \widetilde{D}_t(x)\le c_6\left(\frac{1}{x^{B_3-\varepsilon}}+\frac{1}{x^{1+\eta_1-\varepsilon}}\right) \tag{7.17}$$

for all $x > x_3$. The constant B_3 in (7.17) depends on t and the model parameters and is given by (7.13). The constants c_6 and x_3 in (7.17) depend on ε, while the constant c_5 does not.

Theorem 7.4. *Let $\varepsilon > 0$. Then there exist $c_7 > 0$, $c_8 > 0$, and $x_4 \in (0, 1)$ such that the following estimates hold for the distribution density $\widetilde{D}_t$ of the stock price $\widetilde{X}_t$ in the perturbed Stein-Stein model:*

$$c_7\left(x^{B_3-3}+x^{\eta_2-1}\right)\le \widetilde{D}_t(x)\le c_8\left(x^{B_3-3-\varepsilon}+x^{\eta_2-1-\varepsilon}\right) \tag{7.18}$$

for all $0 < x < x_4$. Here B_3 is the same as in Theorem 7.3. The constants c_8 and x_4 in (7.18) depend on ε, while the constant c_7 does not.

Remark 7.1. Theorem 7.1 is also true for the stock price distribution density in a perturbed correlated Heston model (see [13]). This generalization of Theorem 1 is used in [13] to characterize the asymptotic behavior of the implied volatility for large and small strikes in the case of a perturbed Heston model.

We will prove Theorems 7.1–7.4 in Sect. 7.4. In the remaining part of the present section, we compare the tail behavior of the stock price distribution density before and after perturbation of the model by a compound Poisson process.

Let us begin with the Heston model. It follows from Theorem 7.1 that if $1+\eta_1 < A_3$, then

$$\frac{\bar{c}_1}{x^{1+\eta_1}} \le \widetilde{D}_t(x) \le \frac{\bar{c}_2}{x^{1+\eta_1-\varepsilon}}$$

for large enough values of x. Therefore, formula (7.9) shows that that if the condition $1 + \eta_1 < A_3$ holds, then the tail of the distribution of the stock price in the perturbed Heston model is heavier than in the original model.

On the other hand, if $1 + \eta_1 \ge A_3$, then Theorem 7.1 implies the following estimate:

$$\frac{\tilde{c}_1}{x^{A_3}} \le \widetilde{D}_t(x) \le \frac{\tilde{c}_2}{x^{A_3-\varepsilon}}$$

for large enough values of x. Now formula (7.9) shows that if $1 + \eta_1 \ge A_3$, then there are no significant changes in the tail behavior of the distribution density of the stock price after perturbation. Similar assertions hold for the Stein-Stein model. This can be established using Theorem 7.3 and formula (7.10).

Next, suppose $x \to 0$. Then we can compare the behavior of the distribution density of the stock price in unperturbed and perturbed models, taking into account Theorem 7.2, Theorem 7.4, formula (7.9), formula (7.10), and the symmetry condition (7.14). For instance, if $\eta_2 < A_3 - 2$ in the perturbed Heston model, then

$$\bar{c}_3 x^{\eta_2 - 1} \le \widetilde{D}_t(x) \le \bar{c}_4 x^{\eta_2 - 1 - \varepsilon}$$

for small enough values of x. On the other hand if $\eta_2 \ge A_3 - 2$, then

$$\tilde{c}_3 x^{A_3 - 3} \le \widetilde{D}_t(x) \le \tilde{c}_4 x^{A_3 - 3 - \varepsilon}$$

for small enough values of x. Similar results hold for the Stein-Stein model.

Remark 7.2. For the Hull-White model, there are no significant changes in the tail behavior of the stock price distribution after perturbation. This statement follows from the assumption $\eta_1 > 1$ and from the fact that the stock price density in the unperturbed Hull-White model decays like x^{-2} (see formula (7.11)).

7.4 Proofs of the Main Results

The proofs of Theorems 7.1–7.4 are based on an explicit formula for the distribution density $\widetilde{D}_t$ of the stock price $\widetilde{X}_t$ in perturbed Heston, Stein-Stein, and Hull-White models (see formula (7.22) below). Note that the stock price process $\widetilde{X}$ in the perturbed Stein-Stein and Hull-White models is given by

$$\widetilde{X}_t = x_0 \exp\left\{\mu t - \frac{1}{2}\int_0^t Y_s^2 ds + \int_0^t Y_s dW_s + \sum_{i=1}^{N_t} U_i\right\}, \tag{7.19}$$

while for the perturbed for Heston model we have

$$\widetilde{X}_t = x_0 \exp\left\{\mu t - \frac{1}{2}\int_0^t Y_s ds + \int_0^t \sqrt{Y_s} dW_s + \sum_{i=1}^{N_t} U_i\right\}. \tag{7.20}$$

Formulas (7.19) and (7.20) can be established using the Doléans-Dade formula (see, for example, [19]). Put $T_t = \sum_{i=1}^{N_t} U_i$ and denote by μ_t its probability distribution. Clearly, we have:

$$\mu_t(A) = \pi_0\delta_0(A) + \sum_{n=1}^{\infty}\pi_n \int_A f^{*(n)}(u)du \tag{7.21}$$

where $\pi_0 = e^{-\lambda t}$, $\pi_n = e^{-\lambda t}(n!)^{-1}(\lambda t)^n$ for $n \geq 1$, A is a Borel subset of $\mathbb{R}$, and f is given by (7.5). The star in (7.21) denotes the convolution.

The distribution density D_t of the stock price X_t in uncorrelated models of our interest is related to the law of the following random variable:

$$\alpha_t = \left\{\frac{1}{t}\int_0^t Y_s^2 ds\right\}^{\frac{1}{2}}$$

for the Stein-Stein and the Hull-White model, and

$$\alpha_t = \left\{\frac{1}{t}\int_0^t Y_s ds\right\}^{\frac{1}{2}}$$

for the Heston model (see [10] and [11]). The distribution density of the random variable α_t is called the mixing distribution density and is denoted by m_t. We refer the reader to [10, 11], and [20] for more information on the mixing distribution density.

The next lemma establishes a relation between the mixing distribution density m_t in the uncorrelated model and the distribution density $\widetilde{D}_t$ of the stock price $\widetilde{X}_t$ in the corresponding perturbed model.

Lemma 7.1. *The density $\widetilde{D}_t$ in perturbed Stein-Stein, Heston and Hull-White models is given by the following formula:*

$$\widetilde{D}_t(x) = \frac{1}{\sqrt{2\pi t}x}\int_0^{\infty}\left(\int_{-\infty}^{\infty}\exp\left\{-\frac{(\log\frac{x}{x_0 e^{\mu t}} + \frac{ty^2}{2} - u)^2}{2ty^2}\right\}\mu_t(du)\right)m_t(y)\frac{dy}{y},$$

where m_t is the mixing distribution density and μ_t is defined by (7.21).

Proof. We will prove Lemma 7.1 for the Heston model. The proof for the Stein-Stein and the Hull-White model is similar. For the latter models, we use formula (7.19) instead of formula (7.20).

For any $\eta > 0$, formula (7.20) gives

$$\mathbb{P}\left(\widetilde{X}_t \leq \eta\right) = \mathbb{P}\left[\int_0^t \sqrt{Y_s} dW_s + T_t \leq \log \frac{\eta}{x_0 e^{\mu t}} + \frac{t\alpha_t^2}{2}\right]$$

$$= \mathbb{E} \int_{-\infty}^{z*} \int_{-\infty}^{\infty} \frac{1}{\sqrt{2\pi t}\alpha_t} \exp\left\{-\frac{(z-u)^2}{2t\alpha_t^2}\right\} \mu_t(du)dz,$$

where $z_* = \log \frac{\eta}{x_0 e^{\mu t}} + \frac{t\alpha_t^2}{2}$. Making the substitution $z = \log \frac{x}{x_0 e^{\mu t}} + \frac{t\alpha_t^2}{2}$, we obtain

$$\mathbb{P}\left(\widetilde{X}_t \leq \eta\right) = \mathbb{E} \int_0^{\eta} \int_{-\infty}^{\infty} \frac{1}{\sqrt{2\pi t}\alpha_t} \exp\left\{-\frac{(\log \frac{x}{x_0 e^{\mu t}} + \frac{t\alpha_t^2}{2} - u)^2}{2t\alpha_t^2}\right\} \mu_t(du) \frac{dx}{x}$$

$$= \int_0^{\eta} \int_0^{\infty} \int_{-\infty}^{\infty} \exp\left\{-\frac{(\log \frac{x}{x_0 e^{\mu t}} + \frac{ty^2}{2} - u)^2}{2ty^2}\right\} \mu_t(du) \frac{m_t(y)}{\sqrt{2\pi t} y} dy \frac{dx}{x}.$$

It is clear that the previous equality implies Lemma 7.1. □

Remark 7.3. It follows from Lemma 7.1 that

$$\widetilde{D}_t(x) = \frac{\sqrt{x_0 e^{\mu t}}}{\sqrt{2\pi t} x^{\frac{3}{2}}} \int_{-\infty}^{\infty} e^{\frac{u}{2}} \mu_t(du) \int_0^{\infty} \frac{m_t(y)}{y} \exp\left\{-\frac{(\log \frac{x}{x_0 e^{\mu t}} - u)^2}{2ty^2} - \frac{ty^2}{8}\right\} dy. \tag{7.22}$$

This representation will be used below to obtain two-sided estimates for the distribution density of the stock price in perturbed stochastic volatility models.

The next lemma will be needed in the proof of Theorem 7.1.

Lemma 7.2. *Let f be the density of the double exponential law (see formula (7.5)). Then for every $n \geq 1$, the following formula holds:*

$$f^{*(n)}(u) = e^{-\eta_1 u} \sum_{k=1}^{n} P_{n,k} \eta_1^k \frac{1}{(k-1)!} u^{k-1} \mathbb{1}_{\{u \geq 0\}}$$

$$+ e^{\eta_2 u} \sum_{k=1}^{n} Q_{n,k} \eta_2^k \frac{1}{(k-1)!} (-u)^{k-1} \mathbb{1}_{\{u<0\}}, \tag{7.23}$$

where

$$P_{n,k} = \sum_{i=k}^{n-1} \binom{n-k-1}{i-k} \binom{n}{i} \left(\frac{\eta_1}{\eta_1 + \eta_2}\right)^{i-k} \left(\frac{\eta_2}{\eta_1 + \eta_2}\right)^{n-i} p^i q^{n-i}$$

for all $1 \leq k \leq n-1$, *and*

$$Q_{n,k} = \sum_{i=k}^{n-1} \binom{n-k-1}{i-k} \binom{n}{i} \left(\frac{\eta_1}{\eta_1+\eta_2}\right)^{n-i} \left(\frac{\eta_2}{\eta_1+\eta_2}\right)^{i-k} p^{n-i} q^i$$

for all $1 \leq k \leq n-1$. *In addition,* $P_{n,n} = p^n$ *and* $Q_{n,n} = q^n$.

Lemma 7.2 can be established using Proposition B.1 in [17] and taking into account simple properties of the exponential distribution.

The next statement follows from Lemma 7.2 and formula (7.21):

Lemma 7.3. *For every Borel set* $A \subset \mathbb{R}$,

$$\mu_t(A) = \pi_0 \delta_0(A) + \int_{A\cap[0,\infty)} G_1(u) e^{-\eta_1 u} du + \int_{A\cap(-\infty,0)} G_2(u) e^{\eta_2 u} du, \quad (7.24)$$

where

$$G_1(u) = \sum_{k=0}^{\infty} \left[\frac{\eta_1^{k+1}}{k!} \sum_{n=k+1}^{\infty} \pi_n P_{n,k+1} \right] u^k, \quad (7.25)$$

and

$$G_2(u) = \sum_{k=0}^{\infty} \left[\frac{\eta_2^{k+1}}{k!} \sum_{n=k+1}^{\infty} \pi_n Q_{n,k+1} \right] (-u)^k. \quad (7.26)$$

Our next goal is to estimate the rate of growth of the functions G_1 and G_2 defined by (7.25) and (7.26).

Lemma 7.4. *For every* $\varepsilon > 0$ *the function* G_1 *grows slower than the function* $u \mapsto e^{\varepsilon u}$ *as* $u \to \infty$. *Similarly, the function* G_2 *grows slower than the function* $u \mapsto e^{-\varepsilon u}$ *as* $u \to -\infty$.

Proof. We will prove the lemma by comparing the Taylor coefficients

$$a_k = \frac{1}{k!} \eta_1^{k+1} \sum_{n=k+1}^{\infty} \pi_n P_{n,k+1}, \quad k \geq 0,$$

of the function G_1 and the Taylor coefficients $b_k = \frac{1}{k!}\varepsilon^k$, $k \geq 0$, of the function $e^{\varepsilon u}$. We have $a_k \leq b_k$ for $k > k_0$. The previous inequality can be established using the estimate

$$\eta_1^{k+1} \sum_{n=k+1}^{\infty} \pi_n P_{n,k+1} \leq \eta_1^{k+1} \sum_{n=k+1}^{\infty} \pi_n,$$

and taking into account the fast decay of the complementary distribution function of the Poisson distribution.

This completes the proof of Lemma 7.4 for the function G_1. The proof for the function G_2 is similar. □

The following lemma was obtained in [11] (formula (54)):

Lemma 7.5. *Let m_t be the mixing distribution density in the Heston model. Then there exist constants $H_1 > 0$ and $H_2 > 0$, depending on the model parameters, such that*

$$\int_0^\infty \frac{m_t(y)}{y} \exp\left\{-\left(\frac{\omega^2}{2ty^2} + \frac{ty^2}{8}\right)\right\} dy$$

$$= H_1 \omega^{-\frac{3}{4}+\frac{qm}{c^2}} e^{H_2\sqrt{\omega}} \exp\left\{-\frac{\sqrt{8C+t}}{2\sqrt{t}}\omega\right\} \left(1 + O\left(\omega^{-\frac{1}{4}}\right)\right)$$

as $\omega \to \infty$. The constant C in the previous formula is given by (7.12).

In the following subsections we present the detailed proofs of Theorems 7.1 and 7.2. We do not include the proofs of Theorems 7.3 and 7.4, because these theorems can be established exactly as Theorems 7.1 and 7.2.

7.4.1 Proof of the Estimate from Below in Theorem 7.1

We will use formula (7.22) in the proof. Put $z = \log \frac{x}{x_0 e^{\mu t}}$. Then we have

$$\widetilde{D}_t(x) = \frac{\sqrt{x_0 e^{\mu t}}}{\sqrt{2\pi t} x^{\frac{3}{2}}} \int_{-\infty}^{\infty} e^{\frac{u}{2}} \mu_t(du) \int_0^\infty \frac{m_t(y)}{y} \exp\left\{-\frac{(z-u)^2}{2ty^2} - \frac{ty^2}{8}\right\} dy. \tag{7.27}$$

Note that for the uncorrelated Heston model the following formula holds:

$$D_t(x) = \frac{\sqrt{x_0 e^{\mu t}}}{\sqrt{2\pi t} x^{\frac{3}{2}}} \int_0^\infty \frac{m_t(y)}{y} \exp\left\{-\frac{z^2}{2ty^2} - \frac{ty^2}{8}\right\} dy \tag{7.28}$$

(see [11]).

Let ρ be any increasing function of z such that $\rho(z) < z$ and $z - \rho(z) \to \infty$ as $z \to \infty$. Assume z is large enough. Then (7.27) gives

$$\widetilde{D}_t(x) \geq I_1 + I_2, \tag{7.29}$$

where

$$I_1 = \frac{\sqrt{x_0 e^{\mu t}}}{\sqrt{2\pi t} x^{\frac{3}{2}}} \int_1^{\rho(z)} e^{\frac{u}{2}} \mu_t(du) \int_0^\infty \frac{m_t(y)}{y} \exp\left\{-\frac{(z-u)^2}{2ty^2} - \frac{ty^2}{8}\right\} dy \tag{7.30}$$

and

$$I_2 = \frac{\sqrt{x_0 e^{\mu t}}}{\sqrt{2\pi t} x^{\frac{3}{2}}} \int_z^{z+1} e^{\frac{u}{2}} \mu_t(du) \int_0^\infty \frac{m_t(y)}{y} \exp\left\{-\frac{(z-u)^2}{2ty^2} - \frac{ty^2}{8}\right\} dy. \tag{7.31}$$

Throughout the remaining part of the section, we will denote by α a positive constant which may differ from line to line. Since the function G_1 is increasing on $(0,\infty)$ and (7.24) and (7.25) hold, we have

$$I_2 \geq \alpha x^{-\frac{3}{2}} \int_z^{z+1} e^{\frac{u}{2}} e^{-\eta_1 u} du \int_0^\infty \frac{m_t(y)}{y} \exp\left\{-\frac{1}{2ty^2} - \frac{ty^2}{8}\right\} dy, \quad x > x_1.$$

It is known that $\int_0^1 y^{-1} m_t(y) dy < \infty$ (see [11]). Therefore, the second integral in the previous estimate converges. It follows that

$$I_2 \geq \alpha x^{-\frac{3}{2}} \int_z^{z+1} e^{\frac{u}{2}} e^{-\eta_1 u} du = \alpha x^{-1-\eta_1}$$

for $x > x_1$. It is not hard to see using the inequality $\tilde{D}_t(x) \geq I_2$ that the estimate from below in (7.15) holds in the case where $1 + \eta_1 \leq A_3$.

It remains to prove the estimate from below under the assumption $1 + \eta_1 > A_3$. We will use the inequality $\tilde{D}_t(x) \geq I_1$ in the proof. To estimate I_1 we notice that $z - u \geq z - \rho(z) \to \infty$ as $x \to \infty$. Therefore, Lemma 7.5 can be applied to estimate the second integral on the right-hand side of (7.30). This gives

$$I_1 \geq \alpha x^{-\frac{3}{2}} \int_1^{\rho(z)} e^{\frac{u}{2}} G_1(u) e^{-\eta_1 u} (z-u)^{-\frac{3}{4}+\frac{qm}{c^2}} e^{H_2\sqrt{z-u}} \exp\left\{-\frac{\sqrt{8C+t}}{2\sqrt{t}}(z-u)\right\} du.$$

Since the function G_1 is increasing on $(0,\infty)$ and the function

$$y \mapsto y^{-\frac{3}{4}+\frac{qm}{c^2}} e^{H_2\sqrt{y}}$$

is eventually increasing, the previous inequality gives

$$I_1 \geq \alpha x^{-\frac{3}{2}} \int_1^{\rho(z)} e^{\frac{u}{2}} e^{-\eta_1 u} \exp\left\{-\frac{\sqrt{8C+t}}{2\sqrt{t}}(z-u)\right\} du$$

$$= \alpha x^{-A_3} \int_1^{\rho(z)} \exp\left\{(A_3 - 1 - \eta_1)\, u\right\} du.$$

Here we used the equality $A_3 = \frac{3}{2} + \frac{\sqrt{8C+t}}{2\sqrt{t}}$ (see (7.12)). Since $A_3 < 1 + \eta_1$ we get $I_1 \geq \alpha x^{-A_3}, x > x_1$. This establishes the estimate from below in Theorem 7.1 in the case where $A_3 < 1 + \eta_1$.

7.4.2 Proof of the Estimate from Above in Theorem 7.1

Let ε be a sufficiently small positive number. Denote by $\Lambda_t(z,u)$ the following integral:

$$\int_0^\infty \frac{m_t(y)}{y} \exp\left\{-\frac{(z-u)^2}{2ty^2} - \frac{ty^2}{8}\right\} dy,$$

Then formula (7.27) can be rewritten as follows:

$$\widetilde{D}_t(x) = \frac{\sqrt{x_0 e^{\mu t}}}{\sqrt{2\pi t x^{\frac{3}{2}}}} \int_{-\infty}^\infty e^{\frac{u}{2}} \Lambda_t(z,u)\mu_t(du) = J_1 + J_2 + J_3, \tag{7.32}$$

where

$$J_1 = \frac{\sqrt{x_0 e^{\mu t}}}{\sqrt{2\pi t x^{\frac{3}{2}}}} \int_{-\infty}^{0-} e^{\frac{u}{2}} \Lambda_t(z,u)\mu_t(du),$$

$$J_2 = \frac{\sqrt{x_0 e^{\mu t}}}{\sqrt{2\pi t x^{\frac{3}{2}}}} \int_0^{sz} e^{\frac{u}{2}} \Lambda_t(z,u)\mu_t(du),$$

and

$$J_3 = \frac{\sqrt{x_0 e^{\mu t}}}{\sqrt{2\pi t x^{\frac{3}{2}}}} \int_{sz}^\infty e^{\frac{u}{2}} \Lambda_t(z,u)\mu_t(du).$$

The number s in the previous equalities satisfies $0 < s < 1$. The value of s will be chosen below.

To estimate J_2, we notice that if x is large, then $z-u$ in the expression for J_2 is also large. Using Lemma 7.3 and Lemma 7.5, we see that

$$J_2 \le \alpha D_t(x) + \alpha x^{-\frac{3}{2}} \int_0^{sz} e^{\frac{u}{2}} G_1(u) e^{-\eta_1 u} (z-u)^{-\frac{3}{4}+\frac{qm}{c^2}} e^{H_2\sqrt{z-u}}$$

$$\exp\left\{-\frac{\sqrt{8C+t}}{2\sqrt{t}}(z-u)\right\} du.$$

Since the functions $G_1(y)$ and $y \mapsto y^{-\frac{3}{4}+\frac{qm}{c^2}} e^{H_2\sqrt{y}}$ grow slower than the function $y \mapsto \exp\left\{\frac{\varepsilon}{2} y\right\}$ (see Lemma 7.4), the previous inequality and formula (7.9) imply that

$$J_2 \le \alpha x^{-A_3+\varepsilon} + \alpha x^{-\frac{3}{2}} \int_0^{sz} \exp\left\{\left(\frac{1}{2} - \eta_1 + \frac{\varepsilon}{2}\right) u\right\}$$

$$\exp\left\{\left(-\frac{\sqrt{8C+t}}{2\sqrt{t}} + \frac{\varepsilon}{2}\right)(z-u)\right\} du$$

$$\leq \alpha x^{-A_3+\varepsilon} + \alpha x^{-A_3+\frac{\varepsilon}{2}} \int_0^z \exp\{(A_3 - 1 - \eta_1)\, u\}\, du$$

$$\leq \alpha \left(\frac{1}{x^{A_3-\varepsilon}} + \frac{1}{x^{1+\eta_1-\varepsilon}} \right) \tag{7.33}$$

for $x > x_1$.

The function Λ_t is bounded (this has already been established in the previous part of the proof). Therefore,

$$J_3 \leq \alpha x^{-\frac{3}{2}} \int_{sz}^{\infty} e^{\frac{u}{2}} G_1(u) e^{-\eta_1 u} du. \tag{7.34}$$

Since the function $G_1(u)$ grows slower than the function $y \mapsto \exp\{\zeta u\}$ for any $\zeta > 0$ (see Lemma 7.4), estimate (7.34) implies that

$$J_3 \leq \alpha x^{-\frac{3}{2}+s(\frac{1}{2}+\zeta-\eta_1)}, \quad x > x_1.$$

Now using the fact that ζ can be any close to 0 and s any close to 1, we see that

$$J_3 \leq \alpha \frac{1}{x^{1+\eta_1-\varepsilon}}, \quad x > x_1. \tag{7.35}$$

We will next estimate J_1. It follows from Lemma 7.3 that

$$J_1 = \alpha x^{-\frac{3}{2}} \int_{-\infty}^{0-} e^{\frac{u}{2}} \Lambda_t(z,u) G_2(u) e^{\eta_2 u} du.$$

Since $u < 0$, we see that $z - u$ is large if x is large. Using Lemma 7.5, we obtain

$$J_1 \leq \alpha x^{-\frac{3}{2}} \int_{-\infty}^{0-} e^{\frac{u}{2}} (z-u)^{-\frac{3}{4}+\frac{qm}{c^2}} e^{H_2\sqrt{z-u}}$$

$$\exp\left\{ -\frac{\sqrt{8C+t}}{2\sqrt{t}}(z-u) \right\} G_2(u) e^{\eta_2 u} du. \tag{7.36}$$

The function $y \mapsto y^{-\frac{3}{4}+\frac{qm}{c^2}} e^{H_2\sqrt{y}}$ is eventually increasing. Moreover, it grows slower than $e^{\frac{\varepsilon}{2}y}$. Since $z - u > z$ in (7.36), we have

$$J_1 \leq \alpha x^{-\frac{3}{2}} \int_{-\infty}^{0-} e^{\frac{u}{2}} \exp\left\{ \left(-\frac{\sqrt{8C+t}}{2\sqrt{t}} + \frac{\varepsilon}{2} \right)(z-u) \right\} G_2(u) e^{\eta_2 u} du$$

$$\leq \alpha x^{-A_3+\frac{\varepsilon}{2}} \int_{-\infty}^{0-} e^{\frac{u}{2}} \exp\left\{ \left(\frac{\sqrt{8C+t}}{2\sqrt{t}} - \frac{\varepsilon}{2} \right) u \right\} G_2(u) e^{\eta_2 u} du$$

$$= \alpha x^{-A_3+\frac{\varepsilon}{2}} \int_0^\infty \exp\left\{\left(-\frac{1}{2} - \eta_2 - \frac{\sqrt{8C+t}}{2\sqrt{t}} + \frac{\varepsilon}{2}\right)u\right\} G_2(-u)du. \quad (7.37)$$

If ε is sufficiently small, then the integral in (7.37) converges (use Lemma 7.4). It follows from (7.37) that

$$J_1 \leq \alpha \frac{1}{x^{A_3-\varepsilon}}, \quad x > x_1. \quad (7.38)$$

Finally, combining (7.32), (7.33), (7.35), and (7.38), we establish the estimate from above in Theorem 7.1.

7.4.3 Proof of Theorem 7.2

The following formula can be obtained from (7.22):

$$\left(\frac{x_0 e^{\mu t}}{x}\right)^3 \widetilde{D}_t\left(\frac{(x_0 e^{\mu t})^2}{x}\right) = \frac{\sqrt{x_0 e^{\mu t}}}{\sqrt{2\pi t} x^{\frac{3}{2}}}$$

$$\int_{\mathbb{R}} e^{\frac{u}{2}} \mu_t(du) \int_0^\infty \frac{m_t(y)}{y} \exp\left\{-\frac{(\log \frac{x}{x_0 e^{\mu t}} + u)^2}{2ty^2} - \frac{ty^2}{8}\right\} dy. \quad (7.39)$$

It follows from (7.39) and (7.24) that

$$\left(\frac{x_0 e^{\mu t}}{x}\right)^3 \widetilde{D}_t\left(\frac{(x_0 e^{\mu t})^2}{x}\right) = \frac{\sqrt{x_0 e^{\mu t}}}{\sqrt{2\pi t} x^{\frac{3}{2}}}$$

$$\int_{-\infty}^\infty e^{\frac{u}{2}} \tilde{\mu}_t(du) \int_0^\infty \frac{m_t(y)}{y} \exp\left\{-\frac{(\log \frac{x}{x_0 e^{\mu t}} - u)^2}{2ty^2} - \frac{ty^2}{8}\right\} dy, \quad (7.40)$$

where

$$\tilde{\mu}_t(A) = \pi_0 \delta_0(A) + \int_{A\cap(0,\infty)} G_2(-u)e^{-(\eta_2+1)u} du + \int_{A\cap(-\infty,0)} G_1(-u)e^{(\eta_1-1)u} du \quad (7.41)$$

for all Borel sets $A \subset \mathbb{R}$. In (7.41), G_1 and G_2 are defined by (7.25) and (7.26), respectively. Now it is clear that we can use the proof of Theorem 7.1 with the pairs (η_1, p) and (η_2, q) replaced by the pairs $(\eta_2 + 1, q)$ and $(\eta_1 - 1, p)$, respectively. It is not hard to see using (7.39) that for every $\varepsilon > 0$, there exist constants $\tilde{c}_1 > 0$, $\tilde{c}_2 > 0$, and $\tilde{x} > 0$ such that the following estimates hold:

$$\tilde{c}_1 \left(\frac{1}{x^{A_3}} + \frac{1}{x^{\eta_2+2}}\right) \leq x^{-3} \widetilde{D}_t\left(\frac{(x_0 e^{\mu t})^2}{x}\right) \leq \tilde{c}_2 \left(\frac{1}{x^{A_3-\varepsilon}} + \frac{1}{x^{\eta_2+2-\varepsilon}}\right) \quad (7.42)$$

for all $x > \tilde{x}$. The constants $\tilde{c}_2$ and $\tilde{x}$ depend on ε. Now it is clear that (7.16) follows from (7.42).

This completes the proof of Theorem 7.2.

References

1. Alòs, E., León, J., Vives,J.: On the short time behavior of the implied volatility for jump diffusion models with stochastic volatility. Finance Stochast. **11**, 571–589 (2007)
2. Alòs, E., León, J., Pontier, M., Vives, J.: A Hull and White formula for a general stochastic volatility jump-diffusion model with applications to the study of the short-time behavior of the implied volatility. J. Appl. Math. Stoch. Anal. Vol. 2008 (2008), ID 359142, 17 pages.
3. Bates, D.S.: Jumps and stochastic volatility: exchange rate processes implicit in Deutsche Mark options. Rev. Financ. Stud. **9**, 69–107 (1996)
4. Cont, R., Tankov, P.: Financial Modeling with Jump Processes. Chapman and Hall – CRC, Boca Raton (2004)
5. Fouque, J.P., Papanicolaou, G., Sircar, K.R.: Derivatives in Financial Markets with Stochastic Volatility. Cambridge University Press, Cambridge (2000)
6. Friz, P., Gerhold, S., Gulisashvili, A., Sturm, S.: On refined volatility smile expansion in the Heston model. Quant Finanace, **11**, 1151–1164 (2011)
7. Gatheral, J.: The Volatility Surface: A Practitioner's Guide. Wiley, Hoboken, NJ (2006)
8. Gulisashvili, A., Stein, E.M.: Asymptotic behavior of distribution densities in models with stochastic volatility: The Hull-White model. Comptes Rendus de l'Academie des Sciences de Paris I **343**, 519–523 (2006)
9. Gulisashvili, A., Stein, E.M.: Implied volatility in the Hull-White model. Math. Finance. **19**, 303–327 (2009)
10. Gulisashvili, A., Stein, E.M.: Asymptotic behavior of distribution densities in models with stochastic volatility I. Math. Finance. **20**, 447–477 (2010)
11. Gulisashvili, A., Stein, E.M.: Asymptotic behavior of the stock price distribution density and implied volatility in stochastic volatility models. Appl. Math. Optim. **61**, 287–315 (2010)
12. Gulisashvili, A.: Asymptotic formulas with error estimates for call pricing functions and the implied volatility at extreme strikes. SIAM J. Financ. Math. **1**, 609–641 (2010)
13. Gulisashvili, A.: Asymptotic equivalence in Lee's moment formulas for the implied volatility and Piterbarg's conjecture. Submitted for publication; available at arxiv.org, arXiv:1007.5353
14. Heston, S.L.: A closed-form solution for options with stochastic volatility with applications to bond and currency options. Rev. Financ. Stud. **6**, 327–343 (1993)
15. Hull, J., White, A.: The pricing of options on assets with stochastic volatilities. J. Finance **42**, 281–300 (1987)
16. Keller-Ressel, M.: Moment explosions and long-term behavior of affine stochastic volatility models. Math. Finance, **21**, 73–98 (2011)
17. Kou, S.: A jump-diffusion model for option pricing. Manag. Sci. **48**, 1086–1101 (2002)
18. Kou, S., Wang, H.: Option pricing under a double exponential jump diffusion model. Manag. Sci. **50**, 1178–1192 (2003)
19. Protter, P.: Stochastic Integration and Differential Equations, 2nd ed. Springer, Berlin (2005)
20. Stein, E.M., Stein, J.C.: Stock price distributions with stochastic volatility: An analytic approach. Rev. Financ. Stud. **4**, 727–752 (1991)

Chapter 8
Maximizing a Function of the Survival Time of a Wiener Process in an Interval

Mario Lefebvre

Abstract Let $X(t)$ be a controlled one-dimensional Wiener process with positive drift. The problem of maximizing a function of the time spent by $X(t)$ in a given interval is solved explicitly. The process is controlled at most until a fixed time t_1. The same type of problem is considered for a two-dimensional degenerate diffusion process $(X(t), Y(t))$.

8.1 Introduction

We consider the one-dimensional controlled Wiener process $\{X(t), t \geq 0\}$ defined by the stochastic differential equation

$$dX(t) = \mu\, dt + b\, u(t)\, dt + \sigma\, dB(t), \tag{8.1}$$

where $\{B(t), t \geq 0\}$ is a standard Brownian motion, $u(t)$ is the control variable, and μ, b and σ are positive constants.

Assume that the process starts from $X(0) = x < d$ (> 0), and define

$$T(x; d) = \inf\{t > 0 : X(t) = d \mid X(0) = x\}.$$

That is, $T(x; d)$ is the first time the controlled process $X(t)$ takes on the value d.

Next, let

$$T_1(x; d) = \min\{T(x; d), t_1\},$$

M. Lefebvre (✉)
Département de mathématiques et de génie industriel, École Polytechnique,
C.P. 6079, Succursale Centre-ville, Montréal, Québec H3C 3A7, Canada
e-mail: mlefebvre@polymtl.ca

M. Zili and D.V. Filatova (eds.), *Stochastic Differential Equations and Processes*, Springer Proceedings in Mathematics 7, DOI 10.1007/978-3-642-22368-6_8,

where $t_1 > 0$ is a constant. Our aim is to find the control u^* that minimizes the expected value of the cost criterion

$$J(x) := \int_0^{T_1(x;d)} \frac{1}{2} q u^2(t)\, dt + k \ln[T_1(x;d)], \tag{8.2}$$

where $q > 0$ and $k < 0$. Notice that the optimizer seeks to maximize the time spent by $\{X(t), t \geq 0\}$ in $(-\infty, d)$, taking the control costs into account. Whittle ([7], p. 289) has termed this type of problem *LQG homing*. Actually, the termination cost is positive if $T_1(x;d) < 1$, whereas it becomes a reward if $T_1(x;d) > 1$.

A number of papers have been published on LQG homing problems; see, for instance, Kuhn [1], Lefebvre [2] and Makasu [5]. In [1] and [5], as well as in Whittle ([8], p. 222), a risk-sensitive cost criterion was used.

LQG homing problems can be considered in n dimensions. Then, the random variable T_1 is the moment of first entry of $(X(t), t)$ into a stopping set $D \subset \mathbb{R}^n \times (0, \infty)$. In practice, though, explicit solutions were only obtained so far in special cases when $n \geq 2$, and with T_1 defined only in terms of $X(t)$.

In the next section, the optimal control problem will be solved in a particular case. In Sect. 3, the same type of problem will be studied for a two-dimensional degenerate diffusion process $(X(t), Y(t))$ for which the derivative of $X(t)$ is a deterministic function of $X(t)$ and $Y(t)$. Finally, a few concluding remarks will be made in Sect. 4.

8.2 Optimal Control in One Dimension

First, the uncontrolled process $\{\xi(t), t \geq 0\}$ that corresponds to the process $\{X(t), t \geq 0\}$ in (8.1) is defined by

$$d\xi(t) = \mu\, dt + \sigma\, dB(t).$$

Because the drift μ is strictly positive, we can write (see [3], for instance) that

$$P[\tau(x;d) < \infty] = 1,$$

where

$$\tau(x;d) = \inf\{t > 0 : \xi(t) = d \mid \xi(0) = x\}.$$

It follows that, even if $t_1 = \infty$, we have:

$$P[\tau_1(x;d) < \infty] = 1,$$

where $\tau_1(x;d)$ is the random variable defined by

$$\tau_1(x;d) = \min\{\tau(x;d), t_1\}.$$

Moreover, because the quantities b, σ and q in the problem formulation above are all positive constants, we can state that there exists a positive constant α such that

$$\alpha \sigma^2 = \frac{b^2}{q}. \tag{8.3}$$

Then, we may appeal to a theorem in Whittle [7] to assert that the optimal control u^* $(= u^*(0))$ can be expressed in terms of the mathematical expectation

$$G(x) := E\left[\exp\left\{-\frac{b^2}{q\sigma^2} k \ln[\tau_1(x;d)]\right\}\right]. \tag{8.4}$$

Namely, we can write that

$$u^* = \frac{\sigma^2}{b} \frac{G'(x)}{G(x)}.$$

Thus, making use of the theorem in [7], the stochastic optimal control problem set up above is reduced to the computation of the function $G(x)$.

Assume next that we choose the constant $k = -q\sigma^2/b^2$. Then, the mathematical expectation $G(x)$ simplifies to

$$G(x) = E\left[\exp\{\ln[\tau_1(x;d)]\}\right] = E\left[\tau_1(x;d)\right].$$

Now, if we assume further that the constant t_1 is such that

$$E[\tau_1(x;d)] = c\, E[\tau(x;d)], \tag{8.5}$$

where $c \in (0,1)$, then (see [3], p. 220)

$$G(x) = c\, \frac{d-x}{\mu}.$$

Hence, the optimal control follows at once.

Proposition 8.1. *Under the hypotheses above, the optimal control u^* is given by*

$$u^* = \frac{\sigma^2}{b} \frac{1}{x-d} \quad \textit{for } x < d. \tag{8.6}$$

Remarks. (i) Let $\{X^*(t), t \geq 0\}$ denote the optimally controlled process. We deduce from the proposition that

$$u^*(t) = \frac{\sigma^2}{b} \frac{1}{X^*(t) - d},$$

where $\{X^*(t), t \geq 0\}$ satisfies the stochastic differential equation

$$dX^*(t) = \mu\, dt + \frac{\sigma^2}{X^*(t) - d}\, dt + \sigma\, dB(t).$$

(ii) An application of the model considered in this paper is the following: suppose that $X(t)$ is the wear of a machine at time t, and that this machine is assumed to be worn out when $X(t)$ reaches the value $d > 0$. Then, the optimizer tries to maximize the lifetime of the machine, taking the control costs into account. By choosing to stop controlling the process at most at time t_1, with t_1 being such that (8.5) holds, it is understood that the aim is to use the machine for a certain percentage of its expected lifetime, in the case when no control is used at all. A reason for doing so is the fact that a machine can become obsolete before it is actually worn out. Also, when it is old, the maintenance costs can increase rapidly.

(iii) Notice that the optimal solution (8.6) depends neither on the constant c, nor on the drift μ. However, if we modify the cost criterion (8.2) to

$$J_1(x) := \int_0^{T_1(x;d)} \frac{1}{2} q\, u^2(t)\, dt + k \ln[1 + T_1(x;d)],$$

so that the termination cost is always negative, then we find that

$$u^* = -\frac{\sigma^2}{b} \frac{c}{\mu + c(d - x)} \quad \text{for } x < d,$$

which implies that

$$u^*(t) = -\frac{\sigma^2}{b} \frac{c}{\mu + c[d - X^*(t)]}.$$

If $t_1 = \infty$, so that $c = 1$, we obtain that

$$u^*(t) = -\frac{\sigma^2}{b} \frac{1}{\mu + d - X^*(t)}.$$

(iv) The probability density function of the first passage time $\tau(x;d)$ is given by (see [3], p. 219)

$$f_{\tau(x;d)}(t) = \frac{d - x}{\sqrt{2\pi\sigma^2 t^3}} \exp\left\{-\frac{(d - x - \mu t)^2}{2\sigma^2 t}\right\} \quad \text{for } t > 0.$$

When $t_1 < \infty$, we have:

$$\begin{aligned} E[\tau_1(x;d)] &= E[\tau_1(x;d) \mid \tau(x) \le t_1]\, P[\tau(x) \le t_1] \\ &\quad + E[\tau_1(x;d) \mid \tau(x) > t_1]\, P[\tau(x) > t_1] \\ &= \int_0^{t_1} t\, f_{\tau(x;d)}(t)\, dt + t_1 \int_{t_1}^{\infty} f_{\tau(x;d)}(t)\, dt. \end{aligned}$$

From the previous formula, we can compute the value of c for any choice of t_1:

$$c = \frac{\mu}{d - x} E[\tau_1(x;d)].$$

If we look for the value of t_1 that corresponds to a given proportion c, then we can use a mathematical software package to estimate the value of t_1 from the above formula.

8.3 Optimal Control in Two Dimensions

We mentioned in Remark (ii) above that a possible application of the model considered in Sect. 2 is obtained by assuming that $X(t)$ is the wear of a machine at time t. However, because wear should always increase with time, a better model is obtained by defining the two-dimensional controlled diffusion process $(X(t), Y(t))$, starting from $(X(0), Y(0)) = (x, y)$, by the system (see Rishel [6])

$$\begin{aligned} dX(t) &= f[X(t), Y(t)]\, dt, \\ dY(t) &= m[X(t), Y(t)]\, dt + b(t)u(t)\, dt + \{v[X(t), Y(t)]\}^{1/2}\, dB(t), \end{aligned} \tag{8.7}$$

where f (> 0), b $(\neq 0)$, m and v (≥ 0) are real functions, and the random variable $Y(t)$ is a variable that is closely correlated with the wear.

Let

$$T(x, y) = \inf\{t > 0 : Y(t) = 0 \mid X(0) = x > 0, Y(0) = y > 0\}$$

and define, as in Sect. 2, the random variable

$$T_1(x, y) = \min\{T(x, y), t_1\},$$

where $t_1 > 0$ is a constant.

We want to find the control u^* that minimizes the expected value of the cost criterion

$$J(x, y) := \int_0^{T_1(x,y)} \frac{1}{2} q\, u^2(t)\, dt + k \ln[T_1(x, y)],$$

where q is a positive function, and k is a negative constant.

Assume next that

$$f(x, y) = m(x, y) \frac{x}{y}$$

and that

$$v(x,y) = 2xy.$$

Moreover, let us choose

$$b(t) \equiv b_0 \ (\neq 0) \quad \text{and} \quad q = \frac{q_0}{xy},$$

where $q_0 > 0$ is a constant. Then, there exists a positive constant α for which the relation

$$\alpha v(x,y) = \frac{b^2}{q} \quad \Longleftrightarrow \quad \alpha = \frac{b_0^2}{2q_0} \tag{8.8}$$

holds. Furthermore, at least when t_1 is finite, we can write that

$$P[\tau_1(x,y) < \infty] = 1,$$

where $\tau_1(x,y)$ is the same as $T_1(x,y)$, but for the uncontrolled process obtained by setting $u(t) = 0$ in (8.7). Similarly for the first passage time $\tau(x,y)$.

Then, making use of the theorem in [7], we deduce that

$$u^* = \frac{2xy}{b_0} \frac{G_y(x,y)}{G(x,y)}, \tag{8.9}$$

where

$$G(x,y) := E\left[\exp\{-\alpha k \ln[\tau_1(x,y)]\}\right] = E\left[\tau_1(x,y)^{-\alpha k}\right].$$

Remark. Let $H(x,y) = G_y(x,y)$. We can write that

$$u^*(t) = \frac{2X^*(t)Y^*(t)}{b_0} \frac{H[X^*(t),Y^*(t)]}{G[X^*(t),Y^*(t)]},$$

where $(X^*(t), Y^*(t))$ is the optimally controlled process defined by

$$dX^*(t) = m[X^*(t),Y^*(t)] \frac{X^*(t)}{Y^*(t)} dt,$$

$$dY^*(t) = m[X^*(t),Y^*(t)]dt + 2X^*(t)Y^*(t) \frac{H[X^*(t),Y^*(t)]}{G[X^*(t),Y^*(t)]} dt$$
$$+ \{2X^*(t)Y^*(t)]\}^{1/2} dB(t).$$

Now, it can be shown (see [4]) that the probability density function of the random variable $\tau(x,y)$ is given by

$$f_{\tau(x,y)}(t) = \frac{y}{xt^2} \exp\left\{-\frac{y}{xt}\right\} \quad \text{for } t > 0.$$

We find that $\tau^{-1}(x,y) \sim \mathrm{Exp}(y/x)$. Moreover, we have:

$$P[\tau(x,y) < \infty] = 1.$$

Therefore, even when t_1 is infinite, the condition $P[\tau_1(x,y) < \infty] = 1$ is fulfilled.

Because we deduce from the density of $\tau(x,y)$ that

$$E[\tau(x,y)] = \infty,$$

we cannot set $E[\tau_1(x,y)] = c\,E[\tau(x,y)]$, as in Sect. 2. Nevertheless, we find that

$$E\left[\tau^{1/2}(x,y)\right] = \left(\frac{\pi\,y}{x}\right)^{1/2} \quad \text{for } y/x > 0,$$

which yields the following proposition.

Proposition 8.2. *Under the hypotheses above, if we choose*

$$k = -\frac{1}{2\alpha},$$

then the optimal control is given by (8.9), in which

$$G(x,y) = E\left[\tau_1(x,y)^{1/2}\right] = \left(\frac{\pi\,y}{x}\right)^{1/2}\left\{1 - \mathrm{erf}\left[\left(\frac{y}{x\,t_1}\right)^{1/2}\right]\right\} + t_1^{1/2}\left(1 - e^{-y/(xt_1)}\right).$$

Proof. We obtain the formula for $E\left[\tau_1(x,y)^{1/2}\right]$ by conditioning on $\tau(x,y)$, and by making use of the formula

$$\int \frac{y}{x\,t^{3/2}}\,e^{-y/(xt)}\,dt = -\left(\frac{\pi\,y}{x}\right)^{1/2}\mathrm{erf}\left[\left(\frac{y}{x\,t}\right)^{1/2}\right]$$

for $x, y > 0$. ■

Remarks. (i) If $t_1 = \infty$, we find that the optimal control simplifies to

$$u^* = \frac{x}{b_0}.$$

(ii) In this problem, the relation between the constants t_1 and c is the following:

$$c = 1 - \mathrm{erf}\left[\left(\frac{y}{x\,t_1}\right)^{1/2}\right] + t_1^{1/2}\left(1 - e^{-y/(xt_1)}\right).$$

8.4 Concluding Remarks

We solved explicitly two optimal control problems involving diffusion processes, making use of a theorem that enables us to reduce them to purely probabilistic problems. In each case, in order to obtain an exact and explicit expression for the optimal control, we needed the probability density function of a first passage time for the corresponding uncontrolled process. In Sect. 2, we used the well known density function of the time it takes a one-dimensional Wiener process to hit a given boundary. In Sect. 3, we made use of the density function computed in [4] of a first passage time defined in terms of a degenerate two-dimensional diffusion process. There are few such explicit expressions in the case of n-dimensional processes, with $n \geq 2$.

As mentioned above, the models considered in this note can be applied to reliability theory. The model presented in Sect. 3 is such that the wear of a certain device always increases with time, as it should. We could try to generalize the model by incorporating a compound Poisson process in (8.7). However, then we could no longer appeal to the theorem in [7] to transform the stochastic optimal control problem into a probability problem. Similarly, we could consider discrete processes, instead of diffusion processes. Again, we would have to solve the optimal control problems differently.

Finally, here we solved problems for which the optimizer tries to maximize a function of the survival time of a diffusion process in a given continuation region. In the application presented, we assumed that a machine is used at most until a fixed time t_1. We could also assume that there is a warranty period, so that it will be used at least during a given period of time.

Acknowledgements The author is grateful to the reviewers for their constructive comments.

References

1. Kuhn, J.: The risk-sensitive homing problem. J. Appl. Probab. **22**, 796–803 (1985)
2. Lefebvre, M.: A homing problem for diffusion processes with control-dependent variance. Ann. Appl. Probab. **14**, 786–795 (2004)
3. Lefebvre, M.: Applied Stochastic Processes, Springer, New York (2007)
4. Lefebvre, M.: Controlling a two-dimensional diffusion process at most until a fixed time. (Submitted for publication)
5. Makasu, C.: Risk-sensitive control for a class of homing problems. Automatica J. IFAC. **45**, 2454–2455 (2009)
6. Rishel, R.: Controlled wear processes: modeling optimal control. IEEE Trans. Automat. Contr. **36**, 1100–1102 (1991)
7. Whittle, P.: Optimization over Time, Vol. I, Wiley, Chichester (1982)
8. Whittle, P.: Risk-sensitive Optimal Control, Wiley, Chichester (1990)

Appendix A
SAAP 2010 Scientific Committee

Edward Allen (Texas Tech University, USA)
Jean-marc Bardet (University Paris 1, France)
Viacheslav Belavkin (University of Nottingham, United Kingdom)
Kevin Burrage (University Queensland, Brisbane, Australia)
Zhen-Qing Chen (University of Washington, USA)
Nathalie Eisenbaum (University Paris 6, France)
Vyacheslav Leonidovich Girko (Institute of Mathematics Kiev, Ukraine)
Abdelhamid Hassairi (Faculty of sciences of Sfax, Tunisia)
Shuya Kanagawa (Tokyo City University, Japan)
Peter Kloeden (J.W.Goethe-Universitt Frankfurt, Germany)
Sana Louhichi (University Paris- South, France)
Habib Maagli (Faculty of sciences of Tunis, Tunisia)
David Nualart (Kansas University, USA)
Habib Ouerdiane (Faculty of sciences of Tunis, Tunisia)
Youssef Ouknine (University of Marrakech, Morocco)
Henri Schurz (Southern Illinois University, USA)
Marta-Sanz Solé (Barcelona University, Spain)
Constantin Tudor (University of Bucharest, Romania)
Tusheng Zhang (University of Manchester, United Kingdom)

M. Zili and D.V. Filatova (eds.), *Stochastic Differential Equations and Processes*, Springer Proceedings in Mathematics 7, DOI 10.1007/978-3-642-22368-6,

MIX
Papier aus verantwortungsvollen Quellen
Paper from responsible sources
FSC® C105338

If you have any concerns about our products,
you can contact us on
ProductSafety@springernature.com

In case Publisher is established outside the EU,
the EU authorized representative is:
Springer Nature Customer Service Center GmbH
Europaplatz 3, 69115 Heidelberg, Germany

Printed by Libri Plureos GmbH
in Hamburg, Germany